THE PURSUIT OF NEW PRODUCT DEVELOPMENT:

The Business
Development Process

THE PURSUIT OF NEW PRODUCT DEVELOPMENT:

The Business Development Process

Marc A. Annacchino, P.E.

AMSTERDAM • BOSTON • HEIDELBERG • LONDON
NEW YORK • OXFORD • PARIS • SAN DIEGO
SAN FRANCISCO • SINGAPORE • SYDNEY • TOKYO

Butterworth-Heinemann is an imprint of Elsevier

Butterworth–Heinemann is an imprint of Elsevier
30 Corporate Drive, Suite 400, Burlington, MA 01803, USA
Linacre House, Jordan Hill, Oxford OX2 8DP, UK

Recognizing the importance of preserving what has been written, Elsevier prints its books on acid-free paper whenever possible.

Library of Congress Cataloging-in-Publication Data
Application submitted

British Library Cataloguing-in-Publication Data
A catalogue record for this book is available from the British Library.

ISBN 13: 978-0-7506-7993-0
ISBN 10: 0-7506-7993-X

For information on all Butterworth–Heinemann publications
visit our Web site at www.books.elsevier.com

Printed in the United States of America
06 07 08 09 10 11   10 9 8 7 6 5 4 3 2 1

CONTENTS

8 *MANUFACTURING DEVELOPMENT, 333*

9 *THE PRELAUNCH CHECKLIST, 399*

10 *THE PRODUCT LAUNCH, 439*

11 *THE PURSUIT AND PRODUCT MANAGEMENT, 457*

To Terrie, with whom I share the future,
To Angel Ashley and Alecia who are our future
To my Mother and Father who taught me how to embrace the future.

And

Also to the men, women, and families of the
United States Armed Forces
who work every day to
secure our future.

PREFACE

WORLD COMPETITIVENESS AND LIFELONG LEARNING

The Objective of this Book

The objective of this book is to provide the reader with a thorough understanding of the business development process and how to execute a product development program to grow the business.

This book serves to outline the complexities in the planning, execution, timing, and problem solving skills required to manage a program. We are in a fast changing world where change brings adversity as well as opportunity. We all must practice continued learning to be world competitive, which allows us to participate in these new opportunities.

The Evolution of Employee

The learning process, as we all know, takes place every day during our life. More formalized learning occurring during school gives way to updates, seminars, and possibly more formal training. The first positions out of school allow us to gain experience. As we work longer we begin to amass more experience.

Initially, when we enter the workforce, we are on a quest for experience. This cements our learning. As we work longer, the many experiences both positive and negative shape our opinions and judgment. At some point in our career we reach a point where opinions are solidified.

There is a fine line between extensive experience and preconceived notions. If we are not careful, we cross the line into jaded thinking. A balance must be established.

The marketplace reestablishes the reference each day, forcing us to cope with the realities of energy level, absorption of information, and desire. Each day we can choose to reinforce and harden our own paradigms, or open our minds to new learning. The new learning allows us to remain on top of our game and be a formidable gladiator in the arena of business.

The Evolution of Firm, Corporation, Company, and Industry

The firm goes through a parallel process as it matures: Initially the young entrepreneurial organization meets the market head on with energy and intellect and drive. The corporation, as a unit, is intimate with the marketplace and is an active member of the business. Initial success allows the corporation to grow and expand.

With continued growth comes organizations, structure and corporate momentum. Left unmanaged, bureaucracy begins to set in. Lack of customer intimacy may begin to affect orders and ability to secure orders. Not affected by any one item, the lack of progress on the following perspectives contributes to a loss in world competitiveness.

- The march along the pathway to intellectual competitiveness
- Absorption of external information
- Ability to synthesize solutions
- Ability to analyze situations objectively and form perspectives outside the organization
- New talent mentoring
- Arenas and industry segment activity
- Functional displacement of products and processes
- Leadership's role
- Diffusion index of new information
- Assimilation index

To prevent, or even reverse the natural evolutionary process, the corporation, by virtue of its employees, must take personal responsibility for remaining world competitive by life-long learning. This book's intent is to provide a basis for that learning.

Marc A. Annacchino

THE BUSINESS OBJECTIVE

BACKGROUND

1. New Product Development and the Economy

New product development is an integral part of a healthy, growing economy. This chapter will start out with a review of the various means for economic development, including manufacturing and how product development plays a role. The various types of product development and how they draw on and pay out to the economy are also included. In addition, we will look at our world and what it would be like without some famous new product developments. Finally, there is a review of what happens to a company without new product development, from a purely financial perspective.

A. Economic development

New product development contributes to the economy by generating revenue and profits to a corporation that otherwise would not have been generated. The revenue then is paid out to vendors (other manufacturers). The vendors themselves pay out to their sub-vendors and personnel, or retain the earnings. Salaries are paid to personnel and they, in turn, spend funds to purchase goods and services from other profit-making enterprises. The retained earnings fund the long-term growth of the enterprise and increase the value of the business. The profits are taxed and that goes into the pool of funds to govern the community and provide for the common good.

This is the role that private investment plays in the economic development arena. The public sector contributes to the economic development by providing incentives to encourage manufacturers to establish their businesses in their locale. They provide the means for funding business expansion and growth. A collateral activity is to network with other manufacturers on your behalf for future business. Figure 1-1 illustrates the typical funds flow in a manufacturing enterprise. As will be seen later, the service sector of the economy has a funds flow that is different from that in manufacturing.

Service businesses are not characterized by the leverage associated with manufacturers. Instead, they are characterized by a smaller investment in capital equipment, a smaller investment in each revenue cycle, and a generally faster revenue cycle; as such, they generate incremental profits from incremental investment with low fixed costs. This is shown in Figure 1-2.

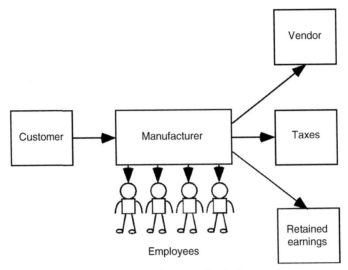

Figure 1-1. Manufacturer's funds flow model.

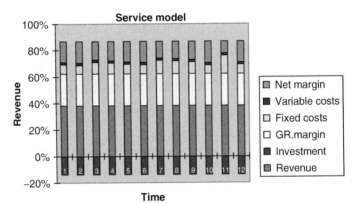

Figure 1-2. Service business funds flow model.

As shown, the service model has relatively constant net profit. This is because there is no major initial investment to absorb and there is a small incremental investment with each order. The other costs are somewhat variable and track with the incoming order rate. Incremental profits can track with incremental revenues. Service businesses generally have a lower barrier to entry than do their manufacturing counterparts.

Manufacturing businesses differ markedly from service businesses in that leverage can occur. This leverage is characterized by larger returns for their investment. A larger investment is required at the outset, but revenue is generated through a stream of returns from year to year through the products' life cycles. This assumes there is sufficient volume to offset the fixed expenses and the absorption of the initial investment. There is significant investment in capital equipment and processes, along with significant payoff if you have hit on the right opportunity with the right product.

The manufacturing financial model (Figure 1-3) shows that net margin increases at a disproportionate rate as revenue increases. This is because the fixed costs are already absorbed

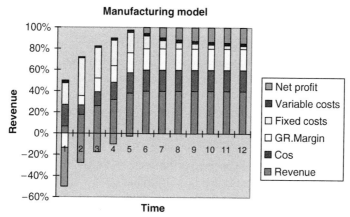

Figure 1-3. Manufacturing financial model.

by some base level of business and the incremental revenue does not require incremental investment with each order. Fixed costs decrease as a percent with increasing revenue.

However, there is a large initial investment to design the product, which can run the length of the diagram in time. Figure 1-4 illustrates how a manufacturing enterprise's funds flow as a result of new product development.

An idea is conceived and qualified in terms of opportunity and overall business sense. There are costs associated with these activities, such as market planning costs, surveys, customer visits, and demographics data analysis. There are also costs in taking the opportunity and the market data fed back and coalescing them into a product opportunity. The scope of the target market is accounted for in this stage, and the product platform must be laid out to reach the market at a cost-effective price. These costs are generally low compared to the other costs in the product development arena.

The next phase is the investment and development phase of the program. This phase takes the product concept and creates the intellectual property required to take a concept and reduce it to bills of material, manufacturing processes, and define a manufacturable product that the market will purchase. There are technical, labor, development, tooling, and other capital equipment costs in this phase.

The next phase in the model assumes that all the development is complete and manufacturing takes place. Here, there are set-up costs, material, labor, and overhead costs required to produce the units. The product moves out of manufacturing and into the sales channel for

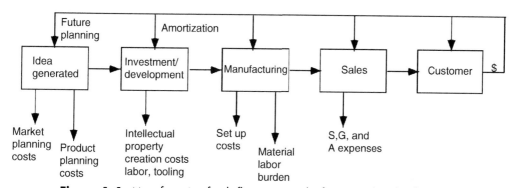

Figure 1-4. Manufacturing funds flow as a result of new product development.

placement at the customer. In this stage there are a host of sales, general, and administrative (SG&A) costs incurred.

Finally, the product is at the customer and funds for the product have changed hands. The manufacturer now can begin to enjoy the profits. These profits come after the funds are distributed appropriately. Dollars must go back to fund the sales expenses, general expenses, and any other expenses previously incurred. Manufacturing must be reimbursed for the materials labor, manufacturing expenses, and inventory carrying costs.

The development of the product must be amortized so as to be able to fund future developments. In addition, funds must also be channeled back to the product planning and market planning function to allow the proper expenditures to verify product viability, market program viability, and future enhancements and products.

B. Types of new product development and their contribution

There is a variety of different types of new product developments in existence today. Each is used for a different reason, and each has its own objectives and dynamics for execution. The following is a common list of the different types and their attributes and contributions.

1. "New to the world" products

These are somewhat revolutionary in the marketplace in that the marketplace never had exposure to the product directly—perhaps only as a concept or prediction from a futurist. They generally create entire new markets that never before existed. An example would be the cellular telephone. Predicted only vaguely by the Dick Tracy cartoon strip and personified as a "communicator" in *Star Trek*, the cellular telephone has revolutionized person-to-person communications in modern-day society. One such product looming on the 10-year horizon today is mass-market fuel cells.

These product development programs generate entire new markets that were not previously there. They enable true growth in the economy by generating revenue to the enterprise. They also have a multiplication effect in the economy by generating requirements for parts and subassemblies that need to be developed and supplied by the vendors. In many cases, they generate new channels of sales and new routes to market.

2. New product lines

These new categories of products allow entry into newer markets not previously participated in by the manufacturer. By adding the categories, manufacturers must be careful to protect the positioning of their existing products, which generate the existing business. Failure to do so will place them in danger of converting loyal customers away from one already successful product to a new one, with no net gain in market share. Perhaps a good example of this type of product would be the Hyundai Azera, a higher-end automobile offering from Hyundai. Here a large manufacturer with many product lines generated an entire new category of car to serve a more discerning customer base. Careful not to jeopardize their existing base, this new car company initially launched its product in the same value-focused dealerships selling the existing products. This is different from the Acura offering distancing itself from Honda.

The new product lines generate incremental revenue to the manufacturer by leveraging the market's familiarity with the manufacturer into new categories of products. In many cases, the market's familiarity with the manufacturer paves the way for new categories of products. Sometimes these products go into new markets, but can also be an alternative to existing ones.

3. Additions to existing product lines

These efforts support existing product lines by creating line completers to extend the influence of the original products' brand to larger audiences or extending range, power, and scope. All are done in the attempt to secure more of a market. An example of this type of product would be the M&M Candy Company extending their product line to M&M peanut and M&M almond and seasonal M&Ms for Christmas and Easter. Another example would be tomato sauce versions—hearty, traditional, roasted garlic, Alfredo, and vodka tomato. By taking the basic product and modifying it, a wider market share is realized.

The addition to existing lines has a similar effect on the company's revenue as the new product lines. They generate incremental revenue by leveraging the existing product familiarity rather than the company familiarity. These programs generate incremental improvement in the economy, but generally fall short of the contribution made by the totally new products.

4. Improvements and revisions of existing products

As time marches on, customers have higher expectations of your product and the competition adds features to their offering. It becomes necessary to improve your company's offering to increase market share or to retain it. By redesigning the product or repackaging it, your company can offer a greater value or satisfaction to the customer. It is possible to temporarily affect this by enhancing perceived value; however, an ever-more-informed customer base will respond to actual value increased in the long run. An example of this type of product development is the automotive companies adding features to their base models each year as standard.

Generally, the improvements to existing products do not generate additional revenue to speak of. They are simply a means to retain the market share or to slightly improve it. They are defensive in nature and in many cases are stopgap measures until a new product program can be introduced. These programs do little to generate a vitalized economy in the long run, but can provide time and revenue to pursue the development of a replacement.

5. Repositioning

Another means of increasing or maintaining market share is through repositioning. A repositioning is an exercise in changing the perception in the mind of the consumer. It generally can happen with products that are lower in value (dollar amount), or the consumer spends little time evaluating the actual data. For high-dollar decisions, the consumer will generally take the time to evaluate the facts and make his or her own decision. Repositioning is truly a marketing activity rather than a development activity. An example of this is a change of advertising by focusing the audience on a possible linkage drawn between certain brands of cereal and a high-fiber, lower-cancer-risk diet.

Repositioning is another stopgap measure for generating revenue from an existing product. It does not generate overall growth in the economy per se; rather, it is similar to an improvement or a revision except that it doesn't even necessarily require a product change. It simply repositions the product in the mind of the consumer.

6. Cost reductions

These programs are strictly a means for reducing the cost of products to offer similar value. They generally are the result of a competitive initiative, either generated internally or from external forces. In many cases, it is simply a means to generate more volume, which will generate less incremental profit (but perhaps more overall profit). Whatever the motive, a cost reduction is generally meant to increase unit volume through the channel. This becomes easier with capital equipment costs and development costs absorbed, and the manufacturer wants to capitalize on the sales channel.

Cost reductions are helpful to the organization by generating additional margin from the existing product. This margin can absorb development costs and manufacturing set-up costs. In many cases, they enable a period of time to continue with the product, generate the revenue, and allow the organization to position itself with a new product. They do not, however, generate any real growth in the economy.

Figure 1-5 summarizes the different types of product developments in terms of (a) time required to develop (b) the revenue to the economy; (c) the revenue to the company; (d) the company's positioning; and (d) the margin impact.

C. Narrative and financial review of a nondeveloper contribution

Figure 1-6 illustrates the dynamics of the income statement of a company that has little new product development. As shown in the financial analysis, a company can grind to a slow halt by not participating in the dynamics of the ever-changing marketplace.

Along the top are the years under study. Along the left side are the income statement categories. The following summary discusses the perspectives and how they affect the overall operation.

1. Cost of sales

The cost of sales has a natural tendency to increase over time. This is a result of the increases in direct labor costs in manufacturing the unit as well as the effect of vendor increases in pricing. For most manufacturers, the product maintenance function of development has to initiate cost reduction wherever possible, just to stay even. Therefore, to progress down the learning curve, significant initiatives must be made to effect cost reductions because these changes must offset the increases already built-in.

Type of development	Time to introduce	Potential revenue contribution to economy	Revenue contribution to company	Company positioning strategy	Potential margin impact
New to the world	Longest	Highest potential	Highest potential	Market development	Highest
New product lines	Long	High potential	High potential	Market development	High
Add to existing	Medium	Medium potential	Medium potential	Line complete	Medium
Improve or revise	Short	Little potential	Medium potential	Market share	Medium
Repositioning	Shortest	Little potential	Medium potential	Market share	Medium
Cost reductions	Shorter	Little potential	Medium potential	Raise margin	Medium

Figure 1-5. Types of product development programs.

	Base Year ($)	Year 1 ($)	Year 2 ($)	Year 3 ($)	Year 4 ($)	Year 5 ($)
Net revenue	100,000	101,000	102,000	103,000	104,000	105,000
Material	45,000	45,950	46,400	46,850	47,300	47,750
Labor	10,000	10,600	10,700	10,800	10,900	11,000
Burden	25,000	25,250	25,500	25,750	26,000	26,250
Total cost of sales	80,000	81,800	82,600	83,400	84,200	85,000
Gross profit	20,000	19,200	19,400	19,600	19,800	20,000
Engineering expenses	4,000	4,200	4,410	4,630	4,862	5,105
Administrative expenses	3,500	3,675	3,858	4,051	4,254	4,467
Sales expenses	3,000	3,150	3,307	3,472	3,646	3,828
Total expenses	10,500	11,025	11,576	12,155	12,762	13,401
Net profit	9,500	8,175	7,823	7,444	7,037	6,599
Provision for tax	4,275	3,678	3,520	3,350	3,166	2,969
Net profit after tax	5,225	4,496	4,303	4,094	3,870	3.629
Percent return on sales	5.23	4.45	4.22	3.98	3.72	3.46

A

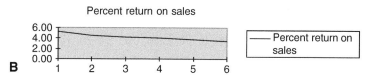

Percent return on sales

B

Figure 1-6. (A) Income statement, nondeveloper; (B) Percent return on sales

2. Profits

Assuming the expenses stay at a constant amount or grow slightly, this increase in materials and labor and manufacturing expenses (burden) creates enormous offsets in the net profit of the company. Salaries and benefit expenses increase with time because the manufacturer must remain competitive with the other, more dynamic companies in the industry. Sales expenses will increase as more concessions, incentives, or travel must be done to achieve the same result from year to year.

3. Strengths

The financial strength of a company degrades with a lack of successful new product development. Unfortunately, the effect is not known immediately. The financial picture initially looks better because the expenditures normally spent on a development are saved. This is, however, a short-term gain, because the market will soon bypass the manufacturer with new products from other firms, leaving the manufacturer with little to sell in the evolved marketplace. In later analysis, the strength of the enterprise diminishes.

4. Market share

If the manufacturer cannot keep up with the marketplace's appetite, eventually the package of values it offers does not meet the customers' needs and it begins to lose market share. By losing the revenue from lost market share, a dangerous downward spiral begins that is difficult to reverse.

5. Response to changes in environment

If the manufacturer cannot keep up the pace established by its industry by evolving product lines on a normal evolutionary path, it is at great risk from the negative effects of changes in the playing field, or legislation changes, or salient attacks from nontraditional competitors. This weakness will manifest itself as a financial problem over time, which will eventually place limitations on the organization's ability to respond in order to survive.

D. A look at our world without some new product developments

1. A world without Apollo

Not all new product development comes from private investment. The U.S. government regularly conducts research in a variety of areas in order to generate understanding, research, and analysis of data. One of the more scrutinized and high-profile development programs in the 1960s was the project to land a man on the moon and safely return him to Earth.

Aside from the sheer magnitude of the program, the raw objective was to develop methods, produce machinery and hardware, and execute a mission to land and return. The task represented enormous technological hurdles in numerous disciplines. The achievement of this objective in July 1969, must be historically underscored by the numerous spin-off technologies that have changed our world.

The Apollo program accelerated the development of technology in an unprecedented manner. The need drove the course of action: If it existed, use it; if it didn't exist, invent it. Time was literally of the essence. Money was no object, and resources were granted on the basis of how much time it would save, not how much it cost.

There were several technologies that experienced accelerated development during that era. The term *accelerated* is the operative one in this case. Most technology eventually gets developed as people become interested in its potential and entrepreneurship drives these elements into society. However, the Apollo program literally rocketed development of certain technologies that would have taken many more years to implement without the driving requirement that necessitated its invention.

Rather than listing and discussing each one, it may be more effective and interesting to look at our technological and societal world without the Apollo program. Let's examine the latter half of the 1990s as we look at "a world without Apollo."

You wake up at 6:00 A.M. The mechanical radio alarm clock clicks on with the local news. The clock is mechanical because a digital alarm clock is too expensive to build out of discrete transistors. The integrated circuit showed a lot of promise in the 1960s as an idea but was never funded, and so its widespread use is just beginning. However, that is not on your mind.

What is on your mind is the temperature outside and the chill you feel inside. There's a 20°F cold snap outside. Back in the 1970s, oil prices skyrocketed, and so Americans living in North America needed to hold down expensive oil and natural gas usage. Your home is insulated with 1960s technology and cannot hold in the heat very well. The less-than-optimal insulation used is due to lack of funding to develop the new technology in the 1960s. The Apollo program hardware utilized specialized insulation and polymerized film materials, and solved its heat retention and heat rejection problems by the use of advanced insulation systems.

You are a jet aircraft pilot and have to fly to two cities and return home early this evening. You are planning to go to the baseball game tonight to see your favorite team play, so you hope it warms up later in the day, upon your return.

Today, your children are going to perform in the school play and your wife is planning to go food shopping and investigate a new hairstyle at the beauty shop.

You are pressed for time because you forgot to buy movie film to record the kids' school play. You can only film three minutes of the play and have no means for sound recording to coordinate with the movie film. The video camera has not been invented yet. Because there was no need to develop the technology for portable television cameras used on the Apollo missions, there was no spin-off to personal use.

When your wife gets to the food store and is ready to check out, there is a long line at the checkout. Because all groceries are checked out by hand, it takes longer to process the customers. Bar code scanners haven't been invented yet, nor is the raw technology in a cost-effective package for use in the commercial sector because integrated circuitry has not progressed far enough down the learning curve, so she waits.

Now, it's off to the beauty shop to experiment with that new hairstyle. Your wife will be taking a significant risk in experimenting because she will have her hair cut very short for the new hairstyle. In addition, the new style requires a permanent-wave treatment. Both items are a one-way commitment to the new style for two to three months. The image processing technology developed for the Apollo program has not yet been invented, since there was no need. Consequently, no spin-off technology allowed the beauty shops a means for taking your wife's photo and digitizing it to try out different hairstyles before cutting or waving.

Your wife now starts to plan the budget for the next month. Unfortunately, she is calculating and adding bills manually and makes a mistake, allowing more expenses than can be covered by the checking account's current balance. The handheld calculator has not been invented yet. There has been no market need for integrated circuits or cost reductions, so there are no calculators yet. Eventually you will have one as the world progresses into the acceptance and the implementation of these technologies at the normal, nonaccelerated pace.

Meanwhile, you are at the airport and ready to start your day. You climb into the flight deck and see the traditional stick and rudder used to fly the airplane. There are very few avionics and no inertial navigation system. Everything in terms of flying the plane is done by manual calculation. There are no major systems, such as navigation or communications, in place. The Carousel system for inertial navigation and fly-by-wire technology have not been invented or implemented in the fleet of planes yet. There was no requirement for inertial navigation for long distances like in the Apollo program, so it never made its way into the commercial marketplace. This system, combining computer technology, electromechanical physics, precision machining, inertial navigation algorithms, and a worldwide infrastructure, will be available through natural, incremental evolution around the time you plan to retire.

You land in the first city of your journey and listen to the television in the airport terminal. There is a news story about an entire family who perished in a home fire. How sad and tragic, you think, visualizing your own family. This family, like many others, died because the fire consumed them while they were asleep. There was no means for early fire detection or warning; the smoke detector was never invented for home use. The Skylab program in the early 1970s used mostly Apollo hardware, and the National Aeronautics and Space

Administration (NASA) needed a means for early detection of smoke and fire for America's first space station.

Your attention shifts back to the weather, and you are thinking about that game tonight and want to know whether it will be too cold to go. You do not have any way of planning for the weather because modern forecasting methods are limited to predictions only—satellite photos using image enhancement have not yet been invented. There was no requirement for it on Apollo, so no spin-off.

Your children go to the school library to do research for a term paper. They comb through the card catalog and cannot find any article or reference material for their papers. The library's focus has not yet shifted from collection of books to access of information.

Your kids will have to wait for the reference material to be sent through the mail because there is no Internet with universal access yet, no low-cost computers affordable enough for the municipal budget, and no network of other users.

You finish your flights, arrive back home, and plan to go to the game. Your wife asks you to hang a new picture she bought today. You go down in the basement to get your electric drill to drill the hole required for the anchor. Unfortunately, you loaned the extension cord to a neighbor. You have no means for electrical power in the area where you will place the picture—cordless tools have not been invented yet. They will be an eventual spin-off from the drill used to core lunar samples on the Apollo missions. You will have to wait until tomorrow for your neighbor to return the extension cord.

You are off to the game at last. It never did warm up, nor did you ever get any advanced warning. It would be nice on a night unseasonably cold as this to have a cover over the stadium to retain the heat. Traditional roofs over stadiums are expensive and require huge means for support. Lighter, more affordable fabric used in domed roofs for stadiums and airports hasn't been developed yet. (They use spin-off fabric technology from the Apollo Moon suit developed for the astronauts to traverse the lunar surface.) It's so cold in the stadium that you get some hot coffee rather than a cold beer!

This short vignette represents only a few technologies developed for the space program that are in the everyday products we now take for granted. It has been often said that the space program was wasteful in terms of spending the funds on space travel rather than "curing" the problems on Earth. By conservative estimates, there are more than 30,000 products developed and launched as a result of the space program.

The reality of these results are that the tax on profits alone from the revenues generated from these products and technologies have paid for the government's investment in the entire Apollo program many times over each year.

Here are a few examples where continued product development has benefited the customers with better, safer products:

Automobiles: The Model T progressed to the Lincoln Town Car.
Aircraft: The Ford Tri Motor progressed to the Boeing 777.
Computers: The Electronic Numerical Integrator and Computer (ENIAC) progressed to the Cray supercomputers in technological integration and their pervasive use in numerous mass market products.
Vacuum Tubes: The vacuum tube amplifier progressed through discrete transistors to integrated circuit amplifiers on a chip.

Incandescent Lights: The original incandescent light spawned fluorescent tubes and bulbs with numerous multiples of extended life. Fluorescent technology applied to bulbs using embedded electronics enable fluorescent light in a standard bulb socket.

The importance of continuing the product life cycle preserves the customer base. It drives competition between manufacturers in the marketplace. Competition is desirable and necessary. It prompts everyone to action to improve the products that they offer, so the customer and the manufacturer both benefit.

1. Differentiating Research and Development

A. Research expands core technology, development implements core technology

There is a unique and interesting relationship between research and development (R&D) in an organization. The two disciplines are diverse and separate, although they are often thrown together in conversation and lumped as one entity. In actuality, the two must function like a relay race, with research establishing the lead position and handing off the intellectual property and know-how to the development people to apply and create new products.

Research can be thought of as a strategic element of the organization, whereas development can be thought of as more operational in nature. Research is science-oriented. It translates phenomena into deterministic events, removes uncertainty, and is a resource tool for the development group. Often there is a disconnect between the week-to-week efforts of researchers and their respective development groups.

The operational issues with research include loose definitions of residual uncertainty; loose definitions of the delivered development tools; a loose definition of *deterministic*; and often a disconnect with the end-user of the technology, resulting in so-called projected customer acceptance.

Many problems in new product development occur when researchers have not completed their effort in totality. Time lines are not met, costs increase, and they forget that they are an integral part of the revenue-generating organization that must produce results on a specification and a time line to hit a window of opportunity.

Development, on the other hand, must focus on the creation of a product that can be manufactured and sold at a profit and achieve actual customer acceptance. Development engineers are charged with creating an accurate bill of materials, enforcing change management, creating documentation for manufacturing and product support, and testing. They also must identify and cement the product usage boundaries to ensure safety. Finally, they are charged with the human-factor engineering of a new product. Development has operational issues in establishing and maintaining linkages with research, ever more pressure on time lines, full absorption of uncertainty, funding constraints, compromises, and decision management between safety and cost. In total, they must translate customer needs into certified hardware.

For fast-moving technology companies, even research needs to be considered as operational, meaning it needs to have usable results, to be on time, to be well-defined, and to be easily transferable to development for industrialization and commercialization.

There are three basic types of R&D used in one form or another in industry today: incremental, radical, and fundamental.

Incremental research is best characterized by a small amount of research and some amount of development. This development is manifested by small, incremental improvements in the product, manufacturing processes, and cost reductions. Not much research is required; however, continuous development is. These efforts generally have large, aggregated payoffs over time but can be unsung heroes of the corporation's profitability index.

The *radical type* is best characterized by a large amount of both research and development. This is the type of development that requires new knowledge to be discovered and understood, as well as development of the means to embody the know-how into a product for a specific purpose. Huge expenses in development follow huge expenses in research. The organization launching into this type of development requires immense amounts of capital to see the programs through to profitability.

The *fundamental type* of program is best characterized by a large amount of research and little or no development. Many times these programs degenerate into gathering knowledge for the sake of knowledge rather than gearing the research toward a specific goal. They may be strictly strategic in nature and won't have a payoff for many years. The senior management must have a great amount of faith and long-range vision because the rewards of the decision and subsequent sacrifice will accrue to the next generation of management.

B. Research expands core technology

Research and development have three major strategic goals in an organization. Their purpose is to contribute to the existing businesses by allowing the organization to defend market and product positions, support sales growth, and expand existing business.

Research fundamentally generates intellectual property for the organization. If the activities of the research element are directed properly, the intellectual property will have tangible value on a balance sheet. This asset will be eventually used to develop new products.

It is therefore incumbent on the senior management to carefully select the programs to fund. They need to select programs that are tied to operational objectives, have a time table and an end, and result in development generating revenue for the organization through product sales. Management needs to think through the vision of the future in today's time frame and assign research work that will generate timely results for the next product developments.

C. Development contributes and makes money through product sales

Development, on the other hand, takes the research, intellectual property, and know-how and generates products that can be sold at a profit sufficient to offset the other expenses of the organization. Figure 1-7 is a simplified illustration of the role of development.

It can best be characterized as a machine that has marketing input, operational input, and core technology available, and generates salable products that can be manufactured at a profit.

D. Development is much more measurable

The actual progress of development can be easily measured, since the development team is charged with the responsibility of committing to a schedule to produce results that are quantifiable. With the fundamental technology understood and made deterministic by the research function, development must concern itself with items such as bills of materials, performance, functionality, manufacturability, and cost. Milestones can be established and tracked with action items, responsibilities, and completion dates. Work content can be managed by manpower hours and loading in general, and results of expenses and time invested can be evaluated.

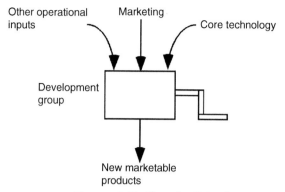

Figure 1-7. The role of development.

E. The need for research to be tied to operational goals

Because research generates the knowledge and is an integral part of development pro-
ductizing the technology, it is very important for the research expenditure to be carefully
made. Costs of progress from research to development increase dramatically, so diligence is
required at the onset. Simply stated, ensure that research function is tied to operational
goals. This means that the area of study must directly source the needs of development, and
it, too, must occur on a time line. If not, development ends up cultivating the core technology
and the project becomes irretrievably delayed.

F. The model for research and development

The following is a model for the research function and how it fits into the organization.
Although organizationally the reporting relationship may vary, Figure 1-8 represents, in
a general way, the flow of information and knowledge. It shows that information comes
into research from several sources, including cooperative agreements with other companies' shared development in a consortium, from university sources, or from other sources

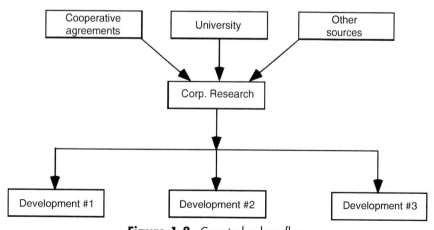

Figure 1-8. Core technology flow.

as required. The corporate research department then generates the core technology to lay into the development programs so products can be developed to hit a market opportunity.

On a strategic basis, Figure 1-9 illustrates how profits are derived from the efforts of research. The goals and objectives identified for the organization's future drive a strategy to achieve them. Research contributes to the development process, and profits are realized through operations. It is a critical link in the chain, and results are required in sequence and on time to be effective.

On a project basis, the research function must contribute knowledge at various points in the planning and development process. Figure 1-10 highlights this contribution of core technology. The project starts with a concept. The marketing analysis defines the customers' needs. Research then contributes the core technology—the necessary knowledge to determine project feasibility. It must answer the question: Can we as a company absorb and cultivate core technology and produce it cost-effectively to create a profit? At the point that the affirmative conclusion is reached, the core development of the technology begins. Research then transfers the technology to development. An integral part of the development is the industrialization of the product, and then commercialization. As shown in Figure 1-10, research contributes at two critical junctures in the process: at the feasibility stage to determine if the technology can be brought to product status, and at core development when research transfers the technology to development.

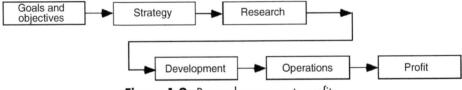

Figure 1-9. Research sequence to profits.

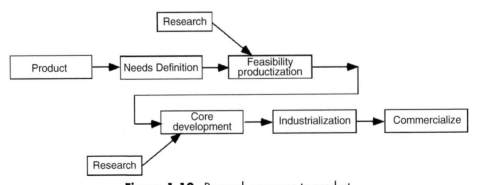

Figure 1-10. Research sequence to product.

PERSPECTIVES INTERNAL TO THE ORGANIZATION

1. The Product Fit

A. Selecting the new product opportunity

Opportunities abound in this world of ours! The challenge is to select and focus on the ones that will yield the desired corporate result. However, what is the desired result for the company? Is it revenue generation, cash flow, profit, or is it tangible, long-term growth for the company, strategic positioning, and satisfying operational requirements? Is it achievable, desirable, and agreed-to? All of these questions must be addressed to determine if the opportunity in question is the right one for your company at this time.

Not all opportunities are truly opportunities for a company. Although some of them represent good prospects from a pure market perspective, they may not fit with the company that is evaluating the opportunity. It is very important to factor into the decision all of the company attributes that will contribute to or detract from success of the new product. This exercise can be referred to as the *fitability* of the new product into the company.

B. Understanding the company's capabilities

A new product opportunity must easily fit into the operational culture and temperament of the firm. A company evolves with several interlocked disciplines that must work together to effectively assimilate, integrate, and execute a new product program. This must happen from a sales and marketing perspective, an engineering perspective, and manufacturing, service, and quality control perspectives.

The new product opportunity must fit with most of the disciplines to be pursued effectively. If there is an area that is not initially compatible or leverageable with the opportunity, then steps can be taken to correct or bolster the weakness. If, however, there are too many areas that are inconsistent with the new product, then the chances of developing the product within the market window of opportunity are reduced, especially while effecting changes in the organization and supporting the existing business. The new product idea must be tested not only in the marketplace, but also within the firm to determine its fitability with the sales organization, the marketing organization, the current and anticipated capabilities in development engineering, and the integratability of the product into the manufacturing organization.

C. How does this opportunity fit into the goals and agendas of the company?

Every company has goals and objectives that generate its agenda. Even if the company doesn't have any stated goals or objectives, there is still an agenda that factors into the new product analysis. It may be as simple as: The objective is to maintain the status quo. In that case, the agenda is to do nothing in the way of product development. All new opportunities are evaluated and summarily rejected for one reason or another. Others are evaluated against agendas, such as target growth rates of 15% to 20% per year. Still others are evaluated against requirements demanded by customers, other parts of the organization, or that are purely financially motivated.

There are also the ownership issues that affect the company's ability to conduct a healthy product development program. A publicly held company with significant resources will evaluate an opportunity differently than will a privately or closely held company. A subsidiary operating from a remote location will act differently than will a division that is part of a larger group within the same venue. A set of poorly prepared objectives for a general manager of a relatively autonomous organization will dictate new product development activity differently than will ones given by an on-site senior manager.

The current financial status of the organization will also determine the risk and size of the development program. In the case of a subsidiary or a division, the status of the parent company must also be examined.

All of these issues must be factored when evaluating the opportunity. The operative point, simply stated, is that a new product opportunity must be evaluated not only from the external factors affecting the success, but also from the equally important internal factors that must contribute to its success.

D. Who makes the sacrifice?

As the saying goes, "No pain, no gain"; so, too, with new product development. Who in the organization will make the sacrifices? Is it the sales team that will make the extra effort to introduce the product? Is it the marketing team that will make the extra effort to prepare an adequate specification that will stand the test of development time? Will it be the development team that will shed the internal conservative agenda and break new ground in terms of costs, functionality, and manufacturability? Or will it be the service team that will drive the product to success even through initial quality and reliability problems? Each and every part of the organization needs to contribute the extra effort to ensure success of the new product, in spite of the pitfalls it encounters.

If the fit is right, the team has a chance at success. The fit alone will not guarantee success; success also requires the *commitment* of the team members and the company as a whole.

2. The Product Strategy

A. What is the company's strategy in pursuing a market?

To effectively evaluate the fit of a new product, it is necessary to understand the company's overall product strategy. What has been its track record for embracing new technologies and product concepts? What were the expectations of each, and what were the results? Often, many companies do not have a clear objective and thus cannot generate a clear strategy.

The previous section discusses leveraging the organization's segments as part of the fit of the product. The product strategy should attempt to leverage the available product opportunities to create a steady stream of products to create a new, sustainable business unit for the company. This forces examination of the interrelationship of the product as well as individual contributions.

Regardless of how lucrative an isolated product opportunity looks, few products can succeed as a one-time hit without subsequent market and product development. It simply puts the organization in a market position that is not defensible. In addition, a single success in an uncontested market will not remain so for very long, necessitating the need for continued development.

Again, the strategy is dependent on several factors, including the structure of the company. A product or service that is in its infancy and has "a lot of runway" may be a good choice for a company that will eventually be taken public. It has the longevity and growth potential required to attract investors in the initial public offering. A product or developing technology that may be required to complete someone else's product line may be chosen to groom the company as an acquisition target. A privately held company will tend toward market niches that are narrow enough to be defensible and wide enough to be worthwhile. Consequently, a well-thought-out strategy executable toward an objective is invaluable to the organization.

B. Why does the business development team think this opportunity will take off?

New product development is a business venture. It must be evaluated like a banker or venture capitalist evaluates any business venture. Some developments may have very interesting technology and be very fun to work on. However, each must result in tangible value to the organization funding the development activity.

As will be discussed later, the development team consists of several interdisciplinary elements of the organization. Therefore, the development team must make the case for the development program. Why will it be successful? What is unique about the product or technology that will be desired and accepted by the marketplace at this particular juncture in time and space? These are the hard questions that must be answered satisfactorily very early in the process.

The strategy, the timing, the functionality, and the company's plans must be instilled into the development team and must be eminently clear to all involved. The team must initiate, champion, and carry the torch of the new product business plan.

C. Where is the connection between the market and company?

At the risk of sounding like P.T. Barnum, I would stress that every product should have a distinct marketing advantage that it brings to the marketplace. Very few products are accepted that are "me, too" in nature, so there must be a clear marketing advantage at introduction. This is what starts the momentum that will carry future sales.

The marketing advantage is generally, but not limited to, the integration of features within the product that differentiates it in the marketplace. Longer, lower, wider, less expensive, more functional, easier to use, and more available are all product design tactics designed to make the product more marketable.

If the so-called hook exists, it is because the team placed it there, and it is there to generate sales. If the product is a "new to the world" product, its novelty alone goes a long way. However, it is then a sales and training job to gain market acceptance.

D. Think the sales scenario through with contingency planning

Each new product must have a vision of how it will play out in the marketplace. The scenarios for success and for failure must be thought through, and a recovery procedure for each worked out. This contingency planning is similar to conducting battles in a war. Competitive battles resemble the tactics and strategies of warfare. For example, if you introduce a new product and it has existing competition, you are the invader and the competitor will attempt to defend its territory. If it has uncontested reign of the market, and it is a worthwhile market to go after, the competition will enter the race with an offering of its own. Then the competition is the intruder. Each attack has its own set of measures and countermeasures as well as scenarios that must be developed by each participant to survive. It is best to attempt to think through the interchanges beforehand to establish contingencies.

E. The benefits of strategic planning

The term *strategic planning* has been a very overused term and, in many cases, is under-implemented as long-term activity. It needs to be specific enough so that its results define the product development and market development direction.

Strategic planning has been traditionally used to prescribe desired market position, dollar amounts of business segments, growth targets, means for justification of acquisitions, and more. It seems to be employed to reset a company's focus and direction initially; however, it soon seems to lose focus in the third, fourth, and fifth years.

Unfortunately, strategic planning has the most value as a learning tool for the organization by setting direction at the beginning of a period—executing a strategy or program and evaluating its success within the strategic framework initially laid out. Most important, the success or failure of the first year must be factored in when setting second-year and subsequent goals.

Consider a strategic plan that calls for accelerated growth by internal development of a steady stream of new products. Also consider that all-important review after that first year—showing the initial target and missing that target by some amount, for example. It is an imperative requirement for the accuracy and legitimacy of the plan that this be factored into the subsequent year's planning. If your company exceeded the first-year objective, what have you learned? What worked and what didn't work? Draw the correct conclusions from successes as well as failures.

If the ground rules of strategic planning are well understood and the company has the discipline to effectively execute them, then the planning process offers several benefits to the organization. As a review, here are the benefits of the strategic planning process:

- It improves profitability by focusing the organization on two levels of profit: gross margin and net profit. It improves profitability regardless of the measurement criteria: return on investment, return on net assets employed, or returns on total capital employed.
- It yields a higher growth rate for the enterprise with it than without it. Without a plan for growth, it becomes difficult to determine the best opportunities to pursue. This compromises the overall growth rate. The organization may happen to stumble onto a great opportunity; however, consistent increase in business and market share is not likely without integrated planning.
- It reduces wasted time, effort, materials, and resources of the organization due to lack of focus. Following on the previous tenet, chasing loose opportunities can consume time and effort.
- It provides focus on the marketplace and examines the business from the customer's perspective. Every business is about customers. It is they who make the business possible; it is they who comprise the marketplace. Strategic planning, although an internal goal-setting exercise for the organization, forces the management to focus on the customers, their problems, and the opportunities to resolve their problems.

- It articulates and details future plans to the workforce to allow early buy-in to a program. It also serves as an early check for opposition to the future directions of the company. This point is not to be underestimated. Strategic planning is an excellent means to trial balloon issues and initiatives to test the waters of the organization. People make up the organization, and people can facilitate or destroy initiatives.
- It fosters greater personnel commitment to the agreed-upon goals. As an essential part of the management commitment process, a personal commitment must be made and followed-through. The strategic planning process allows the participants to contribute and commit to the goals established in the process.
- It provides for a cohesive management team. One of the most divisive elements in an organization is the fractionalization of the management team. Strategic planning allows for galvanization of this team. If it is not evident that the team can be galvanized in this fashion, it allows for change of the players, if necessary.
- Strategic planning provides the framework to understand the competition in terms of competitive advantage, workforce, value added, position in the marketplace, and ability to react. This is one of the most lucrative elements of the entire process because it requires the organization to evaluate the competition in a dispassionate, objective way and to determine how the organization intends to compete and prevail in the markets it chooses.
- It provides high visibility on quality from a holistic perspective: quality of the product, the process, and the entire revenue cycle. The strategic planning process forces quality into the organization. By its very nature of long-term goal and objective setting, the process dictates that the moves must be well considered and execution cannot be sloppy. The philosophy permeates all of the programs generated from the process and drives a higher level of quality onto the organization.
- It yields greater customer satisfaction. Higher quality, an organization that knows where it is going and how to get there, and a realistic customer focus allow a greater level of customer satisfaction.
- It provides a framework for continuous improvement. Because the process is long-term and annually reviewed, the organization can continuously improve through introspection, modification, and feedback.
- It directs a day-to-day vision that is real and achievable. Although strategic planning is long-term, it affects the organization on a day-to-day basis by reinforcing goals and schedules. It keeps the organization focused and on track, rather than reactionary to every diverse opportunity that comes along.

F. The framework for planning

The planning process has a general framework that must be followed to generate the required results. A well-executed process has three distinct parts.

The first is *introspective* in nature. It encourages examination of the company itself to determine strengths, weaknesses, and historical perspectives.

The second part is *creative* in that the company must determine its future and its desired position in the marketplace.

The third part is *pure planning*. It defines the means to the end described in the second part. The following is a framework for the strategic planning process that can be used as a general format for initial discussion.

1. Introspection: Where we are

Initiate a search of records to create a five-year history of the company's performance. This would include financial information and a narrative about the company. (Figure 1-11 presents a suggested format.) Analyze and draw initial conclusions from the obtained data. Review trends in sales engineering and manufacturing expenses, in addition to the product-related performance.

Use a narrative to describe the general health of the business. Interview a cross section of the employees to find where "the bones are buried" as far as product line performance, results of management attitudes, predisposition to new initiatives, and acceptance criteria. Do not focus on individuals; rather, focus on the results of the organization.

Next, create a product line analysis (as detailed as possible) to describe the performance of individual products. Examine the costs, features, benefits, trends, and driving forces for each product. It is also important to include the SG&A expenses for each because costs may not be allocated on an activity-based cost basis.

Product maintenance is what absorbs time and slows down new product development, so it is important to examine, from a historical perspective, how the company handled earlier products and supported them. In some cases a product appears profitable from a margin analysis, but in actuality may break even and absorb a disproportionate amount of the organization's resources.

Figure 1-12 is a framework for this analysis. Be sure to allow for narrative for some of the intangible elements of the analysis.

Here is a review of the information required for this analysis. The product should be replaced with the description of the product. Consider individual products that are part of families as separate. Next, enter the price that your company is getting in the marketplace for the product. Cite important features and benefits that the customer has given as feedback.

Category	Year 1	Year 2	Year 3	Year 4	Year 5
Sales					
Gross margin					
Net profit					
Expenses					
Warranty					
Inventory					
Return on investment					
Return on Capital					
Return on net assets employed					
Accounts receivable					
Cash reserves					

Figure 1-11. Analysis of business.

Category	Product 1	Product 2	Product 3
Price level in marketplace			
Feature/benefit			
Customer base			
Technology base			
Cost			
Market driving forces			
Percent of total sales			
Contribution to profit			
Life cycle stage			

Figure 1-12. Product line analysis.

Do not use sales and promotional literature for this part; use actual feedback from customers who have used the product.

In the next row, describe the customer base for the individual product segments. Next, discuss the technology employed in your company and describe any differences from the competition. Also include manufacturing costs, fully burdened. Look at the current and near-term market driving forces that will affect the acceptance of your technology, product feature/benefit set, competition, and demographic preferences.

Use accounting data to generate the row of sales data contribution to profit data for each one. Finally, enter the best assessment at the life cycle stage for each product. This may be numeric or a narrative. It will be used to gauge product development efforts throughout the overall framework.

2. Future: Where are we going?

This section of the strategic planning process identifies where the organization wants to go strategically in the future. This section is one of the most important of the three sections because it is the goal that must be set for the organization. It defines the call to action, the point in the future that the organization wants to attain. The following are the basic elements of this section and represent the basic requirements for defining the organization's goals.

Theme or mission statement development: This is an often-overused term, and is frequently invoked by senior management as the only means of reinforcement of long-term plans. Simply stated, the theme statement is the short, concise description of the business of the company and its objectives. It describes the company's identity and role in the marketplace.
Narrative of the vision: Create a detailed narrative of the vision of what you want your company to be in the future. It should discuss the time frame, expected results, financial position, market position, product position, and competitive comparison. It should also outline the ownership structure at that time, and whether any changes are expected.

Narrative of how the plan and reality will fit together: Create another narrative that weaves together all of the requirements for the future dream and the present and future operational issues. This is done for the sole purpose of outlining the plausibility of the plan and goals, and to ensure that they are consistent and compatible.

Five-year product line catalog: This is an interesting and enjoyable exercise because it allows management to sculpt the future offerings of the organization by outlining the five-year product and services catalog. You most likely will be required to go through a detailed product and services evolution flow chart to generate the complete catalog. That process will be outlined later. However, at this time it is sufficient to list the products as they would appear in the catalog.

Product scope definition: The product scope definition is a means to describe the product portfolio in a detailed manner. It should show how the products fit together and leverage off of each another, and where the boundaries are. Also, take time to establish low, medium, and high unit volume scenarios for each of the products. This will become helpful for the identification of the core requirements of technology that the company must have.

Specific market segments: Next, detail the specific market segments for each product. Do some market research to determine the trends in the marketplace for these areas and plan scenarios as to how your effort will fare competitively. Management should, as part of this exercise, question why this strategy will work and why this product segment will succeed in the dynamic marketplace.

Industry trends: Finally, the team needs to look at overall industry trends to determine whether they will be active through the planning cycle. If basic natural resource industries, such as the energy industry, are not active from a capital spending point of view for the near term, be aware that you may position the organization for the end of the planning cycle; however, it will be dry run through the planning cycle. Strategically, it's a good plan; however, from a cash generation point of view, it may not be fiscally allowable.

3. How will we get there?

This section is the pure planning section. It specifically outlines how the organization will move from point A to point B in the planning time frame. Whereas the previous narratives describe what the organization will look like, this section outlines what specific steps will be taken. Each area of the organization will be examined and affected in the transition. There will be marketing issues, development issues, manufacturing issues, service issues, and financial issues.

The following outline lists the elements of this section of the plan that must be addressed.

Segment strategy: This portion of the plan views the market strategy of each segment. It should start with a current market assessment of the segment. The purpose is to find out what is going on in the marketplace and to view the trends, both near- and long-term. It is also imperative to cite critical success factors. These two elements then drive the strategy—how the company will position itself and what role the company will play in the marketplace. The plan also needs to define how the company will play this role. By what means will it prevail in the competitive environment: innovation, price availability, or training? These are the questions that must be answered to complete the strategy.

Implementation: Since not all growth is internally generated (especially in a five-year scope), some of the businesses and product lines may be acquired or brand-labeled, or be a

IMPLEMENTATION MATRIX		Year 0	Year 1	Year 2	Year 3	Year 4	Year 5
Product 1							
	Internal development	X					
	Joint venture						
	Brand label						
Product 2							
	Internal development						
	Joint venture		X				X
	Brand label						
Product 3							
	Internal development			X	X		
	Joint venture						
	Brand label						
Product 4							
	Internal development					X	X
	Joint venture						
	Brand label		X			X	

Figure 1-13. Implementation Matrix.

result of a joint venture development. In this part of the plan, each segment is identified and a strategy is developed. In addition, an implementation plan accompanies each one to determine whether it is an internal development or some external mechanism. The chart in Figure 1-13 illustrates the plan.

Investment/Return: In order to add financial validity to the plan, it becomes necessary to identify the scope of funds required to carry out the plan. Each investment or market segment that has been identified, each strategy that has been established, and each company organizational requirement that has been outlined must now have a financial picture attached to it. An investment/return money line should be established for each. This is illustrated in Figure 1-14.

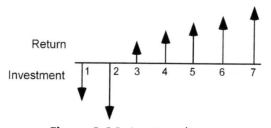

Figure 1-14. Investment/Return

Be sure to include all investments and both the development expense and the capital equipment expense. Although this is not completely pure from an income statement/balance sheet point of view, you are looking for the scope of funds to be expended. This is most often executed with cash, and therein lies the requirement.

The return should be calculated as follows: Take each individual segment and determine the number of projected unit sales. Then generate a projected gross profit based on revenue and costs, and post the product of the gross profit and unit sales to the return stream of payments. It is best not to post revenue only as a percent of market share obtained or as total dollars (independent of cost and units) because this often lacks substantiation. Also, the intent here is to evaluate the opportunity of the segment. As such, you want to compare real dollars invested against real dollars realized.

Next, calculate a growth rate for each segment and establish a trend measurement for each. In this way the opportunity can be evaluated against a desired financial profile and also against each segment.

Finally, aggregate all of the investments and all of the returns and post these results to a money line.

In this way, the financial pro forma accompanies the narrative of the plan, and senior management will get a financial *feel* for the plan and the scope of the expenses.

Requirements: There will be two sets of requirements for the organization and the plan. One will be internally focused and the other will be externally focused. Both sets of requirements must be addressed and cannot be glossed over.

Internal: There may be infrastructure changes as a result of the plan. These should be outlined and specifically listed (with timing) if they are critical to the success of the plan.

External: There may also be external requirements (e.g., in the sales organization, marketing channel, or vendor links). These need the same treatment as the internal changes to the organization.

Recommendation: The recommendation is one of the most important parts of the strategic plan. All of the data observations and opportunities must be coalesced into a firm recommendation for the company to follow. It also determines the acceptability of the plan from a management and commitment point of view. There are several elements in the recommendation section that should be included for the plan to be complete and to increase chances for acceptance. They are as follows:

Trial balloon the plan with internal people from different elements within the organization. This can be done in stages, in parts, or both. By doing this, you can get initial feedback from the organization and a measure of the acceptability to certain groups.

Next, *trial balloon the plan with the sales channel to get their perspective*. This will afford you the opportunity to see who in the channel will support the plan and who is not on board with it. This information will be helpful at a later time when evaluating constituent players as the plan is executed.

Finally, *visit a cross section of the customer base and present it to them*. This feedback is the most helpful because it is the customers who will eventually support the plan by continuing to purchase existing and new products from the company.

It now becomes necessary to *evaluate the feedback from the internal organization, the channel, and the customer base* to factor their input to any modifications that may be required to ensure the plan's success. Any modifications made should be communicated back to the people who suggested them to enhance acceptance and buy-in.

Learn the dynamics and the biases of the organization and retreat from confrontation in the recommendation section of the plan. The objective is to create agreement, not to teach or preach to the organization.

Next, *leave the plan alone for several weeks*, if possible, and work on something else. For example, visit some customers, manage other projects, or simply distance yourself from the plan, if possible.

Then, *reread the plan and examine to see if it's reasonable*. Does it still read like a plausible plan? Does it still seem achievable? If so, proceed.

Finally, *make the recommendation to management and get their agreement.*

Action plans: After you have agreement of the principals, it is necessary to generate an action plan. Without it, the plan becomes a conversation piece and none of it gets executed. Worse yet, senior management may assume that action is taking place when it may not be. This will impact credibility and eventual funding. The action plan should be very specific with times, assignments, and completion dates. Even if it is not totally complete, it is important to establish momentum early and set the pace for the organization. If funds are necessary to be appropriated by senior management, put them on the action plan with an assignment and a completion date. You will find out early if this is a real plan or just wishful thinking. The final element in constructing the action plan is to refrain from taking half-steps. If the plan is reasonable and the recommendation is well-considered and achievable, then do not compromise at this stage by taking half-measures. Execute the plan as designed.

3. Dead-Ended Single-Product Programs

A. A new product should fit into: A family supported by a strategy

There has been significant discussion regarding strategy, vision, and operational planning and all of the steps associated with it. The reminder here is to refrain from one-time-hit programs. The single-hit programs are characteristically seductive, lucrative, and have huge upside potential. These programs also have generally longer lead times, significant uncharted development waters to navigate, and often require huge capital outlay. Long-term success of an enterprise relies on the need to stay on strategy, and nurture and execute a plan quickly. Focus on solid strategies; then execute tactical and operational planning to carry them off effectively.

B. Product development initiatives should not be easily copied by competition

Take care in selection not to choose programs that are all alone, like an island. An enterprise must have significant financial resources to embark on a development program with no previous core technology or background. These firms are at risk when embroiled in these types of programs because they can consume tremendous amounts of capital and personnel. Progress may be slow and obstacles might require disproportionate effort to overcome. They generate characteristic revenue curves, making them easy targets to knock off. They are financially stretched; they have no depth to the core technology; they have not productized it before; and their slow progress allows a competitor or consortium of competitors to leapfrog them technologically with their product offering. The marketplace is very dynamic and fast-paced. Loyalty is as strong as your last product advantage. Alliances between companies can be fleeting, and your ally today may be your opponent tomorrow.

C. The family of products retains momentum and therefore strength

As discussed previously, a family of related products under development, in which the technology employed is one of the core technologies in-house, serves the best long-term prospects of the organization. It retains momentum, offers flexibility, and fosters financial and competitive strength for the company.

D. Practicing the concept of leverage

This is the basis for practicing the concept of leverage. Combine efforts to serve a strategic goal or use a single effort to serve related, multiple goals. This philosophy leverages the product development resources in an organization. The managers of several programs should spend their time looking for and creating this leverage. Start with modest goals that are achievable and build on the successes. Along the way, gather ideas and execute selected ones that contribute to an overall portfolio.

4. Consistency

A. Is this a reasonable opportunity?

As part of determining if the prospect in question is a real opportunity, one needs to conceptualize and visualize the product offering as part of the company's future standard offering. You need to visualize a sales call with your sales team working the channel. Evaluate the plausibility of the product. Can you see your team assimilating and effectively promoting it?

Do not assume that the organization will step up to the plate to promote the new product. Take care to strike a delicate balance between a new product stretch for the organization and something that is unachievable given the resources and mix of players.

Given the development cost, resource allocation, and lost opportunity cost incurred in the creation of the product, a certain amount of introspection and due diligence are required to ensure that the organization can be effective.

B. Setting the criteria for evaluation

To provide some consistency in the evaluation process, it is helpful to have a criterion by which to evaluate these new product prospects. A framework could be as simple as a list of criteria that must be satisfied in order for the product to fit in the organization. This has little to do with the market fit or opportunity for market success; rather, it deals with the internal limitations of the organization.

The benefit of this type of evaluation is that it removes the emotionalism in the decision-making process. It also limits the effect of personal preferences and agendas, making the selection more objective and driven by a corporate decision-making process. To make this even more effective, factor in the individual company's time to develop into the criteria, to determine the ability of the company to capitalize on the market window of opportunity.

Perhaps the best way to accumulate the criteria is to establish a rating system that assigns weight to each of the criteria and allows for an arithmetic means for selection. In addition, a single opportunity may not have to satisfy all of the criteria; in some instances it may only need to satisfy 70% to 80% to qualify. However, a few of the criteria may be mandatory. The following are examples of criteria for a company that has changed ownership and is highly leveraged. In this case, opportunities taken must be deliberate and must have large payoffs to offset the risk in valuable capital:

- Is this a finite opportunity or a new market?
- Are the projected sales at least $2.5 million per year for four years?
- Can our manufacturer's reps sell this in most of the territories?
- Can we design and build it?
- Do we have to install it?
- How much after-sale support is required?
- How many people will we have to add to headcount to support this?
- Can we sell it to other customers?
- Can the design be leveraged from another design or to another design?
- Can we develop it in less than a year?
- Is the gross margin more than 50%?

As an example, for the highly leveraged organization that must make every dollar spent pay off, projected sales, corporate capability, channel leverage, and development leverage may be absolutely mandatory, whereas there may be some latitude in the remaining criteria.

C. How much of a stretch is this for sales, engineering, and manufacturing infrastructure?

It is important to assess how much of a stretch the new product will be for the organization on all of its operational fronts. As discussed earlier, one cannot blaze new trails on every front and expect any leverage to take place. A degree of stretch is desirable in each of the operational elements to keep the organization fresh and competitive. A visual way to look at this is to generate the graph in Figure 1-15.

Radially place lines from a center point as shown. Have each line equally placed at the same angle. The length of each line indicates the measure of difficulty to implement the

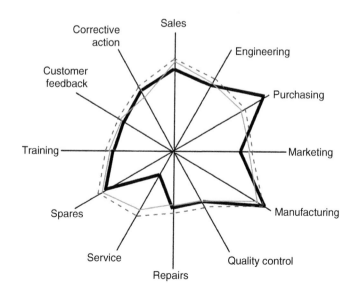

The haeavy line is the company
The light line is the market
The dased line is the new product

Figure 1-15. Fitability graph.

new product. A separate line is used for each of the critical operational areas that contribute to the product's success. There is one line each for sales, engineering, manufacturing, quality, purchasing, service, and repairs. Assess where the organization is with respect to the competition and plot on each of the operational areas. Also plot where the new product will fall with respect to the present internal capabilities of the organization. In this visual way it is easy to assess the degree of difficulty in implementing the new product.

The black line represents the company's capabilities in the various disciplines. The red line represents the present capabilities of the marketplace, and the blue line represents the present requirements of performance for the new product or business. By looking at the relative comparison between these various assessments, one can evaluate the company's capabilities to execute the new product development program.

As can be seen, the company is a little short in the sales and marketing discipline and significantly short in service. The diagram also shows that the product requires service capabilities even ahead of the market's expectation for service; therefore, the company will have to make some changes.

D. Ensuring that all of the business stakeholders are committed

As stated earlier, it is crucial to the new product development's success that all essential stakeholders of the organization are in agreement and are committed to the execution of the program. It is therefore incumbent on the person entrusted with the leadership role to assess this commitment as early as possible. Interview the key players in the organization to assess their levels of commitment to the program. Correct or alter any attitudes that will stand in the way of success. If you cannot, change the players. The energy needs to be spent in executing the program to position the product favorably with respect to the marketplace. Do not expend the energy trying to convert nonbelievers within in the organization.

E. The danger of "promised commitment" or getting commitment early

There is a danger in getting what seems to be commitment early from several people in the organization in word, but not necessarily in deed. Consequently, it is good practice to involve all the players in the organization at the outset. Require their participation early in the process and observe their attitudes. Don't shy away from confrontation; in fact, invite some of it. In this way the truth comes out and it can be resolved early on. Also, there is a benefit to some confrontation in that some good ideas and observations may be brought to light (which otherwise would have not been discussed) and integrated into the program. In addition to inviting the confrontation, it is the manager's responsibility to drive it to a conclusion for the team. In this way the group can see how the objections, discussions, and disagreements can be a positive force in bettering the program.

F. Taking the "basal temperature" reading

As the program progresses, it becomes necessary to periodically take the basal temperature of the organization with regard to the program. This serves as a reaffirmation of the commitment of the key managers and the key players to impart a proper sense of urgency to the project. Do not let apathy or negativism begin to permeate the group, because this begins to feed on itself and eventually will slow progress. In the leadership role, your responsibility is to smooth over internal speed bumps and focus on the competitive- and market-driven issues.

5. Continuity

A. The continuity of development is as important as the development itself

The continuity of development can be considered as important as the development itself. What this statement means is that a company must become proficient at recognizing an opportunity, initiating action to capitalize on it, and driving a program through to success with it. If an organization can achieve this with the determination to weather the obstacles along the way, it will eventually be able to get to anywhere it desires to go.

B. On again/off again

Conversely, the developer's nightmare is a management structure that issues on again/off again orders and constantly changes priority on projects. In these cases, the bulk of the effort that needs to go into the development of the product actually goes to winding down a program, only to later reinitiate it.

C. Building momentum

Figure 1-16 illustrates this point. If there are interruptions in the development processes, the returns in revenue cannot be additive, according to the model for leverage discussed

Unleveraged revenue					
	Year 1	Year 2	Year 3	Year 4	Year 5
Product 1	Delayed	1	1.2	1.5	1
Product 2		Delayed	1	1.2	1.5
Product 3			Delayed	1	1.2
Product 4				Delayed	1
Product 5					
Total	0	1	2.2	3.7	4.7

A

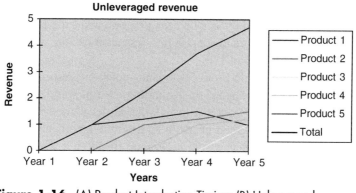

B

Figure 1-16. (A) Product Introduction Timing; (B) Unleveraged revenue.

earlier. According to the leveraged revenue model, each successive program builds the revenue and profit curve toward a more exponential shape. Interruptions will therefore alter the time scale and shift the revenue (missed market opportunity) and profit-generating potential scale away from the desired characteristic shape.

D. Geometric progression of rewards

The objective, therefore, is to reap a geometric progression of rewards for the invested development. The operative word here is investment, in that a company desires to harvest as much revenue and market share as possible from an investment made in product development. Future revenue dollars are discounted by inflationary pressures, and development dollars are invested presently at full value and generally after tax (outside of R&D tax credits used from time to time). Therefore, almost by definition, a certain amount of leverage is mandatory.

E. The pursuit is as important as the attack

The important point of this section is to understand that continuity is critical in new product development. The company must initiate a program, execute it in a timely manner, and harvest the reward of its investment. To that point, the pursuit of the business is as important as the original attack. Each program and action must be deliberate and have a payoff. There can be no voids in the continuum because this wastes time and dollars; hence, the pursuit is as important as the initial attack on a market opportunity.

6. Leverage

A. Each new product development effort should leverage the organization

The key to effective new product development is to create leverage in the development arena similar to creating leverage in the manufacturing and sales arenas. It simply is unaffordable for most companies to continuously start from ground zero on every new product development; consequently, the desire is to leverage past development efforts when starting new programs. Figure 1-17 illustrates the issue.

As shown in the figure, successive development programs can be thought of as developing concentric rings around a base of core technology. As the programs become more complex and involved, the core increases. Each activity then pushes the outer envelope in terms of technology, purchasing, sales and marketing, and manufacturing systems. In this way the company develops expertise in all areas with each program and experiences real growth in sales, margin, and the infrastructure to support it. A less abstract and more operational way of looking at this is shown in Figure 1-18.

Figure 1-18 shows that a single product demands company resources in the procurement of materials, the engineering of technology employed, the manufacturing processes, and the sales channel to get it to the marketplace. As new opportunities are evaluated, a portion of the evaluation needs to address the leveragability of the product within the organization for the successive programs. Building a business using similar manufacturing processes, engineering and development, and routes to market contributes to the leverage discussed.

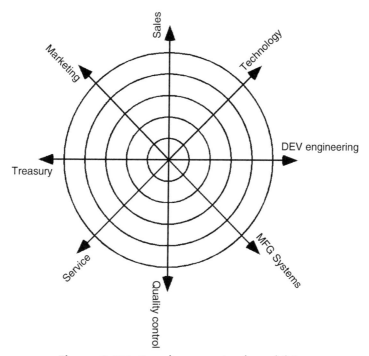

Figure 1-17. Extending operational capabilities.

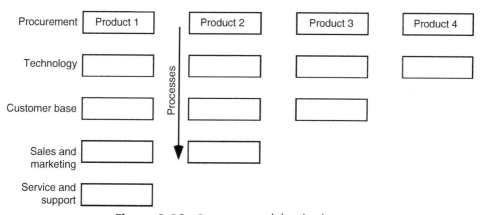

Figure 1-18. Company capability development.

B. The result should be 1, 2, 4, 8 versus 1, 2, 3, 4

By leveraging the SG&A elements of the business, a successful series of programs will generate financial leverage in the form of growing revenue and profit. Because the knowledge is being cultivated in-house and applied to the programs, the leverage shows up as faster program execution times and lower costs. Faster time puts the company in a better market position, and lower costs make the program easier to amortize. Revenue- and profit-generating programs then sum up to a result that generates a profit curve that begins to look exponential rather than linear. The difference is the leverage that is created. This is illustrated in Figure 1-19.

Normalized sales results					
	Year 1	Year 2	Year 3	Year 4	Year 5
Product 1	1	1.5	2.5	4.375	7
Product 2		1	1.5	2.5	4.375
Product 3			1	1.5	2.5
Product 4				1	1.5
Product 5					1
Total	1	2.5	5	9.375	16.375

A

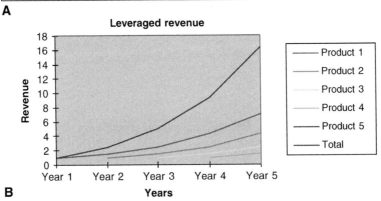

B

Figure 1-19. (A) Normalized sales results; (B) Leveraged revenue.

Each product generates growth for the company, starts out small, and builds momentum in the marketplace. As the company then structures the programs in succession (and does this for several product lines), the revenue and profits grow disproportionately. This is the leverage that is desired; however, there is also a collateral benefit, as we shall see in the next section.

C. More flexibility in the event of failure

This philosophy will allow the organization to recover from a single failure more easily than with other strategic philosophies because there is overlap in all areas, and each obstacle does not result in navigating uncharted waters. This philosophy also takes more time and effort to initiate, but results in a more solid momentum after you get it going.

D. Improving the chance of success

Consequently, this will better the company's chance of success in a fast-paced, changing marketplace. Most articles on new product development will discuss the percentage of failures in the marketplace and the historical reasons for failure. This philosophy represents tangible, positive steps that can be taken to ensure a higher percentage of successes. It also becomes important when discussing platform development and recovering a program that is in trouble.

7. Flow

A. Take and execute manageable chunks of development

To keep the flow of a new product development program smooth and timely, each step of the process needs to be identified and broken down into component parts for analysis and execution. It is important to refrain from selecting huge chunks of activity to complete without detailing the specific parts. By taking small elements, the team can use trial-and-error in managing uncertainty and essentially practice its problem-solving techniques. As will be discussed later, the smaller the component part, the easier it is to populate the team and create a timely flow of progress.

B. Get great at executing small, incremental improvements

The company needs to get highly proficient at executing these small, manageable chunks of development activity so as to remove the uncertainty associated with the huge project at the outset. By doing these small chunks effectively, repeatedly, and predictably, one can then manage a large program within a budget in terms of time and money. The only way to confront the uncertainty hidden in a large program is to break it down to its component parts for execution. This also has a collateral benefit of ensuring completeness by including all of the component parts required for the program.

C. Smoothing the progress and making it measurable and predictable

The benefit here in terms of project management is to make these component parts measurable in terms of effort and talent required, and to be in a position to plan and complete activities in a predictable and repeatable fashion. This takes practice and several programs to become proficient. To achieve this means that almost any program can be broken down into its component parts for analysis and a measurable amount of effort can be applied, and the result can be predicted within a time frame.

8. Pursuit

A. Becoming relentless at improvement and execution wins the business

The company should evolve a culture of consistent improvement, incremental product development, and timely execution. The competition does not maintain the status quo, either in isolation or collectively; this mandates your firm to initiate consistent, predictable actions that result in increased market share and better competitive posture.

B. Key vocabulary: Small, incremental, definable, measurable, predictable, and fast

Once the strategic vision is locked-in, the company must perform. The watchwords of this performance are small, incremental changes that are definable, measurable, and have leveraged payoffs. Any actions that the company takes need to be fairly predictable at the outset, executable in a short period, and relentless in the succession of these steps. The firm needs to maintain the strategic vision and strive to relentlessly execute steps to achieve it.

PERSPECTIVES EXTERNAL TO THE ORGANIZATION

1. Defining the Marketplace

A. Who is the customer?

Always understand and know who the customers are. They are the ones generating the revenue through acceptance of your solution to their needs. Continue to solicit their opinions, their wants, their changing customer needs, and their pressure points. This will ensure that as changes in the landscape take place, your company will be positioned to respond appropriately.

B. Demographics

Do not be lulled into a false sense of security by demographic summaries. Because so much of our lives is governed by lumping people together and treating their issues as common, keep in mind when serving an original equipment manufacturer (OEM) market, the *customers* are trying to differentiate *themselves* to create a competitive advantage. Consequently, they need your company to customize your solution for them. In trying to serve a consumer market, keep in mind that there are different classes of customers with different needs that have to be met by the product or service. Products cannot necessarily be lumped together and mass-marketed with success, as a demographic summary may indicate.

C. Users, OEMs, resellers, and agents

Remember that a market is a collection of customers, not a uniform, homogeneous purchasing machine that can be pushed, prodded, and calculated with decimal-place accuracy every time. There are intermediaries in the process who have their own problems, issues, and agendas. Leveraging the initial product development and the launch requires real-time monitoring and action to build and preserve momentum.

D. What are the customer's alternatives to your solution?

Always remember that the customer has alternatives, including competitive offerings at a lower or higher cost, and functional displacements. In some cases, there may be personal apathy on the part of the customer. The operative point here is that the customer your company is approaching to sell to may not be predisposed to the solution you are offering. To understand this better, place yourself in the customer's frame of mind and evaluate his problem and look at all of the industry solutions. Then measure your solution against *his* alternatives.

E. How is your competitive advantage embodied in the product?

Often this is evident in examining competitors' solutions to customer needs. Many times a unique or lower-cost method may be used to accomplish the same function. The difference may be that the development engineers have biases and mental paradigms, as we all do in addressing and solving problems. The innovative engineers, designs, and companies break these paradigms to offer alternative approaches embodied in their products for satisfying customer needs. That is the business objective in new product development.

2. Understanding the Marketplace

A. It's an ever-changing, complex world

The world is becoming more homogeneous as time progresses. Actions taken overseas affect transactions and business here. Changes in currency affect global trade and individual buying decisions. The marketplace is simply becoming more complex and initiatives and actions need to be multidimensionally evaluated before implementation, or they may be ineffective or disastrous.

As an example, let's assume your company supplies electronics to an OEM in North America for sale to end-users in North America. A sister company to the OEM (half ownership) resides in the Tuscany region of Italy and sells its equipment to the European continent and the rest of the world. The Italian company is the primary mechanical manufacturer for the domestic OEM and has electronics supplied by a privately held Italian electronics manufacturer, supplying the same package near your cost. Do you have a domestic threat from the Italian electronics company here? What are the critical factors to secure continued domestic market share? Can you maintain domestic loyalty here even if the Italian mechanical manufacturer prefers its own supplier? Can you be functionally displaced by the Italian electronics manufacturer who is working in close conjunction with the Italian mechanical manufacturer on a new platform that will be brought to North America?

All of these factors must be evaluated. You will find that it begins to mimic a chess game with an international flavor. The operative point is that the world is a complex place.

B. What are the customer problems to focus on?

The effective way to navigate the marketplace is to focus on the customers' needs. Their problems are the ones that need to be solved in the embodiment of your products and services. By keeping close to the customers' wants and needs by resolving their problems, you position yourself as a problem solver and a consultant to be trusted and relied on (i.e., a business partner), a position that every supplier aspires to.

C. Understanding marketplace dynamics

Because the marketplace is a fast-paced, dynamic medium, it is most important to understand the dynamics of a market. Your company is attempting to do one of two things. Either you are trying to develop a new market that currently doesn't exist, or you are trying to increase market share by taking it from others. Both *actions* drive *reactions* in the market. The first will draw attention to the lucrative opportunity. The second is basically a catfight to secure the business. Knowing the marketplace dynamics will assist in developing the strategy for each. Failing to understand the dynamics can be fatal to the effort.

D. Understanding customer perceived value purchasing triggers

What does the customer purchase—a product, the satisfaction of a need, or peace of mind? This question and its answer are the essence of the marketing and sales of a new product. The products and services are fundamentally the embodiments of the customer's satisfied need. Many marketers assume that the product they make is what the customer buys (i.e., the customer buys these features). The customer actually buys benefits and solutions to problems. If you fail to understand that or provide for it in the product embodiment, sales will be fleeting.

If you fail to satisfy the need, there is no compensation that will offset what is lacking. Meet the need well and there are many things that will overlooked, all else being equal. Remember that the product you sell and get profit from is simply the embodiment of a satisfied need and is crucial to formulating any program. Failure to understand and accept this fact may cause wasted time and effort for the company. A basic example of this is an insurance policy: Is the customer purchasing a document with legalese and clauses to satisfy a need, or are they purchasing peace of mind from the insurance company?

E. Positioning the product

This is why it is so important to position the product in the mind of the consumer. This positioning must tell the story in a convincing manner so that the buyer's perception of the product satisfies the need. It is important to understand that, at the outset, it is the customer's perception that is making the decision. Subsequent buy decisions after the first sale will be based on product performance in meeting the need. In other words, no matter how good any product is, fundamentally, the customer must have the perception that the product will satisfy their need at first sale or you will never get the opportunity to let performance satisfy the need. Positioning the product creates that initial perception.

Positioning the product means different levels of difficulty for different industries and products. For example, repositioning a product by repackaging the same basic elements is easier than redesigning a product platform for an added or different feature set.

F. What triggers the purchase order

A study of business transactions would indicate that, during the exchange between the company representative and the customer, there is a point at which the customer has all or most of his perceived needs met and decides that this is the best deal he can reasonably get. This is the precise point where a closing needs to take place. This is what will trigger the purchase order. Asking for one before this point is reached is pointless, and waiting past this point is dangerous. It becomes a critical success factor in a sales transaction to seize the opportunity to create this point and act on it.

G. Practice, learn, adapt

They say that practice makes perfect. That is generally true because inherent in practice is scientific methodology. Try, evaluate, learn, modify, try, evaluate, and so forth. This is the value of the lost sales report—it is a key element in the analysis of the transaction and is integral to any improvement change to become ultimately successful.

H. Testing the effectiveness of different sales styles

There are different sales strategies and styles for different products, and it is important to match the sales style to the product type. If, for example, you have a technological leadership position, the sales style would be different than if you were supplying a "me, too" type of product. The difference is a value-based style versus a price-based style. This is an area that must be identified and coordinated with the product launch and that is integral to the marketing plan. If it is not, you may be wasting the sales personnel's time and risking their disillusionment. Many companies have failed in product launches because they assumed that a brand new, highly technologically advanced product could be marketed by an existing sales force with no background in the technology.

I. Modifying the sales approach for effectiveness

Marketing and selling a product is similar to telling a story to the customer, at least from the promotional aspect. Ideally, the interaction needs to symbiotic but, for purposes of this discussion, the marketing "story" needs to be consistent with the launch objective. If it is observed that, for whatever reason, the presentation needs to be modified, it should be reevaluated in conjunction with the entire product plan and product launch.

J. Tiered market structures: High, medium, low

Also, different tiers in the market demand use of different sales strategies and tactics. If the product line consists of vehicles that traditionally had status and utility associated with them, there are several strategies to use to be successful with each tier of luxury, value, and low-cost transportation. Parts of the promotion need to appeal to the different wants and needs of the buyer. The operative point is that the multiple-tiered market must match up with the multiple product line to be effective.

K. How can it be sold?

Product marketing personnel need to constantly evaluate the effectiveness of the promotional programs and continuously ask how the product can be sold. Because the landscape is constantly changing, the positioning must be modified to keep pace. This can be quite difficult to do during an initial launch because it requires introspection and openness in assessing sales levels and degrees of success.

L. Where is the value created and exchanged?

To effectively market and sell products, the company must determine where the value is created and where it will be exchanged. This is the essence of the transaction. The customer is looking for the solution to a need at a good price. The manufacturer must add value at a profit to continue operations and to fund future developments. The value must be created and exchanged for revenue so that both parties come out to the good. Anything less than this is temporary and fleeting.

3. Global Product and Business Development

A. Small world, isn't it?

Disney Corporation dedicated an entire Magic Kingdom attraction to it, the Internet has made it an information reality, and Boeing's 777 made it easier to circumnavigate. It is indeed a small world and getting smaller every day. What does this mean for new product development? It simply means that isolationism in one market, country, or continent is no longer possible. What your company does affects, to a degree, the global marketplace for that particular market. What the rest of the world does may indeed affect your company. Product development and market development no longer occur in a vacuum. The downside is more and more intense competition. The upside is instant access to information and data.

B. Typical foreign content of today's products

If it is not obvious that the world is getting smaller from a communications aspect, consider the foreign content of many of today's products. The foreign content has become so commonplace that several products publish it in terms of a percentage. Consider the Hyundai automobiles manufactured in the United States: they are manufactured by Americans for a Korean company. It would be an exercise just to document in detail the funds flow in terms of raw, manufactured cost and intellectual property amortization. A similar split in content occurs in cameras, digital video discs (DVDs), cellular telephones, and computers.

An interesting exercise would be to document the mileage on shipped parts in a sub-assembly through manufacturing to its final destination. Consider the following route a power transistor takes in getting to its final destination: The power transistor will be used in the final output stage of an electronic drive for controlling the speed of an electric motor, which is part of an elevator drive system in Algeria. The power transistor is manufactured in Japan and is shipped to a distributor in North America. That distributor ships it to a drive manufacturer in the Midwest for integration into a drive. The drive is shipped to the drive manufacturer's subsidiary in the United Kingdom as part of a stocking order. The drive is pulled from stock and shipped to Bologna, Italy for integration into the elevator control panel. The elevator control panel containing the drive is tested and shipped to the end-user, a building in Algeria!

C. Analysis of exchange rate sensitivity on foreign content

The presence of foreign content in products also adds a dimension to product costing. The actual value paid for foreign content components is subject to the changes in currency from one country to another. This can place pressure points on a product design from a cost sensitivity point of view. In addition, the supply of these products can be affected by global competition, export compliance, and political turmoil.

D. Venues for procurement

This means that the venue for procurement of the component must be carefully selected, in addition to the specification and the cost. Failure to consider these ramifications may cause loss of supply midstream or intense cost pressure. Later in the text (see Chapter 8), we will examine a vendor profile that goes beyond parts specifications and costs.

E. Venues for manufacturing

The same can be said for manufacturing; however, an additional dimension occurs here. What happens if the local lawmakers pass legislation that affects the plant? Can your capital equipment be idled by changes in law, political pressure, or local labor initiatives? In this case, your company is at greater risk than in the procurement case because your manufacturing capacity is affected.

F. Local laws

Local laws can place greater demands on new product development because some require adherence to international standards. These may be stricter than U.S. laws and standards; consequently, the development may have to be more complex and costly.

G. Export compliance

The United States requires manufacturers to adhere to export compliance legislation. This is mostly a marketing effort to be selective in targeting certain locations to exclude from sales and shipment. Failure to comply may result in revocation of an international license, or worse. The export compliance guidelines exist to limit shipment of U.S. technology to foreign countries that may be a security threat. It can significantly affect your company's marketing effort.

H. Technology transfer

Technology transfer is generally a company decision, although the U.S. government can play a role regarding certain technologies. Some foreign governments mandate a transfer of technology to their country as part of a sale. This can adversely affect your company's market position and strength elsewhere, and the decision to transfer technology must be carefully evaluated.

I. Joint ventures

Joint ventures are complex partnerships, with differing cultures, profit motives, and taxes—know your partner's agenda. Such partnerships can be very successful or very costly. Watch where the money changes hands to determine allegiances, agendas, and motives. This will be discussed in detail in Chapter 2.

J. Cost roll-up spreadsheet with sensitivity

Figure 1-20 summarizes the effect on product factory cost impact from foreign content. As can be seen, the effect can be favorable or unfavorable depending on current market conditions. The object point is that your company may enter into an agreement with the conditions being favorable, and lock into terms that will force cost pressures and eventual market pressure if the conditions go unfavorable. This puts your effort in a precarious position. In addition, the manufacturing venue can cause greater problems depending on the percent content of labor, local laws, and relative numerical difference in wages. This also assumes that the hours required to produce the product are uniform and homogeneous across the globe; however, they are not, because different venues have different workforce skill levels.

BOM	with labor adjustment							
•••								
Description	P/N	Qty	Cost each	Extended cost	Domestic	Foreign	Currency	Cost impact
							Favorable/unfavorable @Standard cost	
Part 1	654321	2	43.25	86.5	Yes	No	1	86.50
Part 1	654320	2	0.23	0.46	Yes	No	1	0.46
Part 1	654319	2	0.48	0.96	Yes	No	1	0.96
Part 1	654318	6	4.30	25.8	Yes	No	1	25.80
Part 1	654317	1	6.20	6.2	Yes	No	1	6.20
Part 1	654316	4	7.90	31.6	Yes	No	1	31.60
Part 1	654315	8	12.30	98.4	Yes	No	1	98.40
Part 1	654314	10	8.75	87.5	Yes	No	1	87.50
Part 1	654313	3	3.40	10.2	No	Yes	0.99	10.098
Part 1	654312	7	1.25	8.75	No	Yes	0.98	8.575
Part 1	654311	5	1.65	8.25	No	Yes	1.12	9.24
Part 1	654310	2	2.40	4.8	No	Yes	1.12	5.376
Part 1	654309	1	2.75	2.75	No	Yes	1.12	3.08
Part 1	654308	3	3.45	10.35	No	Yes	1.12	11.592
Part 1	654307	4	6.45	25.8	No	Yes	1.12	28.896
Part 1	654306	5	4.12	20.6	No	Yes	1.12	23.072
Part 1	654305	7	2.34	16.38	No	Yes	1.12	18.3456
Part 1	654304	8	45.70	365.6	No	Yes	1.12	409.472
Labor		12	35.5	426	Yes		1.05	447.30
Burden		1.75		745.5	Yes		1	782.78
Miscellaneous expenses		100		100	Yes		1	100
			Total cost	2,082.40			Adjusted cost	2,195.24
							Impact	112.84
							% Impact	5.42

Figure 1-20. Foreign content sensitivity analysis.

As can be seen in Figure 1-13, many factors contribute to the product factory cost and, as such, it is the responsibility of the manager to account for these issues. The architecture of the product transcends the bill of material; it is affected by all of these contributing issues. The product needs to be planned out from all aspects to ensure continuity in sales and manufacturing.

4. Market Investigation

Market investigation is so important to the long-term growth of a company that it is included in the following presentation. It is presented here from a "fit" perspective to provoke the strategic thought and planning of new product development early in the process, as it should be. In Chapter 2, and more so in Chapter 3, the mechanics of marketing will be covered in detail. The objective here is to present how the concepts relate to the overall evaluation and selection process, and what one should look for in a new business development program.

A. Primary market research

There are two basic types of market research—primary and secondary. The primary type of research is one of the most direct and accurate forms available because it deals with information obtained directly from a customer. There are a variety of ways to secure primary market feedback, but most fall into the following categories: surveys, interviews, and demonstrations.

Surveys can be administered in a variety of mediums and are structured means for obtaining information. To be conclusive and effective, they need to be administered in the same way each time, with no changes in format. *Interviews* are a face-to-face means of gathering information directly from the respondent. These are very valuable for securing qualitative information, exploratory information, and discovery of latent needs among customers. In fact, this type of sampling can be used as research for determining how to structure a survey. *Demonstrations* are another type of primary market research. This type of research offers information in the form of a demonstration and solicits feedback from the respondent. These are critical to confirming features, benefits, and acceptability of a product to a customer. Each type of primary market research has its advantages, and each needs to be used in different aspects of the information-gathering process.

B. Secondary market research

Secondary market research is characterized by an indirect means for gathering information from readily available sources. Typically, research of this type consists of, but is not limited to, the following means for obtaining information:

Trade shows: The trade show is an excellent means for obtaining product information as well as determining the scope of companies in any one field of endeavor. In addition to enabling personal interviewing, observation, and literature search, trade shows typically publish a program that lists companies by product type or process. These are typically segmented several ways to allow a quick view of the market.

Trade literature search: The literature search is an ongoing activity to keep in tune with the marketplace to see who is promoting what to whom. It is an excellent means for gathering information for a future competitive comparison.

Articles: Articles are excellent for documenting trends in the marketplace. They are, by virtue of their authors, expert sources for further information and perspectives on the marketplace. They are very useful to the industry novices who need to familiarize themselves with a product area.

News pieces: News pieces are timely information about the actions of the leaders in the marketplace in a given area. They tell you who is active in an area and to what extent, and what boundaries are being pushed.

Observation of trends: This is an internally generated activity that looks at what is happening in the marketplace from period to period. By looking at the historical activity, one can extrapolate trends to the future. This will be an invaluable exercise when producing a product evolution flow chart.

Financial reports: Financial reports offer a wealth of information about competitors and, as such, collectively about the market. These can be monitored for information and financial information to get a picture of the finances, the principals, and the capitalization of a competitor. Monitoring of the financial markets and their news can provide insight into acquisitions and mergers affecting business development.

Textbooks: Textbooks offer the novice background information on products or services and technology. These sources can be useful to understand elements of the technology.

Patent search: Patents are available for search to determine the state of the art in a field of endeavor, as well as to determine your company's position in the state of the art. It can also be a good source of competitive information on technology.

Lost sales reports: These internally generated documents can be helpful in determining a competitor's ability to use their product to secure business. By evaluating their technology and your company's performance against them, these reports can be an invaluable source of information.

It almost goes without saying that the Internet is of the most prolific means of information-gathering about companies, products, competitors, pricing, trends, and countless other data. It can be an invaluable resource for immediate data-gathering if the data is available and accurate.

C. Interviewing the customer

Although previously mentioned as one of the primary means for market investigation, interviewing is a valuable technique because, if approached properly and under the right conditions, the customer is the most accurate means for assessing your company and its standing in the marketplace. If you listen and attempt to internalize what the customers are telling you in interviews, they will communicate most (if not all) of the qualitative data you will need.

When conducting an interview, it's best to first review the customer's company operations. This will give you a feel for the operation and how this customer makes money. It will also give you a perspective on how your company and its products will fit into the customer's overall plans, as well as determine the legitimacy of the respondent's comments by materiality. The interview will allow you to discover latent needs that the customer may not readily know they have. This generates opportunities for new product ideas.

It's best to have samples, models, examples, and props to enhance the exchange and draw out responses. Do not be averse to broaching issues that may be controversial (pertaining to the subject matter at hand), because this will give perspective on the attention level of the respondent and his or her biases, and will yield perspective on other comments.

D. Engaging the customer

As part of a customer visit, you will want to engage the customer in a frank discussion about their business, its driving forces, its vulnerability, and how they navigate the competitive threats to find success. This discussion should result in symbiosis: to have a real appreciation for the customer and their needs, it becomes necessary to understand their business.

The process of customer engagement consists of three basic themes:

- First: listen to the customer. You have created the venue for an exchange, so take the time and energy to really listen.
  - The customer has an issue or problem to resolve. Recognize it!
  - This is why *you* are here—to pick up on subtleties and act accordingly.
  - The customer is not predisposed to your solution, so do not try to convince them at this time.
  - The best evaluation of competitive standing is to remove yourself from the sales arena and place yourself in the customer's perspective.
- Second: move the interchange to a process of discovery.
  - Conduct a dialogue, not a sales presentation!
  - The interchange should be a mutual process of discovery.
  - Probe for conditions, customer alternatives, acceptance scenarios, and prerequisites.
  - Understand and frame the customer's business, alternatives, decision breakpoints, and their product pathways.
  - Determine and reduce the information to a flow chart and breakpoints.
- Finally: look for the hidden opportunity for your company.
  - How can your company create a product or service that enhances the customer's ability to pursue their market more effectively?
  - Discovery should produce the customer's hidden opportunities, competitive situation, and threats to securing the business.

To get started, consider the following as a framework for an initial discussion. These are questions you may want to use to engage the customer. These discussion interrogatories are open-ended and require thought to create mutual understanding.

- What are the areas in your organization that currently fall short of the skill set and performance index required to achieve your business vision? Who, what, where, how, and why? (This allows for an open-ended forum between you and the respondent. It allows you to draw out complaints initially, and then to get specific ideas onto the table for discussion. Share your company's experiences, also.)
- From a historical perspective, what are you doing about them? (This determines the commitment level of the management and the organization to change, and to the strategic vision.)
- What is your business vision? (This is a good test to determine whether the organization has established a vision and whether it has filtered down through the organization.)
- What operational or product changes are you making to achieve this result? (Again, this determines the effectiveness of the management structure to effect change toward a specific goal.)
- What do you need to make your operations easier and more efficient, to increase output, and to exercise better control? (This open-ended discussion allows the respondent to identify any latent needs in the organization.)
- If you were asked to double your output with the same resources, what would you expect to be required to achieve this? (This removes the specter of the status quo and forces the respondent to push performance of the organization. It helps the respondent to identify future needs.)

- What is currently considered impossible or impractical, but if you had access to it, you could increase your effectiveness? (This question allows the respondent to dream of a future workplace and identify latent needs to achieve it.)
- What do you look for in a supplier? (This identifies the customer's expectations.)
- How can we become a preferred supplier to your company? (This determines whether you have a chance at the business and the account.)
- What are your critical competitive pressures? (This allows you to determine a way to partner with them.)

The key to identifying the potential new product is to synergistically combine all of these ideas and requirements into the tangible improvement of the next-generation product.

E. Surveys and customer feedback

Surveys and customer feedback should not be conducted only at the launch of a program. These need to be an integral part of the process of refinement, product evolution, and subsequent product developments. Never get too far away from the customer because this will lull an organization into complacency or, worse, remove the feedback loop in the process.

F. Extracting the opportunity within a customer complaint

The customer grants a significant favor to you by complaining. Your company sold them something that fell short of their expectations, and they thought enough of it to complain. We as market and product development professionals may not like or want to listen to the complaint; however, there is a lot to be mined out of one. In effect, the customer is telling you what you did wrong. If you evaluate the math of the situation, a single complaint is worth quite a bit to the organization if you are going to do something about it. If you view it as a nuisance, then it has no value; however, if you plan on changing to meet the customer's expectations, then it has great value. In other words, how many other customers who didn't complain would have to stop buying from you for you to lose enough market share to notice? How many customers would you have to survey to find out what this one complaining customer is telling you directly? Clearly, it is a valuable means of feedback.

G. Exchange of knowledge and experiences

During your discussion, it is critical to establish an open and free exchange of ideas, problems, and opportunities between the two companies. Establish a spirit of cooperation at the management level and drive it through the organization. Create the atmosphere of going after a market opportunity together, with each entity contributing a portion of the talent and effort to get there. Constantly look for how the two companies can both profit from the market segment, since very few companies are absolute leaders in a market with a position strong enough to dominate and affect control of the segment by themselves.

H. Establishing rapport, expertise, assistance, improvement, and authority

This exchange should establish a lifelong rapport between the personnel of the two businesses that will transcend any minor market fluctuation, upset, or competitive thrust. Your actions need to position you as an indispensable partner who is integral to your customer's plans. Make sure to plug the customer (now your partner) into your organization. Balance the sacrifice you will make in availing the customer of your resources with the positioning of indispensability within their organization.

I. Correlating the data

Examine the responses carefully and group the data in some meaningful way to draw product conclusions from it. Look for product configuration patterns in the summary of data so that product planning can serve the market with the proper product.

Do the data received make sense with other information gathered? Have the data been corroborated in some way to substantiate the results and accurately draw conclusions? Did you notice any conflicting data that would require more research? Summarize the data in some kind of a spreadsheet to allow viewing and sorting. Organize it for eventual presentation to management. The results should tell management the story of the opportunity.

J. Testing the data

Finally, test the results and the data supporting the conclusions by visualizing the product available for sale today. Evaluate how it will sell in the marketplace today. Would it sell? Why or why not? Visualize the product selling in the future, project the future market needs, and (after development time has elapsed) determine whether it *will still* sell or whether you would have lost the window of opportunity. Do the results of this reinforce that you are targeting the right generation of product?

SUMMARY

This chapter had several objectives for the reader to achieve. The first theme was a discussion on how product and business development relate to the economy. The thematic thrust of the chapter was to draw the relationship between product development as the business objective and to examine it from perspectives external as well as internal to the organization. It was also designed to give a qualitative feel for the different steps in the process of new product development. As the chapter unfolded, the reader learned the various aspects involved with the process. There was a discussion of the new product opportunity's fit in the organization and how it ties into the strategy of the organization.

There was also a discussion of the directed purpose of research and the function of development, how to leverage efforts, and create continuity. Business development programs should each contribute to a corporate goal, and the market investigation is used to refine the goals and the opportunities.

The chapter should also give the reader an appreciation of the interrelationship of the various disciplines within the enterprise that must function in concert to successfully implement a program. Some of the material presented here in broad overview will be discussed in a more detailed manner in later chapters. However, it was presented here for continuity in outlining the process of taking an idea from inception to completion. The subsequent chapters will detail the process, starting with Chapter 2: The Market Opportunity.

THE MARKET OPPORTUNITY

THE MARKETPLACE WITHIN OUR WORLD

1. The Faster-Changing World

The world is an ever-changing place, and the changes that we are currently undergoing are occurring at an ever-increasing pace with wider swings and larger differences. The process of new product development must factor in this uncertainty. Product iterations are coming faster and demand for changes is more prevalent than before. Not all of these changes are necessarily good, nor do they necessarily affect a program; however they must be recognized, evaluated, and factored into a program (and possible modifications of that program) for it to be successful in the marketplace.

In terms of evaluating humankind's knowledge base, each discipline needs to be examined for its accumulated knowledge and real-time progress. Many disciplines increase their body of knowledge in very short periods of time. Since this accelerates with the experience and the passing of time, the amount of change in certain disciplines is enormous. Consider, for a moment, some of these fast-changing technologies:

A. Computer technology

1. Wireless technology

The following looks at data that is somewhat representative of the growth of knowledge, so as to flavor the discussion of how new product development must operate in these fast-changing times.

The U.S. patent system can be considered a measure of knowledge by certain standards. All else being equal in terms of novelty, technological innovation, and application, the number of patent applications and issues is a measure of a level of knowledge development. Figure 2-1 is a graph showing the number of U.S. patent activities from the period from 1836 to 2005. This 169-year period in our history shows a steady increase of the number of applications and patents issued.

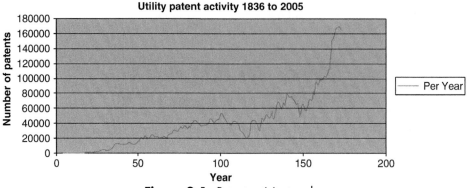

Figure 2-1. Patent activity trends.

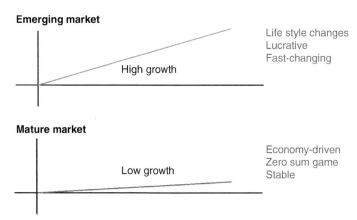

Figure 2-2. Emerging versus mature markets.

2. Population and time

Another measure of humankind's knowledge level is attributable to the sheer growth in the population. More people, more studies, more ideas, more creativity, and more applied knowledge development time ultimately generate a larger knowledge base from which to draw. In addition to these, the advent of computers and Internet communication schemes accelerate information transfer to lightening speed.

B. Effects of faster doubling of human knowledge

1. Faster pace of life

The effects of this increase in humankind's knowledge are creating a faster pace of life. Information flows faster, with more sources of input. However, there is also greater inaccuracy in the information, causing false starts and changes in direction. Many aspects of life in modern times have become competitive from a quantity rather than a quality point of view. People are racing toward an ever-more-mobile target. The result is a faster pace with questionable value added in many of the transactions.

2. Increasing pressure

This faster pace is causing increased pressure on our society. We can see the pressure manifested in loss of patience, wider swings in violence and crime, and loss of reflection and relaxation in people. In many cases this pressure, unnatural by its extreme, is causing much anger. The pace is also accelerating as time progresses. There is a higher level of learning among people, with requirements and demands that are constantly increasing.

3. Disconnection index

With all this excessive movement forward, the danger of becoming disconnected is increasing. This so-called disconnection index is a measure of a person's lack of keeping pace with the changes around himself or herself. Fifty years ago a person could have conceivably left the Earth for 10 years, returned, and essentially picked up where they left off. For example, between 1840 and 1850 the amount of change in life was rather small and not easily discernible, compared to the changes from 1990 to 2000. A person leaving the earth for 10 years in 1990 and returning in 2000 would find sweeping changes in foreign policy, security of the European continent, vast changes in communications, computer power, and availability of and access to information.

4. Aging population

Unfortunately, these changes are being thrust onto the populous as if people were homogeneous and equal in terms of their ability to assimilate, absorb, digest, and accept these "improvements."

For example, how many people of the baby-boom generation can easily program a digital video disc (DVD) recorder without instructions? How many elderly people feel comfortable with the seven-segment readout for numbers? How many people enjoy reprogramming all of the clocks in the house after a momentary power outage? As more products get more sophisticated and the population ages, this disconnection index will grow naturally. From a product development perspective, it will have to be addressed in order to appeal to the mass market with a single product platform.

5. Technical focus versus value focus

Another change that has been taking place is that young people, in their quest for keeping up with these changes, have become technology-focused rather than value-focused. This fact is probably little observed as we lead our busy lives, but it is crucial to recognize this in understanding the demographic changes, the needs, the preferences, and the general attitudes of the different segments of our population. Although it might seem that this is only occurring in the United States (because it is more noticeable here), these changes are occurring all over the world, just at different rates.

2. The Seasons of Change

It would seem that each era spawns a new generation of knowledge that emerges, grows, flourishes, declines, and dies off to give way to the next generation. This closely parallels the business development cycle in that the product emerges, gains acceptance, is copied by competitors (indicating, oddly enough, acceptance in the plurality), flourishes, and then eventually dies off, making way for the next generation of product.

It's important to understand how knowledge levels affect our society and its development. Just as the steam engine was a technological revolution that drove far-reaching changes in industrial society, the technological revolution that facilitated knowledge transmission will revolutionize and accelerate drastic changes in postindustrial society. The caveat for business and product development is that one must understand that change and growth are an inevitable part of the cycle of life as well as the business and product development process.

Development patterns exist in our civilization as time progresses. They are often hard to wrestle with on a day-to-day basis but, when considered within the historical perspective, they are readily apparent. These patterns drastically affect the marketplace and the market's acceptance of new products.

One such development pattern has a time base of approximately 80 years, or a single human being's lifespan. The pattern seems to be broken up into four distinct segments, as follows:

- High Phase
- Awakening Phase
- Unraveling Phase
- Crisis Phase

These four segments can be likened to the seasons of change, namely spring, summer, fall, and winter, respectively. In the high phase (spring), a new beginning emerges. This usually follows a crisis of some sort and, as society emerges, there are unbridled optimism and growth. Society expands, prosperity increases, civic responsibility increases, and advancement occurs in most aspects. This phase is followed by an awakening phase in which society begins to question itself and its direction. Spiritual agendas begin to dominate the social landscape. Scientific agendas begin to lose influence, public order declines, and society's general level of enthusiasm begins to wane. War and conflict, driven by newfound idealisms, break out. The next phase is the unraveling phase. This phase emulates the tone of its title. Society begins to acquiesce, public trust declines, guilt begins to dominate the moral landscape, and general attitudes of pessimism abound. People "feel" but no longer "do." Wars and other conflicts are fought with moral fervor and debates on idealism and morality abound. There are "liberating" forces at work, fragmenting society as they have had their fill of "rebirth and soul-searching." It has low resistance to external forces, placing it in extremis.

As can be expected, this phase leads into the crisis phase. This phase separates the conversation and soul-searching of the previous phases into actual conflict. The threats become real, there is a sense of public urgency, and the family unit strengthens. War and conflict are prosecuted furiously and with purpose. The government governs and extraneous laws are shunted aside to make way to meet the crisis head on. Crime diminishes and optimism increases as the society shifts from individualism to national or civic purpose. This phase then wanes and ushers in the next high phase. Each phase seems to average 20 years in length.

For a historical perspective, World Wars I and II represented international crises where national and international purpose drove many changes in society. The government governed and created a Promethean war machine. Society shifted to a more unified entity with purpose.

After World War II there was a tranquil period of growth or "high" in the 1950s, followed by the beginning of the awakening in the late 1960s and early 1970s. The awakening evolved into the unraveling of the 1980s and that brings us to the eve of the next crisis. The 1950s and early 1960s were a tranquil time when growth and prosperity abounded. This high spawned the moon venture, consumerism, freedom, and enthusiasm.

The late 1960s and early 1970s ushered in the awakening, a period of searching, questioning, and shift to individualism. The unraveling continues until the next crisis. In recent times, elements of all four phases are evident in different locations. This would suggest that many time-shifted phases are occurring simultaneously in different parts of the world and within different parts of a same location. Accordingly, products serving these societies must be targeted to serve the phase at each place and time.

The history of cycles and trends needs to be understood and considered within the framework of new product development. Although some products may be achievable, they are not right for the times. For example, would instant access, connectivity, and the freedom of the Internet have been successful in the 1940s and 1950s with the Red scares and security breaches? Would products promoting individualism and personal freedom be successful in an era when national and civic purposes dominate the society? Or, more basically, would fuel efficiency and environment issues embodied in automobiles be accepted in an era where performance and styling dominate the taste of the consumer? These are just a few examples of why product planners need to understand the periods we are in and factor this into the product and company strategy. In a crisis period, products need to rapidly get practical.

The market is looking for products that are functional, affordable, able to be manufactured with available materials, and reliable. Technology generally needs to be proven because the market will have little tolerance for ushering in new technology. After all, they are navigating crisis. In a period following the crisis or high, the market is much more accepting of new things, implementation of new technology, and new products and markets. Regardless of whether one accepts the model or the time line, one must accept that the products marketed within a phase must fit the acceptance patterns of that phase.

MARKETPLACE DYNAMICS

1. It's a Fast-Changing World

The world economy shows a nonhomogeneity of needs and satisfactions all over the world. There are different peoples at different stages with different politics and different goals. The new world order is a work in process. Some societies are way up on the hierarchy while others are struggling at the base for basic needs. This disparity drives product development directions differently. For example, in South Africa a domicile may require 150 watts of electrical power. In contrast, the average U.S. requirement is 2.0 KW, spiking to 18.0 KW

Disparity in the product evolution and infrastructure implementation will change the assumptions. For example, telecom is evolving to a wireless system in the United States *after* a landline infrastructure had been installed for many years, whereas in other parts of the world telecom will start off with wireless infrastructure immediately and landlines will probably never be implemented. This will change the mix of telecom products, the product evolution, and the obsolescence, and legacy pass-downs of used equipment will evolve.

The same thing may happen with distributed and centralized energy! Once the bastions of a society, centralized electrical power generation may give way to decentralized, distributed electrical power where products enabling an existing infrastructure and support system of fuel delivery allow immediate implementation.

A. Reduced ability to cope with changes

For technology to have a long-lasting benefit to humankind, it must be accompanied by the wisdom to use it. Failure of the wisdom element in the equation results in abuses, misapplication, and eventual rejection. With many of the recent technological advances, we have seen the accompanying moral dilemmas in the medical and health-related fields. Failure to garner the wisdom to utilize the technology can result in an inability to cope with the changes that it brings.

B. Diverse value systems

The diversity of the world results in the diversity of values all over the world. We would like to consider them to be homogeneous, but they are not. Accordingly, the development of business and products must factor in these value systems and how they impact acceptance.

2. Market Characterization

A. Market maturity

One of the most valuable exercises is to determine the level of market maturity that exists in the specific market under consideration. Depending on the degree of maturity, the market can be low-growth or high-growth. A mature market will be flat or no-growth. Conversely, an emerging market will be more growth-oriented. Figure 2-2 illustrates the difference and will be used as a basis for illustrating the market share and profit potential of each.

As shown in Figure 2-2, the differences in emerging and mature markets manifest themselves differently. This affects the business development plan. The high-growth markets have more opportunity associated with them. They generally effect lifestyle changes in the customers. An example of this is the automobile industry in the early 1900s. It was a growth market that caused sweeping changes in the lifestyles of the customers; it was very lucrative and fast-changing, which challenged the manufacturers to keep up with the changing market demand.

A low-growth market has less opportunity, is more price sensitive, and can be thought of as a zero sum game. What market share you increase is obtained only from competitors, not from a growing number of customers using the new product or technology. Company growth is determined by swings in the economy, more than other factors, and there is stability in the marketplace.

Figure 2-3 shows the impact on the companies participating in these markets. The benefit shown in the emerging market is that the company can more easily gain absolute shipment dollar revenue through increasing its market share in an increasing market than in the low-growth markets.

In either case, as part of the business plan there should be a clear understanding of what the market dynamics are and how you intend to participate in them.

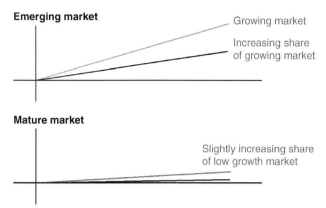

Figure 2-3. Company participation.

B. Life cycle and market cycle implications of profit

Figure 2-4 illustrates the specific impact on a company operating in these diverse markets. In the emerging market, the economic benefit to the firm in dollars of margin far outweigh what can be generated in a mature market.

These examples are for illustrative purposes only, and the absolute values or percentages are somewhat exaggerated for effect in illustrating the difference. Also, the market size and revenues have been scaled for the example. However, the analysis shows the vast difference

Growth patterns					
	Year 1	Year 2	Year 3	Year 4	Year 5
Emerging market					
Available market (000's) Dollars	1,000	2,500	4,500	8,500	12,500
Market share (%)	1	3	7	12	18
Revenue (000's) dollars	10	75	315	1,020	2,250
Cost of sales	0.6	0.65	0.7	0.725	0.75
Margin (000's) dollars	4	26.25	94.5	280.5	562.5
Mature market					
Available market (000's) Dollars	1,000	1,500	2,000	2,500	3,000
Market share (%)	1	2	4	6	8
Revenue (000's) dollars	10	30	80	150	240
Cost of sales	0.7	0.725	0.75	0.775	0.8
Margin (000's) dollars	3	8.25	20	33.75	48

A

Figure 2-4. (A) Growth pattern comparison.

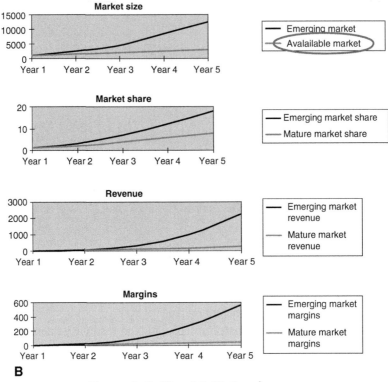

Figure 2-4. (Cont'd) (B) Growth patterns.

in actual gross margin between the two markets. In the emerging market, there is a growing percentage of market share to work with. In the mature market, the behavior is relatively fixed. The graphs show the difference.

Another critical point to remember is that markets are not exclusively either mature or emerging. They start out as emerging and move into mature markets. Therefore, the business plan developed for new products and business has to factor this into the development and account for the transformation.

Figure 2-5 indicates how this may look in terms of market growth over time. The market growth starts out small and increases dramatically, and then remains somewhat constant up to a peak growth rate. The growth rate then diminishes to essentially zero, where the market size remains constant. This assumes, of course, a normal distribution and behavior without overt manipulation by any one dominant competitor who might be large enough to unilaterally effect a change in the marketplace.

C. Low-cost producer

There are certain strategies that a company can use to navigate the changes in markets and use them to their advantage. In the graphs in Figure 2-6, the various behaviors are presented in the framework of the various markets to show why it is desirable—from a survival point of view—to be a low-cost producer by the time the market has matured.

The graphs in Figure 2-6 show a contrast between a mature market and an emerging market implementing older, less cost-effective technology and newer, cost-effective, state-of-the-art technology. The behavior of prices and costs tells the story of company profits.

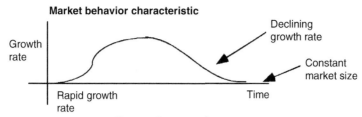

Figure 2-5. Market aging.

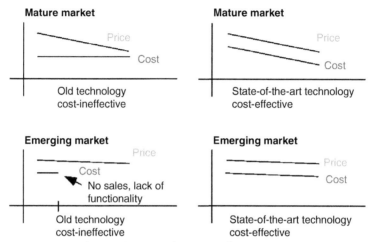

Figure 2-6. Market aging characteristics.

D. Leading-edge technology

In the first scenario (a mature market being addressed with older technology), the cost remains relatively flat while the prices are driven down. In the second example (a mature market with state-of-the-art technology), the decreasing prices are offset by decreasing costs. The third example shows that in a growth market the prices remain stable while the demand for functionality can quickly outstrip the older technology's ability to supply the market demand.. In this example, sales cease after a time because the older technology is limited and cannot satisfy the market needs. The fourth and final example (a growth market with state-of-the-art technology), shows prices slightly declining with costs declining at a greater rate, thus preserving the margin.

As shown in Figure 2-6, the best overall strategy a company can take is to be at the leading edge of technology in order to keep current with the market requirements for functionality and price, as well as cost containment, to maximize profits. It simply puts your company ahead of the competition initially so that you can establish and maintain a market position.

The danger is in allowing the company to be placed in a squeeze situation wherein the market is demanding a low-cost, highly functional design and you have a high-cost, barely functional design to offer. There will be lackluster sales and minimal profits, resulting in thin profit margins and eventual inability to reinvest properly.

PRODUCT CONCEPT SYNTHESIS

1. The Business Concept Embodied on the Product Idea

A. The better mousetrap

"Design a better mousetrap and the world will beat a path to your door." This has been the battle cry for new product developers for many decades. However, there are increasing requirements for success today, such as: Who else makes mousetraps? What must our mousetrap do to induce people to buy it over others? Are we a low-cost mousetrap producer or are we a niche mousetrap player? Have we complied with the new mousetrap standards? What steps have we documented to ensure that humans will not get injured using our mousetraps? Is there a significant influx of foreign mousetrap manufacturers encroaching on our market? Are there new ways to functionally displace mousetraps, such as high-frequency deterrents? All of these issues make the age-old adage now multidimensional.

B. Most systems generate the idea and fail to encompass a plan

Most systems for generating new product ideas focus on the idea, the market opportunity, and the timing. The concept for a new product must encompass the plan to carry it off effectively. The business concept must be entwined with the product concept. Each will draw on the other in the development phase. Each contributes to the other's progress.

C. Tie the business operations to the idea

The manager needs to visualize the business operations as the product is being outlined. This visualization will assist in defining product features and configuration that will aid in future business operations. For example, the demographic data may indicate a number of versions of a product; however, the business operations—manufacturing, procurement, and inventory control—require a finite, practical number of versions that are more manageable.

D. Critically look at the business of the product

It is essential to critically look at the business of the product. Where will the money change hands? Why will the customer be willing to part with money for the product? How will the transaction take place? Provisions and steps must be taken to address these issues to have an effective launch and program. As an example, the electric car is an age-old idea which, with the advent of computerization, modern batteries, electronic drive technology, and improved internal combustion (IC) engine technology, and the hybrid vehicle (a cousin to the electric car), is becoming a market reality. If a hybrid electric car is designed, produced, and introduced into the mass market, there are latent infrastructure issues to deal with. Issues such as safe disposal, battery remanufacturing, financing, and supply/demand changes could affect the price of conventional fuel and thus could affect operating costs. Operating cost changes could affect market acceptance. It is not just a great product that makes great businesses. It is also the infrastructure that supports the product that must be integrated.

E. Tie it together

The saying goes, "To the victor go the spoils." So, too, with the business of business and product development. The program that covers all or most of the bases to ensure success will most likely be successful. This is because the manager asked those hard questions of the program, rather than ignoring the issues altogether only to find the marketplace asking the same questions and displaying little patience for lack of planning.

2. Solving the Customer's Problem in the Product

A. Customers purchase perceived value of the solution to a problem

The buying decision is based, in part, on the perceived value of the solution to the customer's problem. The customer buys a solution; the manufacturer constructs a bill of materials and processes to assemble it into a product. It is therefore a requirement for successful marketing, design, and sales to embody the solution into the product. The value placed on it is generally governed by the customers. They will determine the amount paid based on this initial perception, and subsequent follow-up sales will be influenced by experience factors that may weigh the value differently.

B. How does this new product development solve a customer's need or problem?

This is the basis for a marketing plan. Effectively answering this question will directly tie the customers' needs into the product concept. The pathway is the sales presentation. It relates to the customers and how the solution to their problem or need is embodied in the product your company is selling them. As we know, different issues affect classes of customers differently. What is a critical buying decision for one customer is not necessarily all that important for another. The market must be broken down to effectively create the product's architecture so that the lion's share of the available market can be addressed with a basic platform design. This platform design allows a wide variety of features and a feature gradient to be incorporated in a basic unit, with options to scale up performance or packaging of values for the customer.

C. Under what conditions does the customer commit?

What is the universal model for customer acceptance? The answer to this question is as complex as the number of different customers there are in the marketplace. The arithmetic that takes place in the mind of the purchaser varies with situation, need, time pressure, alternatives, and many other factors.

For example, under normal circumstances a cool glass of water is little more than an act of kindness, worth a generous tip in a restaurant. Under conditions that are much starker and manifested by shortage, a cool glass of water can be worth considerably more if the need is great enough. Certainly, in a desperate situation one would be willing to pay considerably more money for a less-than-cool, less-than-clear glass of water than under normal circumstances. What would be the quality or price gradient in this example?

The operative lesson in the example is to understand and know the customers' conditions and frame of mind when they make that buying decision. Failure to understand and plan for it properly may position your company with product inventory and wondering why customers aren't buying.

An additional means for gathering data on the customer buying-decision dynamics is to monitor competition. How are they selling products? What is their pitch and is it effective? Gather and analyze brochures, data sheets, marketing literature, lost sales reports, and field sales reports to get a feel for their strategy and tactics.

D. How can this new product development fit the demographic conditions?

The marketing assessment outlines the market segments (i.e., that group of customers who have identified features, needs, and preferences that are similar). The manager needs to break down the market opportunities into their component parts so they may be reassembled into product versions to be used to address the market. What version of the basic platform will be used where, and why? Will it offer a competitive advantage? In a more macro sense, what will the product mix consist of? What will it cost and how will it be priced? These decisions need to be made correctly early in the process. A sensitivity analysis needs to be conducted on the cost and the price. If it's expected that the price will lower after introduction and competitive response, this new price needs to be factored in to see if there are any latent pressure points in profitability due to a skewed mix.

3. Idea Evaluation within the Framework of the Business

A. Creating a system for objective evaluation

To begin to objectively evaluate a wide variety of diverse new product ideas and business opportunities, it becomes desirable to have criteria. These criteria remove the emotion from the evaluation and allow different personnel in the organization to do the evaluation. Each idea then must be tested against the criteria in the same way, with the same questions.

B. Setting up the criteria

One of the challenges is in setting up the criteria. What questions are important? What are the company's pressure points? What are things to avoid? For example, is the opportunity dependent on the development of technology that requires a few select people? Are they available within the time frame you desire? If they leave the project, what means for recovery do you have? Given that this is an example of a pressure point as it relates to personnel, there may be a host of other pressure points that affect the selection. As a rigorous part of the business development plan, it is important to develop these criteria customized for your specific business.

C. Testing each idea

Each idea should be tested against the criteria, as previously stated. Figure 2-7 is a format that can be used to evaluate an idea. It is structured with a weighted sum arrangement of criteria. Each criterion is listed as a separate item. Next to each criterion is a preference assessment of that specific criterion. The preference assessment relates directly back to the strategic initiative. Does the business desire that attribute in a new product development? The range is from −1.0 to +1.0. Next to this preference value is the assessment of how the product relates to that specific criteria. The columns are multiplied and summed to get the weighted value.

By observing the individual values posted for each criterion or attribute, one can learn a fair amount about how the organization will absorb and implement the program. As can be seen, the organization has a high desire to select an opportunity that has sales of over $2.0 million over four years. They are highly technical and are not concerned about absorbing

Evaluation criteria for new product development					
Evaluation criteria	Company desired weight (−1.0−1.0)	Actual score (0.1−1.0)	Product 1 weighted score	Actual score (0.1−1.0)	Product 2 weighted score
Sales > $500k per year for four years	1	0.9	0.9	0.9	0.9
Existing sales channel secure orders	0.75	0.85	0.6375	0.6	0.45
Absorb the technology and design the unit	0.85	0.45	0.3825	0.75	0.6375
Training, service, and support in the field	−0.25	0.2	−0.05	0.75	−0.1875
Sold to a variety of customers	0.1	0.6	0.06	0.8	0.08
Already our customer base	0.9	0.7	0.63	0.5	0.45
Leverage to another product line	0.5	0.35	0.175	0.2	0.1
Projected margin over 50%	0.65	0.4	0.26	0.6	0.39
Installation is required	−0.5	0.95	−0.475	0.85	−0.425
Total			2.520		2.395

Figure 2-7. Evaluation criteria.

the technology. They do, however, have an aversion to field service and support, perhaps due to a weak field organization. This criterion shows up as a negative value. Consequently, if the product being evaluated has a high dependence on this for success, multiplying the preference and product assessment will be negative. When summed up with the other criteria, these will detract from the overall weighted value. This analysis simply serves as a numerical evaluation (which is objective) rather than talking yourself into the assumption that the product or business won't need that much field organization support.

Another aspect is to admit that the criterion of field support is not one that your company is good at, and to reexamine your commitment in that area. In addition, perhaps the product and the sales channel could be designed such that a field support organization won't be required.

As with any weighted summation, the absolute value of the sum does not mean much. When two programs are set side by side, the sums can be compared on a numerical basis to select the best alternative. This can be a valuable tool, given two conditions:

1. Be very careful when setting up the company criteria. Make sure they are an accurate reflection of the desire and capacity of the company.
2. Be accurate and honest when evaluating an opportunity against these criteria.

If you are diligent about these two items, this will serve as a useful tool. Figure 2-7 presented here has fixed values and fixed criteria.

D. Consistency

Consistency is one of the key benefits in using this objective type of evaluation. As times and personnel change, it becomes a challenge to establish and maintain consistency in the organization. This exists in all aspects of the organization, not just in the identification of new product opportunities. It is most important to be consistent in the new product area, however, since this area drives all future movements. Wild swings in strategy and tactics will misdirect the organization, thus removing whatever shred of leverage it may have had and rendering it ineffective.

E. Measure of ongoing effectiveness

As the system is put in place and used for some time, it is important to measure its effectiveness. What ideas were brought in, how were they evaluated, and what were the results in selection? What degree of success was obtained in pursuing the market? Are the evaluation criteria still the right ones? Are they consistent with the business development plan? These questions need to be periodically reviewed. Figure 2-8 is an illustration of how the process should work.

F. Objective criteria versus agendas and areas of comfort

The final and summary thought in this section is that new product opportunities must be evaluated with two methodologies in mind. The first is that they need to be evaluated in a consistent manner. The second is that the consistent manner be tied back to the strategic plan and the goals of the company. With these two prerequisites for an evaluation system, new product ideas can be brought in and evaluated in a systematic manner, free from personal agendas and immune to organizational areas of comfort.

PRODUCT INTEGRATION INTO THE BUSINESS

1. The Product as the Vehicle to Profitability

A. How does it use and generate cash, profit, and value?

All products use and generate cash. This becomes an important factor as product development programs become larger and have more impact on the overall finances of a company. Depending on the type of business and the labor and material content, every dollar of added sales requires additional dollars of operating cash to fund it. This cash winds up as inventory either in raw goods or completed assemblies, as well as accounts receivable. The development requires cash for the core intellectual property development and also the capital equipment needed to eventually manufacture the products. Accordingly, the organization experiences cash pressure on several fronts: expenses, capital, inventory, and accounts receivable. This is shown in Figure 2-9, where a negative cash flow exists for the duration of the development program. Depending on the cost of capital employed, the inflation rate, and the time to develop, this negative cash flow can impact the organization significantly if it is not recovered rapidly through profitable revenue.

Some products require enormous amounts of real-time cash that could have been retained earnings. The impact is manifested in the value of the stock and shareholder equity. Consequently, timing is of the essence in recovering the initial investment.

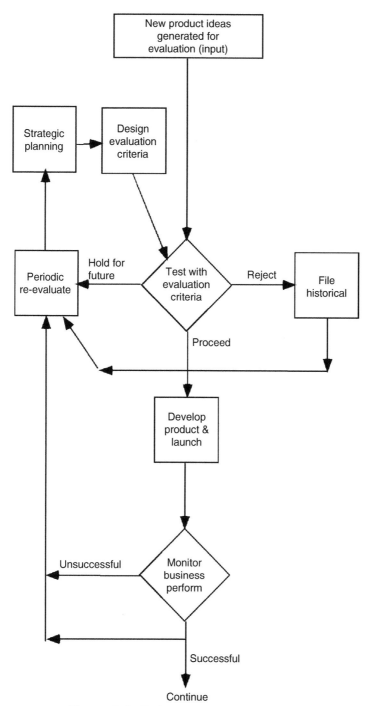

Figure 2-8. Evaluation criteria maintenance.

B. The goal is dollar and value contribution to margin dollars

To use a British colloquialism, at the end of the day the objective is to have the development program financially pay off in the form of margin contribution and value contribution to the organization. It may be a long road to get there, but if the opportunity is real and

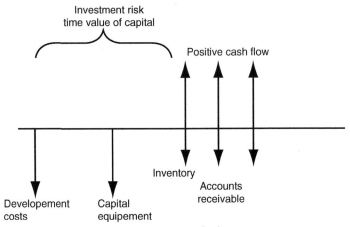

Figure 2-9. Cash Flow.

the plan is achievable the organization must have the faith to see it through. The following example illustrates this point.

In 1948 the Halloid Company paid $25,000 for the right to develop xerography. Their original target for introducing a Xerox machine was in 1950, some two years later. The model, a copier, had 1260 separate components. Fraught with problems, the first real, marketable unit wouldn't be introduced until 1960. The financial drain on the company to develop and market the product was tremendous. The total development cost was $75 million, which amounted to more than twice the operating earnings of the company during that time frame! However, the first unit sold for $29,500 at introduction in 1960, and the financial impact of the revenue was staggering. By 1970, Xerox's stock had grown in value 66-fold, a fine example of big risk, big payoff.

We must mitigate the financial exposure in the process by holding firm on development costs and pursuing the reduction in development time. The preservation of the market window is one of the most important aspects of the enterprise.

D. Integrate the business development team into the business

One of the most positive things a manager can do to improve the efficiency of business development team is to integrate them into the business. Most members of a development team know something about business. They are generally familiar with the cost structures, the material content, and the overhead and burden rates, and have a general product-costing orientation. Very few, however, have an appreciation for the more strategic and tactical aspects of the business. In addition, very few have an appreciation for the sense of urgency required in the product development playing field.

Consequently, training the team in these aspects will have payoffs. The manager needs to outline the impact of slipped schedules and creeping factory costs, and what that means to the revenue line, the recovery of investment, and the loss of competitive advantage. Selected elements of accounting and finance should be reviewed with the team as a group and reinforced on an individual basis where appropriate. Because the team needs to be interdisciplinary, each contributor needs to have broad-based training.

2. Playing the Business Development Game

A. Identify opportunities in which small, incremental changes mean big advantages

If you loosely define the competition of new product development as a game, then consider it to be in the organization's best interest to learn to play the game well. The need to become proficient at incremental changes has been previously discussed. An adjunctive activity is to become proficient at identifying those items that are small and can be accomplished with a minimum of effort, and execute them for incremental gains. A series of these will mean big payoffs in the near and long term. They serve as practice for larger programs and also keep competition at bay by constantly raising the barrier to entry in the specific marketplace.

B. Determine latent opportunities for competitive advantage, and incorporate them

Here is where the engagement of the customer becomes so important. At this interface, if you listen carefully, the latent needs of the customer will become evident. In that latent need the identification of the product features, configuration, or both will be defined and the value posted by the customer. By incorporating these features, the company can gain a competitive advantage with small, incremental effort and protect margin against erosion as a result of competitive pricing actions.

3. Winning the Business Development Game

The marketplace today is characterized by an unprecedented number of versions of similar products. Everyone wants designs custom-tailored to individual tastes, and the products have unprecedented requirements for documentation and support. Just consider the entire infrastructure that has been enabled to support computers, software, and operating systems. Many products generate an enormous potential for errors in an environment where there is unprecedented potential for punitive damages accompanied by an unprecedented lack of consumer responsibility in using the products. This makes creating a product even more difficult, above and beyond the normal commercial pressures and competitive issues.

A. The concept of platform management

Many companies are adopting a product development system that employs platform development as a basis. The product platform serves as a basic building platform on which several versions of a product could be built. In this way, several versions of a product can be quickly implemented on a platform to serve the market rapidly. At a later date, a cost-reduced version could be implemented after volume begins to climb.

An example of a product platform could be a computer tower. The basic system consists of an enclosure, a power supply, a motherboard, and slots for peripheral components such as hard drives, compact disc (CD) and DVD drives, graphics cards, and user interfaces such as monitors and keyboards. Some of these components are used across many product configurations, allowing the manufacturer to configure a wide range of products to meet specific market needs.

Thus, as market needs change, the manufacturer can reconfigure and reprice an offering very quickly. Conversely, imagine the effort that would be needed to create a unique version for each and every customer from scratch, with each component optimized for the use configured!

The problem with platform development, however, is that a disproportionate amount of development must go into the platform as opposed to a specific version. This is because the designers must factor in compatibility with components, obsolescence, re-sourcing of component parts, and so forth. Platform planning and development will be considered in more detail in Chapter 3.

B. No company or market exists in isolation

To give some perspective on the impact of the faster-changing world, consider that information is readily available almost anywhere and to almost anyone. This makes more knowledge available to more people every day.

It would be quite straightforward to institute plans for your company if they could be made in isolation. They could be well considered and every contingency planned out. However, no company really operates in a vacuum, either on the purchasing or the selling side. A network of suppliers, vendors, customers, and employees interconnects each side. Since there is this interconnection, information travels fast and fairly pervasively, albeit not always accurately. Consequently, any advantage is short-lived and information must be acted on rapidly and decisively to capitalize on it.

C. The race for the prize

Firms are interconnected in a race to secure incremental business. What one supplier loses in business, another secures. With a fixed market size, it is a zero sum game. In our world today, many of the markets are fixed in size or are growing at a very slow rate. To create real, large growth in a market, one must enable and cultivate a new set of customers. This is done by ushering them into the world of commerce as participants. By developing new customers, markets can grow, thereby relieving the constant competitive pressure.

Therefore, an individual firm must move rapidly to arrive at the opportunity before its peers do. If there currently is no competition in an uncontested area, there soon will be. Undiscovered or newly discovered niches and segments will attract competitive activity. Lucrative niches developed by one firm will be plundered by another. It is the manager's challenge to carve out the company's place in the market and hold onto it amidst all the activity.

D. Six degrees of separation, and networking

In 1967 a social scientist, Stanley Milgram, at Harvard University conducted experiments that resulted in the "six degrees of separation" theory. The hypothesis was first proposed in 1929 by the Hungarian writer Frigyes Karinthy in a short story called *Chains*. This theory addresses the connectivity that exists in our world, suggesting that we are all interconnected by a vast network of people that essentially can link us with any other person, within six or fewer links. Figure 2-10 illustrates this point.

This is important in that it describes a sphere of influence that surrounds us all. This being the case, we can gain access to people and information, through each other and daisy-chain to reach otherwise unreachable people. The same can be said of organizations. They each have a sphere of influence that extends to other companies, which can allow access to otherwise unreachable information, contacts, customers, corporations, and suppliers. This is illustrated in Figure 2-11.

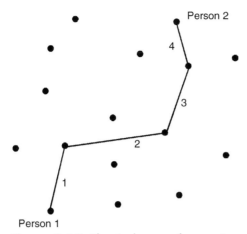

Figure 2-10. The six degrees of separation.

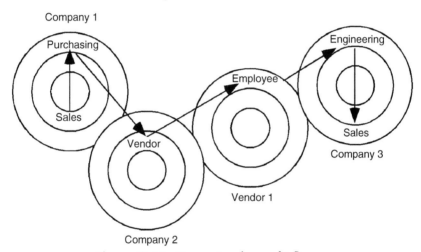

Figure 2-11. Corporate spheres of influence.

Your company initiatives are communicated through your employee's spheres of influence to suppliers, customers, and competitors. Thus, care must be taken in initiating plans and product developments to prevent inadvertent communication of sensitive material. In other words, if you can gain access to competitive data via your network, outsiders can gain access to your data. It is indeed a faster, changing world.

E. More speed, accuracy, and agility before your competitor

How does your company win the competitive game? Everyone seems to be pursuing the same business that you are. How can you differentiate your company from all the others? The perspectives in answering these questions lie in your company's ability and desire to demonstrate speed and agility to the changing market conditions.

It was previously stated that no company or market is an island. It also follows that conditions of advantage and marketability do not remain fixed. Responding to these changing conditions in a timely manner will ensure success.

It might seem from the tone in this and previous sections that winning in the new product arena is a formidable, almost impossible task. It could be, for firms that do not continually strive to improve their products, their marketability, and their position in the marketplace. If, however, your company does execute these continual steps in new product and market development, you will be well-positioned to compete. If you have a strong foothold in the customer base, it will be that much easier to defend it and grow.

F. Consultative selling

"Our customers come to us because they view us as the leaders and trust our expertise to resolve their needs." This statement exemplifies the result of continued consultative selling. The company has positioned itself as the resource that the customer will use in the normal course of their business. It also creates a dependence on the company that results in a competitive advantage. In addition to the normal advantage in a bidding or competitive situation, it also positions the company to be on the customer's cutting edge of new developments, placing them there first. This type of selling allows the customer to become a partner in their future success. Such a relationship requires mutual trust and involvement.

G. Facilitating the customer's dependence

How can you create this dependence between your company and the customer? Your solution needs to provide features and benefit services that the customer would ordinarily need to provide. By providing your solution to them, you can offload this service burden from the customer. If they redeploy or reduce headcount, then you have created the dependence. You need to be selective in this area because if you are not, you run the risk of attempting to be all things to all people. Create the dependence where it does not cost your company too much in time and effort or, better yet, do this in an area where you are exceptionally proficient.

This type of selling, although very successful in the long run, does require a significant amount of energy and funds to develop and cultivate. Your company needs to make a commitment to select a customer and see it through to completion. By effective partnering with the customer's future plans, links between the two companies can be cemented, which will help transcend competitive thrusts.

4. The New Product is the Competitive Weapon

A. New product development is the survival means in a market

Most companies cannot compete on price and delivery alone. In fact, that is quite a dangerous pathway. New product development is often used as a competitive weapon in the marketplace. With most companies, customers and competitors know their history, patterns, and areas of comfort. New product development is a means for a company to gain advantage, secure a position, or win a new customer. It is the ultimate weapon in changing the playing field and should be used as such.

B. Leapfrogging the account and the industry

Examine the industry and the marketplace. Determine the products and the technology to be employed to solve customer problems. Cite advantages and disadvantages of each of these implementations, then forecast the future of each of these technologies. This is where the competition will be at *your* next product introduction. Will you exceed customer requirements, or be at parity with your competitor? Your program needs to leap ahead of where the others will be at that future point. Aim farther as a starting point so that at the time of introduction you will be in a superior position.

C. Attacking your entrenched position

So things are going fairly well. No major problems, volume is reasonable, quality is acceptable, and costs permit a good margin. Time to relax and reap the benefit of all the work, right? Wrong. Now is the time to begin to "attack yourself." Do the proper amount of introspection to determine how long you will have the advantage, what the vulnerable points of the business are, what steps need to be taken to secure future business levels and more. The competition will be doing this, so it is in the company's best interest to do the same, and to do it before anyone else. This activity is healthy because it reduces complacency and rekindles a sense of urgency in continually improving the processes, the products, and the relationships with customers. Now is a good time to revisit that loyal customer and solicit their feedback.

D. The latent opportunity

Are there latent opportunities within your customer base? What are the specific requirements they need to compete better? Find that specific requirement and seize the opportunity to capitalize on it. These are the actions that bind you to a loyal customer. The difficulty is in finding out what the latency is, because it is not always readily obvious. Often, the customer needs to be led through the discussion to obtain this information. This skill is not readily available through most sales channels and requires people who can communicate effectively and solicit information in an unbiased fashion. However, since it is somewhat difficult to obtain for your company, it will also be difficult for your competitor.

E. Raising the barrier to entry

In any situation, it benefits the company to keep the barriers to market entry as high as possible. Surround and imbue the company with expertise and equipment that gives specific market advantage. Investing in capital equipment financed from profits raises the barrier, thus discouraging ill-equipped start-ups from encroaching on your territory. Investment in the business will not only raise the barrier but will also lower costs and improve repeatability and quality. These have a disproportionate effect in terms of competitive advantage and position. Someone chasing your business accounts will find them all that much more difficult to capture.

F. Raise customer expectations beyond the competition's ability to deliver

Your company's actions should continually raise customer expectations of provided products and services. By setting the standard for performance in your marketplace, you make the competition dance to the standards you set, not vice versa. This continual resetting of the customer's expectation level will also raise the barrier to entry from a features and business services perspective—another difficult target for others to reach. This can also catch others off-guard because they won't understand the schedules that the partnership establishes for implementation. Your schedules will chart the pathway and will be known only to you and the customer/partner, making it a guessing game for anyone else to intercept.

5. The Business and Product Evolution Flow Chart

A. Identifying the opportunities that contribute to the vision

Business growth through product evolution is one of the most valuable processes a forward-thinking organization can undertake. It is the essence of the road map to success. It defines where the company plans to go, in what time frame, and how it will get there. The content of a well-considered flow chart can be worth more than the strategic plan itself, since this defines the product path to be taken to the future. The strategic plan then defines the financials and the standing of the firm against its competitors. The first step, therefore, is to identify the opportunities that contribute to the vision of the future and focus on only them. If the vision of the future is agreed to and locked onto, then reject those opportunities that cannot contribute to it, regardless of how lucrative they may first appear.

B. Mix of product and service offerings

The flow chart should not be limited to hard products. It should also include the service aspect of the business. In addition, the mix of products and services should also be projected because this will define infrastructure requirements, finances, and personnel.

C. Contribution to revenue and profit

Where will the revenue and profit come from in the next several years? What percentage will come from new products and new services, and what will the margin structure look like? These questions, although not having to be answered, need to be reflected on to prepare a well-considered flow chart. Figure 2-12 illustrates the concept.

The chart in Figure 2-12 represents an attempt to factor profit contribution centers into the product evolution flow chart. The chart is split into two areas, namely, products and services. The products can be existing and new products to be added in the future. Services, likewise, can be existing and new. The material, labor, and burden components, together with the price and quantity, drive the contribution to margin analysis. By extending this chart, trends can be drawn by loading in historical data, and future mix analysis can occur by forecasting and planning new products and services. This will assist in assembling the flow chart as well as adding financial credibility to it.

D. Product mix analysis, new versus existing, and technology patterns

The real value in this process is getting a feel for the trends and the degree of sensitivity each item has, and how it affects the company as a whole. In a historical analysis, look for patterns or performance behaviors in products and businesses. In a future-planning exercise, decide which products and businesses will be the big contributors and which will create large variations in margin and profit numbers. This is a more objective way of evaluating risk factors associated with new technologies and the products that are generated from them.

E. Mechanics of the product evolution flow chart

It is time to begin the product evolution flow chart development process. As discussed earlier, this chart will serve as a road map for the products that will be developed, acquired, or brand-labeled. The product evolution flow chart, if prepared properly, can be a helpful

Product and services mix										
Products	Material (dollars)	Laboar (dollars)	Burden (dollars)	Cost (dollars)	Price (dollars)	Margin	Qty	Contribution to profit dollars	% Contribution	% of total
1	1	23	4	28	55	0.49	10	270	8.9	5.4
2	2	24	4	30	67	0.55	10	370	12.2	7.4
3	2	23	5	30	68	0.56	10	380	12.5	7.6
4	2	25	6	33	89	0.63	10	560	18.5	11.3
5	2	56	7	65	125	0.48	10	600	19.8	12.1
6	2	45	8	55	100	0.45	10	450	14.9	9.1
7	2	6	8	16	34	0.53	10	180	5.9	3.6
8	2	7	9	18	40	0.55	10	220	7.3	4.4
Services						Subtotal		3030	100	61.0
1	0.5	39		39.5	75	0.47	10	355	18.3	7.1
2	0.5	32		32.5	75	0.57	10	425	21.9	8.6
3	0.5	24		24.5	75	0.67	10	505	26.0	10.2
4	0.5	76		76.5	75	−0.02	10	−15	−0.8	−0.3
5	0.5	87		87.5	75	−0.17	10	−125	−6.4	−2.5
6	0.5	98		98.5	75	−0.31	10	−235	−12.1	−4.7
7	0.5	12		12.5	75	0.83	10	625	32.2	12.6
8	0.5	34		34.5	75	0.54	10	405	20.9	8.1
						Subtotal		1940	100.0	39.0
						Total		4970		

Figure 2-12. Product services mix analysis.

tool in charting the future course of the company. By its very specific nature, it forces identification of the products and services that will be required to support the narrative of the strategic plan. It also forces product planning to the point of determining contribution to the overall financials of the strategic plan. Figure 2-13 is a typical illustration of a flow chart. It shows several things: the pathway that products will take in evolving; how they

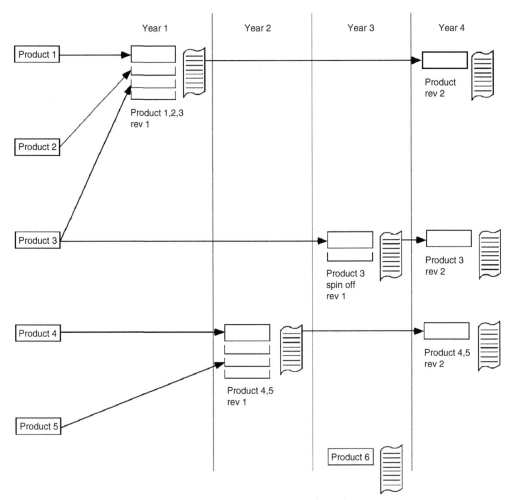

Figure 2-13. Product evolution flow chart.

combine with others in establishing product platforms; and differentiation of products serving different market segments. It also establishes a time frame for when all of these things will happen.

As shown in Figure 2-13, Products 1 through 5 exist in basic form in year 1. The product evolution flow chart shows how these products can be combined from a features and functionality perspective. For example, Products 1, 2, and 3 combine as a result of a development program started in year 1 and introduced in year 2. All of the features in all three products are combined into one product configuration or platform. The basic platform can be configured in one of three versions. This product implementation then runs as a product for years 2 and 3. In year 4 the next-generation product, which combines all the functionality and features of the different configurations, is launched. Newer technology and cost reductions make this condensing of products cost-effective.

In a similar manner, Product 3 will spin off a series of product configurations in year 3 from a basic platform. In year 4 a refined product will also be introduced.

To make the flow chart more meaningful, a narrative should accompany each product implementation, which explains the actions taken and for what reason. Also, the objective of the implementation should be included. The narrative should contain the following components: a description of the product, a definition of the markets served, the degree to which the competitive stance is improved, the product configuration, and the expected level of technology employed. In this example of a flow chart, Product 6 could be added to the portfolio as a result of an acquisition or a brand label arrangement.

As an additional perspective to this product-planning exercise, it may be desirable to prepare several flow charts showing alternative options and pathways. In this way, the best approach can be selected for implementation.

F. Internal requirements, external requirements

Until this point, we have examined the product planning process mostly from an external market perspective. There may be instances in which a requirement for a new product would be a cost reduction or a quality improvement. In such cases, these internally generated reasons for new product development must require a narrative in the flow chart. From a general management perspective, each product must generate cash and profit. Existing programs that are in trouble, either from a prime cost issue or a cost of quality issue, need to be addressed.

G. Future catalog of products and services, features, and technology

What will the future product offering look like? What types of technology will be employed? What level of complexity will it involve? Will training be a marketable commodity or will it be assumed as part of a sales transaction? This process should actually result in a future company catalog with the complete offering included. This 5-year catalog will also drive a thematic narrative of company operations and services in the future. By assembling this catalog, one can see if the strategic story holds together when the products are attached to it.

H. Projection of competition, company, framework, products, technology, and positioning

With all this great planning, your company is positioned to capitalize on the future. The competition is not standing still, however, or at least not all of them. Consequently, it is a good exercise to project what *they* will do based on what you currently know about them. This should be done on the product evolution flow chart and the 5-year future catalog. If for no other reason, it will force consideration of their plans and allow you to evaluate the risk of your plans resulting in future product offering that are only at parity with the competition or, worse yet, nearly obsolete by comparison.

I. Who, what, where, why, and how?

A good plan requires specifics of the mobilization to be clearly articulated. What are the specifics of the product evolution flow chart? This product was developed and introduced by such and such date. This technology was absorbed into the organization by another certain date. Additional channels of sale were established, in place, and functional by still another date. These are the specific statements that actions must follow to accomplish results. The flow chart, the narrative, and the 5-year future catalog must answer the five questions of specificity: who, what, where, why, and how.

J. Organize stages

Finally, organize the action items and the stages of the plan so that it all ties together into a cohesive set of specific objectives and their supporting action items. Look for commonality of purpose and resources to optimize the work needed.

K. Optimize and maximize benefit for dollars spent

Resources for development are always in demand. In addition, the resources for development represent only a small part of the overall investment in a new business opportunity. It is therefore in the company's best financial interest to maximize the results for each dollar spent. Financially sound, well-run small companies consistently deliver the maximum benefit for every dollar spent. They must in order to survive. Larger companies run the risk of failing to do so because of politics, lack of consistent information, and lack of communication flow. In either case, the cost of lost opportunity and the hard cost of research and development demand that the development team maximize the results of dollars spent.

THE COMPETITIVE ARENA

1. The Competitive Game

A. Upsetting the playing field to your advantage

There are several marketing strategies in use today in terms of leadership roles: guerrilla tactics for niche players, defensive positions, and offensive positions. However, all of these strategies become much more valuable when the new product initiative upsets the current playing field. By finding a way to upset the playing field, you improve your position twofold. The first way is to compete with the existing players with a new product. The second way catches them off-guard as to the approach used. Depending on the uniqueness of the approach, it may be all that more difficult for the competition to respond.

The following example illustrates the concept from two perspectives, long-term and short-term. In the 1970s, U.S. automobile manufacturers lacked certain quality features that the American public was looking for in the product. Several foreign manufacturers embarked on a program to improve the quality and service of their products to the American consumer. By the 1980s, the perception had shifted regarding the foreign manufacturer, from that of shoddy goods to a high-quality, consistent product with exceptional service. Their battle cry: American consumers are entitled to exceptional quality! This changed the playing field for Detroit and, ultimately, severely impacted its market share. This was a long-term change in the playing field that led to a distinct market advantage for foreign manufacturers. Tremendous amounts of time and money were required for Detroit to place the competition at bay.

In more recent history, the computer industry exhibited short-term changes in the playing field to create market advantage. Different architectures in operating systems, software, and license issues affect access and market position every day.

B. Faster, more-informed solving of the customer's problem

A key competitive advantage is to mobilize the operation to excel at determining business and product opportunities by engaging the customer and becoming proficient at designing and introducing products with customer solutions embodied in them. The organization needs to bring formidable factors (internal and external to the organization) to bear on these opportunities. The better one is at this, the better one's chances for growth.

C. Traditional market differentiation has changed

The marketplace is different from that of 10 or 20 years ago. We have experienced well-made products at ever-lowering costs, and better service and more availability of product. Out-of-the-box failure rates are becoming more scarce in different industries. Competitive edge won't come from these areas; they are assumed. The competitive advantage will come from mass customization, applying basic platforms to a diverse set of customers, improved availability, application of specific products, and from low-cost, feature-rich innovative producers.

2. The Strategic Difference—Large Versus Small Companies

A. Large companies: Leverage buying, dominate markets, intimidate, steamroll, and cover the market

In a homogeneous market condition, a large company can have an advantage. By virtue of its size, its capabilities, and its staying power, a large company can dominate a market. If it is a dominant player in the specific market segment, it can influence the future of the segment and its general progress. An entrenched player has the capability to eliminate competition through pricing actions, availability, supply logistics, and reduced costs.

B. Small companies: Niche players and guerilla warfare

Small competitors have advantages of their own, but in different ways. A small company can react to changing market conditions quickly, and capitalize on them. The smaller companies will enter a market to exploit an opportunity and be prepared to leave it at a moment's notice. They do not have the staying power or temperament for a long, drawn-out market battle. By virtue of their culture, they will abdicate certain markets under contest and look for new opportunities. "Fast to enter and fast to leave" is often their operational perspective.

C. While large companies evaluate and strategize, small companies execute tactics

There are distinct differences in how large and small firms, plan, strategize, evaluate, and then act. Figure 2-14 illustrates a review of their respective product life cycles and how large and small companies differ in their approach, execution, and longevity in the specific segment. In addition, the financial profile generated in each of the examples and how they differ are examined. Figure 2-14 shows the traditional product life cycle. There are three distinct phases of the product life cycle (i.e., emergence, growth, and decline). The emergence phase is the start of the new product's business. The volume is low and the product is embryonic in its market share. The second phase is the growth phase, where planned volumes are achieved. The third phase is the decline wherein, if left untended, the company will lose market share and volume will drop off. Generally, at this point the company is no longer competitive in that specific market segment.

Ideally, at the time when volume is stabilized and the company is enjoying good profit margin and good volume, the project to replace the product with an updated or next-generation product

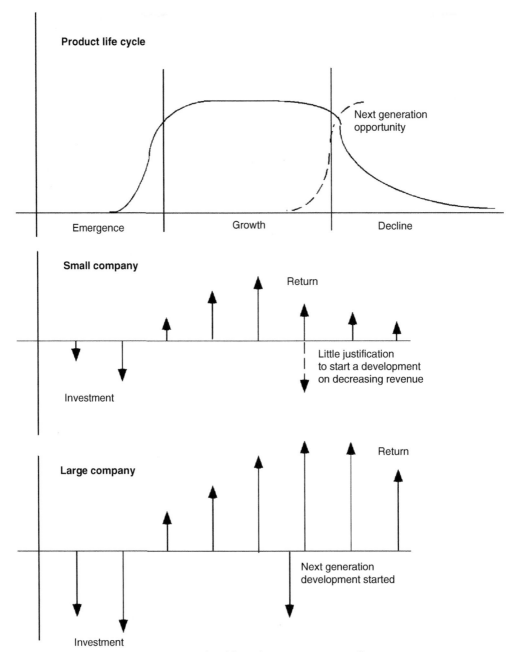

Figure 2-14. Product life cycle: Large versus small company.

is initiated. This is timed so that when this development is complete it will coincide with the reduction in volume of the first-generation product. It is through these follow-on programs, product improvements, and cost reductions that the company will retain and grow its market share. At this time the company may want to reexamine the market and segment the product offering to capture different tiers in the market, thus enhancing market share as time passes.

The difference in the size of the company also determines the differences in implementation. As can be seen from the third diagram in Figure 2-14, larger companies may be slower at the initial implementation but achieve significant volume once the massive selling and manufacturing machine is mobilized. The small company may be lightning-quick at the first-article implementation; however, it will quickly lose momentum compared to the larger company as a result of lack of critical mass. The consequences are clear when the following financial analysis is evaluated, concurrent with the chart in Figure 2-14.

The small company characterization in Figure 2-14 reinforces the assumption that as a product opportunity is discovered, a rapid implementation is generated and put into the marketplace for consumption. The firm makes a small investment to test the marketability. Investment in capital equipment is minimal, and the product is introduced. Initial volume looks promising as it begins to build up in each successive year.

The problem that now results is twofold. The first is lack of sufficient market share initially established to gain hold of the market. The second is cost pressure due to lack of low-cost product platforms. This results from lack of sufficient capital expenditures to keep the costs low. In addition, there is not an adequate barrier to entry in this market from a capital perspective. The development commitment should not be to test the market. It should have been to secure the market.

At the time when the next-generation product should be developed, the outlook for the first-generation product is in decline. Consequently, the company decides that it is not worth pursuing the opportunity, in light of all the other competitors, and seeks another uncontested niche.

In contrast, let's examine a large company's implementation. In this example, the large company researches the opportunity and commits to exploiting it. This involves an interdisciplinary effort within the corporation that focuses on developing the product and the business for the long term. A significant investment is made in development in years 1 and 2. This investment is in developing the product and the manufacturing infrastructure to gain market share. The result is a rapid growth to a favorable market share figure.

As time progresses, the market share and volume increase to a peak. As the volume begins to taper off in subsequent years, the second-generation product has been scheduled for development and is introduced to keep market share stable and growing. The result is a much stronger position from the outset through the product life cycle, thereby generating revenue for profit and subsequent development. This is how a solid commitment in a well-researched program can contribute to long-term growth that can be retained and built upon.

Figure 2-15 is a summary of these differences.

Differences in product execution: large versus small companies			
	Market condition		
	Emerging	**Growth**	**Decline**
Large company **strategy**	Initiate	Growth and market share	Last man out
Characterization	1. Slow to mobilizet	1. Investment in product	1. Retain with other means
	2. Deliberate and methodical actions	2. Interdisciplinary and massive investment	2. Fight to retain and grow market share
Small company **strategy**	Develop and secure	Enjoy differential advantage	Abdicate and find new
Characterization	1. Quick to mobilize	1. Immediate limited growth	1. Growing list of competitors
	2. Lack of long-term planning	2. Missed customer signals	2. Failure to redesign to cost

Figure 2-15. Comparison of large company versus small company product execution.

THE PATHWAYS TO THE PRODUCT

1. Corporate Skills Assessment

A. Business capabilities—you have the heart; do you have the tools?

The honest assessment of corporate skill level in each area of business development is a very difficult thing for most companies to do accurately. The process calls for deep introspection within the organization. In addition, it is very difficult for biased personnel (everyone has biases) to prepare an unbiased assessment. Since there is no absolutely accurate way of performing this assessment, the best we can do is to invite the firm to address the following perspectives in an open, honest, and complete manner. In addition, it may be a benefit to have the customer assist with this assessment.

- What do we do best? (Distance yourself from the sales and marketing literature and focus on the customers' and market's perception of who you are.)
- What do we do worst? (Take an honest look at corporate failures and what the contributory causes were.)
- Who are we kidding? (Take a look again and now come up with the real contributors.)
- How do these align with market opportunities? (Take the time to actually reduce this to paper and look at the comparisons and areas where you fall short.)
- How do we create a sustainable competitive advantage? (Look at each market opportunity to determine how you can create a scenario that allows the company to establish a success in a market and retain the business in that market on a long-term basis.)
- Can we partner with someone and accomplish our goals? (This is often overlooked in an analysis. Many companies cannot partner with another company and accomplish something. Sometimes the team is not aligned; other times, communications systems are not pervasive or do not have appropriate protocols established.)
- Do we have the talent to compete? (Are your firm's sales channel and personnel capable of effectively battling for an order or a piece of business and able to prevail over a competitor? This requires a special skill set; if it is missing, one needs to instill the capability in the responsible group.)
- Do we have the talent to assimilate? (The corporate ego would indicate yes, or that you could soon obtain it; however, in reality, your firm may not be able to assimilate the technology or integrate the technology into a product within the proper time frame.)
- Do we have the temperament for business in today's market? (The acquisition of business in today's climate is fast-paced, highly competitive, and difficult to maneuver in. Accordingly, firms must be up to the task. This will not happen with a fixed mental paradigm or a workforce that lacks some diversity.)
- Do we want the market as badly as our competitor? (Historically, how much has your firm been willing to fight for a piece of business? More important, how much has it been willing to change, adapt, and improve to capture a piece of business? Often, the intent is there in words, but implementation is not there in reality.)
- Are our people motivated to compete? (Addressing the culture of the company, are the employees motivated to compete and prevail? Do they have a winning mentality and act accordingly?)
- How do we view our ability to employ technology? (These next perspectives to consider are a little more difficult to understand since they require both internal examination and external examination. The assimilation of technology and developing the capability

to be conversant with it to the point of creating a product is a long way from the commitment to do so.)

- How do we view our ability in application development? (This is an extremely important corporate skill that is often lacking. It is absolutely necessary to develop the right product and to apply the developed product and create a satisfied customer.)
- How do we view our ability in setting pricing? Are we a cost-plus company trying to operate in a market-price marketplace? Do we know how to design to cost? Do we accept the first cost or do we work the design and procurement until we achieve the desired result?
- How do we view our ability to absorb uncertainty? (Do we have a firm whose employees are averse to digging into details, questioning information, comparing facts and drawing conclusions, and performing tests to clarify data? If so, the firm's ability to absorb uncertainty is a limiting factor in delving into new areas.)
- How do we view our ability to derive the optimized solution? (In the same mode as the cost issue, how does the firm operate in achieving the optimized implementation for a product design? Does it go the extra mile in wringing out all the excess cost? Does it achieve the best performance at the lowest cost and still satisfy customer requirements? These and other items define the company's ability to engage in development in a world-competitive manner.)
- How do we view our ability to profit from an opportunity? (How has the firm historically profited from an opportunity? Is it consistent with the original estimate? Is progress toward profitability measured along the way and steps taken to correct or enhance it along the way?)
- What areas of the organization need remedial assistance or additional infrastructure to be a contributing factor to success? (There are specific tools required to execute the program, and an introspective look at the strengths and weaknesses is in order. The corporate capabilities need to be integrated and coordinated at near-full value. If one of or a set of these required capabilities is weak or missing, it must be addressed before the program begins. Failure to do this will result in a resource-starved program, dooming it to failure of timing, performance, or momentum. Examine each area of the business to ensure it can support and contribute the necessary resources to effect success of the program. If the organization is lacking, fix it now!)
- Can you have access to the skill sets? (Is it possible to obtain the tools for the organization from other sources? In the next section, a variety of ways will be discussed to allow companies to compete in nontraditional markets by partnering for capabilities. The objective is akin to obtaining a cut sheet of lumber, not to necessarily own a sawmill. If this is the case, don't go looking to buy a sawmill, unless you use so much cut lumber that it is in your firm's best interest to invest in capital to become a low-cost producer and have control of the operation. This would only take place when usage drives the make-or-buy decision. Obtain the tools that make most sense to the operation and the new product development.)

B. Partnering for capabilities

1. Why collaborate?

Technology progression is moving at an incredible rate. This places the absorption of uncertainty of business development programs and business operations at an ever-increasing expectation to achieve the required assimilation of technology—the cost and time to develop technology can cause companies to expend vast amounts of resources. Remember the forecasts for business levels due to new products many years ago? The maintenance of these products is also costly to your firm in time and personnel, making it increasingly difficult to devote talent to new business development. These factors have historically driven companies to begin the process of collaborative development. However, there are both commercial and technological issues associated with the collaborative business development process.

Commercial collaboration issues

On the commercial side, business practices and methodologies are being developed rapidly. Also, they are being developed all over the world with different value systems attached to the methods. It is very difficult for a single company engaged in the process of conducting normal operations to be expert or even conversant with these methods.

Having a collaborative partner to assist in tracking the evolution of best business practices and routes to market is a plus. In addition, these collaborative partner's experiences can assist in channel management and how to best navigate market contention issues. It is also quite difficult, given the complexity of our world, for an individual company to select the optimum long-term manufacturing venue and to create manufacturing supremacy.

Will your partner be as willing as you are to make certain sacrifices, given their status? This investigation also extends to ownership structures and motivations that will influence this commitment. Also, what is their degree of sophistication regarding relevant groups and what is their experience with collaboration?

Technological collaboration issues

How will the intellectual property issues be handled in the cooperation? Who will pay and who derives the benefits? Where does the technology reside? Who has the core technology? Who has the right to spin off products? Many times these issues are glossed over in boilerplate contract language and problems can arise later.

Once the products are launched and the business is developed, who governs and provides ongoing product support? If the product requires changes, which partner will provide these changes? If both companies are marketing the product, how do customer-driven changes get incorporated? The ability to adapt to changing market conditions is critical to success. When there are two cooks to the broth, trouble can occur, especially when there has been no previous agreement.

Finally, in the absence of formal intellectual/property protection, trade secret protection can be an excellent way to protect products and processes. The partner company must therefore enforce disciplines to keep trade secrets, secret. Accordingly, this must be a factor in the collaboration negotiation.

Finally, there is the issue of match between the two collaborative partners: Are they equally experienced in the process or does one have far more experience developing and launching products in this arena? Take time to define the complementary technical prowess and experience with collaboration.

C. Don't be proud; be practical and careful

Corporations are not inanimate entities; they are composed of people. People have needs, emotions, and biases. People feel threatened and react to those feelings. Many a logical, forward-thinking, and beneficial idea has been sidestepped rather than embraced because of these factors. Consequently, the notion of securing assistance or know-how from outside of the organization often does not receive wholehearted support. It is the manager's responsibility to ensure that the organization rises above these factors to embrace the tools required for success, wherever they may come from.

D. Getting the tools and remembering the objective

There is an unwritten corporate covenant in business development that exists between senior management and the new product development champion. It is a covenant whereby management will accept the developer's leadership in exchange for a certain amount of due diligence in implementing their program. Consequently, when obtaining the tools needed for a program it is necessary to exercise care in their procurement. Get only the tools needed. Get the tools as a means to the new product end. It is a vocational hazard to quickly get in the habit of procurement on a grand strategic scale and bypassing the critical mass items required for just the program. The operative point is that these tools must be absorbed and amortized by the program, so restraint is best practiced at this time. In the course of procurement it is wise to procure in a forward-thinking, forward-compatible manner. Where it is a small, incremental cost to prevent obsolescence of equipment, expenditures should be made freely. When getting the tools, remember the objective. In the same way you engage the customer to assess where they are, it is a requirement to engage the partner in a more detailed way. Find out their plans, aspirations, and desires. Only in this way can you determine the best deal to strike for both parties.

E. Contingency plans, evaluating with a partner, and future in-house capability

"The best laid plans of mice and men often go awry." This saying implicitly defines the scope of new product development, but especially in the framework of partnering. Knowing this is an accepted fact, make sure to plan contingencies within the partnering arrangement.

An unfortunate fact of life is that unforeseen driving forces can alter a partnering relationship for reasons unrelated to the development, its progress, or its potential. A financial hardship may present itself to the partner of your company and force abandonment of the program and their contribution.

There are also instances in which management focus may change within the partner's management structure. A different agenda may be embraced, or a separate alliance that may potentially overshadow or conflict with the partnership may have been reached. Therefore, it is best to identify alternative and contingency plans at each stage of the process of partnering. As the deal is negotiated, constantly ask yourself, "What if this doesn't work out? What will be my options to achieve the new product development objective?"

One possible alternative may be to plan on absorbing the partner's technology into your organization after the term of the partnership. In many cases, a relationship where this is identified at the outset may eventually be the best for both companies. The initial cooperation is therefore very telling in how the relationships will progress.

Carefully watch the partner's behavior in several situations. Assess their loyalty to the program and to the team of personnel involved. Determine if they act in a long-term, decisive manner or if they operate in a day-to-day, tactical mode with loose focus. It is imperative to assess this quickly in certain programs replete with uncertainty because you will need to evaluate the difference between time lost in overcoming problems or chasing other opportunities before completing the tasks at hand.

Take care to set up a communications infrastructure that preserves secrecy where necessary but encourages a free exchange of knowledge where it enhances the effort, relationship, or the program. Also set up communications at different levels between the two firms, as shown in Figure 2-16.

As the figure illustrates, your firm and the partnering firm should set up a communication protocol that is multilevel. Patterned after a hierarchy, bilateral communications can take place at the operative level. For most issues, if nonconcurrence occurs at the lower level it can be arbitrated at the next level. Senior management preserves and orchestrates the relationship between the two firms. Avoid single-point communications because it is only as strong as the individuals at that point. Multiple communications can also allow information to be brought to light in a nonbiased way, whereas the single-point entry can be prone to one individual's biases.

F. Collaborative communications

The following is a list of the possible levels of communications in a collaborative partnership arrangement that may be typical:

- Executive level communications (product planning and product evolution, product management)
- Development management
- Manufacturing
- Aftermarket

There are also protocols for different stages of a program:

- Beginning of program
- Development
- Problem resolution

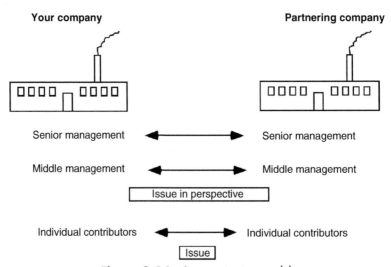

Figure 2-16. Communication model.

- Liability exposure
- Loss mitigation
- Reconstitution after a setback
- Hotly contested competitive areas

H. Focus on the mutual opportunity

The basic concept behind partnering is that, together, two entities can achieve more than either one could alone. The partnership fills in the gaps in the individual players. To reiterate, remember this objective: 100% of nothing is not nearly as beneficial to your company as 50% of something.

The very structure and operational dynamics of a relationship between two companies contributes to diffusion of focus. It will soon be apparent that many interesting things surface from the collaboration, not necessarily related to the program. It is the manager's responsibility to refocus the companies on task. In addition, the partnering relationship may not work out for every company. There has to be a certain corporate appreciation for the differences between the players, their history, and (very important) the relative importance of the project to each firm. As you structure an arrangement, ensure that the players selected have the patience for the partnership.

I. Concepts of corporate strategy

There are differences in various companies' corporate strategies. Although the following model could have been presented in the planning section in Chapter 1, it is more relevant here where it can be recognized in a potential partner during negotiations. Presenting a "hard and fast format" for corporate strategy in the planning section would tend to preempt creative thinking by encouraging comparison and locking onto a specific type prematurely. When contemplating or planning a joint venture, brand label arrangement, or an acquisition, it is critical to determine the seller's corporate strategy. Although most companies espouse some sort of strategy, some do not practice it effectively. In some cases, this depends on the size of the company as well as the ownership structure. For example, a small, privately owned company will execute a strategy differently than a large firm. This is because of three factors:

1. Influence of top management in the process
2. Sensitivity to changing market conditions
3. Inertia of the organization.

A large company sets its pathway and starts execution. It directs its inertia to start in the chosen direction. Changing market conditions generally do not affect this. This is the appeal of the smaller firm, which can react to changing market conditions. This resiliency, however, is a danger to executing a plan because management may change direction more often, thereby negating any momentum. It is incumbent on the acquirer to understand these dynamics.

It would be tempting to create a chart of the various scenarios of acquisitions, joint ventures, and brand label arrangements to summarize the corporate strategies of each, and how to react to them; however, each case is its own set of circumstances that directly affect the dynamics of a deal. Consequently, each must be evaluated and a pathway must be charted through the circumstances.

A large company with a lot of long-term debt may sell off a division at some price not necessarily reflective of its value, and recognize the sale on its financial statement. A smaller firm would execute the transaction totally differently. It is important to investigate the driving forces *behind the deal* as you are *creating the deal*. This can have the most payoff when negotiating.

Some of the factors that come into play are as follows:

- Large versus small
- Public versus private
- Technology developer versus pure manufacturer
- Product offering, which is commodity versus service
- Profitable versus not profitable
- Management under fire versus not under fire
- Availability of cash
- Competitive nature of the deal
- Perception in the marketplace
- Perception of weakness
- Limiting factors on one firm in the deal
- Recoverability index for both companies should a deal go bad

These frameworks also factor into the success of a launch of a new product.

This list outlines the various concepts and practices of corporate strategy. It is desirable to study the philosophy of each of the types and to become familiar with them as you negotiate the partnership arrangement. In this way it will be easier to evaluate the partner's motives, pressure points, and actions.

Figure 2-17 is a summary of corporate strategies and their characteristics.

2. Internal Product Development

Once a firm assesses its capability (overall operational ability) to pursue the business and product opportunity, it must now determine the best way to secure the opportunity and make it a sustainable enterprise. This is often a difficult decision for both senior management as well as the business development practitioners. If the firm has the internal development capability and also is conversant with the next-generation platform of the technology, it can start immediately on the program. This is a rare occurrence for a brand-new product. More often, there are certain elements missing from the organization that must be integrated. Although the next sections present a hard, defining line between the internal development and partnering development, the more common approach may be somewhere between the two ways of approaching the program. They will be presented separately here, however.

A. Internal skills development

Most companies do not have all of the skills required to secure a market with a new product. Therefore, as part of the development program there must be sufficient planning to upgrade corporate capabilities. This means time and funds must be appropriated accordingly. This is a difficult thing to do because most teams are anxious to get started and often understate the time and expense of integrating and assimilating new technology. One need look no further than the time it takes to integrate a new piece of software advertised to cut development time!!! It takes development time to learn!!!

Corporate strategies and characteristics			
	Strategic prerequisites	Organizational prerequisites	Common pitfalls
Portfolio management	Superior insight into identifying and acquiring undervalued companies	Autonomous business units	Pursuing portfolio management in countries with efficient capital marketing and a development pool of professional talenty
	Willingness to sell off losers or sell off winners opportunistically	Small, low-cost corporate staff	Ignoring the fact that the industry structure is not attractive
	Private company or underdeveloped capital markets	Incentives based on business unit's success	
Opportunistic fix-it strategy	Superior insight into restructuring opportunities	Autonomous business units	Mistaking rapid growth or a "hot" industry as sufficient evidence of a restructuring opportunity
	Willingness and capability to intervene and transform acquired business units	A corporate organization with the talent and resources to oversee turnarounds and strategic repositioning of the business units	Lacking the resolve or resources to take on troubled situations to intervene in management
	Broad similarities among the units in a portfolio	Incentives based largely on acquired units' results	Ignoring the fact that the industry structure is not attractive
	Willing to sell off when restructuring is complete and market is favorable		Paying lip-service to restructuring but actually practicing portfolio management
Transference strategy	Proprietary skills in activities important to competitive advantage in target industries	Largely autonomous but collaborative business units	Mistaking similarity or comfort with new business opportunities as sufficient justification for diversification
	Ability to accomplish the transfer of skills among the business units on an ongoing basis	High-level corporate staff whose roles are those of integrators	No practical provision for skills transfer
	Acquisitions of beachhead positions in new industries as a base	Cross-business-unit committees, task forces, and other forums to serve as focal points for capturing and transferring skills	Ignoring the fact that the industry structure is not attractive
Integrative strategy	Activities in existing units can be shared with new business units to gain competitive advantage	Objective of line managers is to include skill transfer	Sharing for its own sake rather than for competitive advantage
	Benefits of sharing outweigh costs	Incentives based in part on corporate results	Assuming sharing will occur naturally without management playing an active role.
	Both start-ups and acquisitions as entry vehicles	Strategic business units that encouraged to share expertise and activitiest	Ignoring the fact that the industry structure is not attractive
	Ability to overcome organizational resistance to business unit collaboration	An active planning role at group, sector, and corporate levels	
	Fostering collaboration, not competition	High-level corporate staff members who see their roles primarily as integrators	
		Incentives based heavily on group and corporate results	

Figure 2-17. Summary of strategies.

Many technologies require capital equipment support and that requires funding and time. It is best to accurately assess and state the requirements and load them into the program schedule. In this way you will be on the right technology platform and be in a technical position to actually execute the development on time.

B. Management development

The lack of experience may not lie with the core development capabilities alone. It may lie with management. Consider a group whose best efforts are put forward in developing software for a product. Each week some progress is made; however, the end is not in sight and lingering bugs keep cropping up. A change in the code to fix one problem introduces another problem. In this case, the development group is too close to the problem resolution and is feeling the pressure to perform. They may have lost objectivity. Even though they are working flat-out to resolve and complete the software, an experienced manager may have to intervene and guide the group to a different way of completing the program. Painful as this may be, the manager must have the experience to recognize the technical impasse and navigate a way around it. Not all mangers have this capability or experience. Failure to have the experience will result in missed deadlines and potential program failure.

C. Funding

As we will see later, there are two important aspects to funding. Obvious as they may seem, they are often minimized and can cause trouble later. The first is the actual amount of funding. This includes the time and money to implement the program. Complete funding is often not budgeted or thought through at the beginning. Management may take a conservative approach by metering out a little funding and watching what happens. This lack of complete appropriation places the program at risk for follow-on funding because another opportunity make look sweeter later on. Get complete appropriation at the beginning, with phased draws, and segregate it from other programs.

The second aspect is where senior management may place one group in competition with another and fund the perceived winner. The cold truth about these situations is that the group does not have the confidence of management to drive the program to a successful completion. If there is no confidence, there will be a lot of waste with this approach. Strive to get complete funding or at least an unencumbered pathway toward it.

3. External Means to Product Development

A. Model for brand label, joint venture, and acquisitions

1. Perspectives on partnering

A brand label, joint venture, or acquisition for purposes of new business development *can* be a successful approach. These relationships allow the company to gain access to markets quickly with products that are on target.

Partnerships, however, do not last forever. They are simply a means to an end. Consequently, initial plans in negotiating a venture should include plans for how the company will absorb the technology, buy the company, or continue the partnership past the term. Failure to adequately plan this at the outset will result in consequences at the end of the term, with minimized alternatives and options. After all the effort to initiate a development and create a business, it can be devastating to lose control of the business because of a lack of transition planning at the beginning of the negotiation.

This section will present a model for brand labels, joint ventures, and acquisitions from a product development perspective. It can be used as a planning guide for initial discussions and as a check-off list for implementation.

2. Levels of investigation and discussion

There are four basic levels of investigation required to understand another partner's business. Depending on the framework of the partnership (i.e., brand label, joint venture, or acquisition), the more in-depth the investigation should be. An acquisition will require more information than will a brand label arrangement, for example.

The levels of investigation are therefore arranged in increasing depth into the partner organization. Level 1 is cursory, whereas level 4 delves into the core of the company—its finances and structure. The following is a summary of the levels:

Level 1

General

The general concept behind level 1 is to investigate if it is feasible to create a deal between the two companies. This section allows a summary of the opportunity to be documented.

Background

The background outlines the market driving forces and events that led up to the identified opportunity.

Market information (secondary)

The market information cited here is secondary and is obtained from trade literature, periodicals, and industry-published data. It is differentiated from primary market data by the fact that it is information obtained indirectly rather than through primary customer contact. It is useful as trend data and initial substantiation, but cannot be used as verifiable market data. That must come from primary customer contact.

1. Size, demographics
2. Market position
3. Market plan

Contacts (Who are the company contacts?)

1. Hierarchy (What is the organizational hierarchy?)
2. Key players

Structure of the deal

1. Diagramatic
2. Product flow
3. Support

Level 2

Relation to overall business development plan

Establish how the proposed partnership arrangement fits into your company's overall strategic plan. Is this a means to gain expertise and immediate entry to a marketplace? If so, what is the plan for after the term of the agreement?

Commercial

Take care in outlining the commercial implications of the arrangement. If a careless analysis is performed, the entire objective of the program could be missed.

1. *Costs and pricing*
   Who are the key players in both companies and who are the deal-makers and -breakers? What is the basic structure of the deal? This should be constrained to a single-page explanation that is understandable to all concerned. Diagram the relationship between the two companies and show what goes where, and where everything comes from. If this cannot be done easily and understandably, recheck the thinking of the opportunity and the partnering relationship. Something is not right. Establish where the product comes from and how it gets through the infrastructure to the customer. Place notations on the diagram to note key points, areas of concern, and potential failure areas. Show how the product will be supported during the relationship. If you can complete this in a relatively short amount of time without major conflicts, the basic tenets of a deal are in place to build on. What are the projected costs and prices? Is there enough margin for each of the players? What will happen if the marketplace puts price pressure on products, creating margin pressure?

2. *Feature/benefit mix*
   Carefully outline the feature and benefit matrix and tie it to price. Do you have the properly configured platform to take to market with your partner? Is the product properly positioned in the marketplace at the time of product introduction?

3. *Ramp-up plan, volumes (reasonable and achievable?)*
   Check to make sure that the two firms are capable of the market-driven ramp-up. Is there a manufacturing infrastructure to support the projected business levels? Are the business-generated unit volumes required to justify the program reasonable and achievable from a manufacturing perspective?

4. *Primary market information to check secondary market information*
   Be diligent in performing primary market research to verify feature acceptance, pricing, value, and trends in the marketplace. Does the product need to be changed to fit the updated market criteria?

5. *Best-case scenario and effects*
   One of the best ways to determine feasibility and reasonableness is to outline a best-case scenario and the assumptions needed to achieve it. Also project this best-case success onto the organizations (both yours and the partner organization) to determine the effect it may have. Will success change the philosophy of the agreement and the principals? Will the contributing organization feel well-served in the transaction, or will they tend to regroup and exercise an alternative strategy?

6. *Worst-case scenario and effects*

   Likewise, create a worst-case analysis to determine downside effects. Also include the effect of missed opportunities because of time pressure. This can be used to create a sense of urgency in situations that may need arbitration between the parties. Also take time to examine effects of a worst-case analysis in terms of actions that may be taken by both companies. If the contributing company senses that the venture will not work out to plan, they may have a tendency to regroup and refocus effort elsewhere, which will further jeopardize the venture.

7. *Most likely scenario and effect*

   Finally, determine the most likely scenario that could occur. Hopefully, this scenario will generate the best-desired result for both parties. Take time to also project your plans and proposals for after the term of the agreement, based on the success. This can provide the backdrop for future negotiations.

8. *Product line analysis: second-generation*

9. *Order flow*

Technical

1. Meets functional description
2. Meets relevant standards
3. Meets brand-labeled documentation
4. Manufacturability study and cost reduction
5. Quality system assessment

It is also helpful to project the second-generation product analysis at this time. It forces a realistic, critical examination of the features and benefits to determine market fit. If a fit is lacking certain features at this time, the second-generation analysis will highlight it and allow a judgment to be made. At this time it is also beneficial to generate a flow chart of product and order flow. Design the process by which product orders will be entered and processed within the organizations and delivered to the customer. If there is partnership beyond design and manufacturing through to the sales channel, then it becomes necessary to outline the protocol and flow of product here.

This area deals with the technical aspect of the arrangement. There are a host of elements that must be orchestrated between the two firms to effectively meet the customer's requirements, and this section will outline those requirements. Does the product, as designed, meet the functional requirements of the marketplace? Does it meet the basic features and performance criteria at a deliverable cost? Does the product, as designed, meet the relevant standards that govern the marketplace, or is redesign necessary to incorporate measures to meet these standards? Has the required documentation been effectively prepared to be acceptable to the marketplace? Is it consistent with the corporate requirements of the brand labeler? How about the manufacturability of the product? Are the manufacturing strategy and product platform open-ended or close-ended from an enhancement and cost-reduction point of view? What is the level of quality system in the contributing and utilizing corporation? Are the two systems consistent (because this is required if one corporation is to design and the other is to produce)? Are the systems capable of continuous improvement and tracking each other? This is important to resolve at this level because it is virtually impossible to reconstruct one in the midst of a product problem or recall situation.

Risk analysis

Finally, as part of a level 2 investigation it is desirable to perform some numerical financial risk assessment in. Also, what is the qualitative risk in terms of reputation and positioning in the marketplace? Look at the partner company's income statement to see trends and business cycles. Cyclical activity generates cyclical performance and rearrangement of priorities over time. This reprioritization can affect progress on a project. The balance sheet can indicate the structure and financial strength of the potential partner. Lack of financial strength may cause delays and possibly require additional contribution of cash on your company's part to complete the program.

Level 3

Review partner's business plan in relation to original estimates and management track record

A level 3 assessment will evaluate the partner company's ability to set plans in motion and execute them to completion. It is important to know this perspective, since the proposed venture will depend on their ability to be decisive in their actions. As an individual manager, it is also a good idea to do a self-assessment at this time in order to evaluate your *own* company's ability to carry out a plan.

Detailed review of financials and performance data

A level 3 assessment will involve a detailed analysis of the financial performance of the company. This is the level where a joint venture would require a form of due diligence investigation. Because a joint venture would require significant deployment of intellectual and hard assets, your company must perform enough of an investigation to protect these assets.

1. Income statements
2. Balance sheets
3. Funds flow
4. Structure of organization, number of shareholders and who they are—the interrelationships, the length of investments, interviews, and profile of investors.

    Get a feel for the ownership structure of the firm. Who are the key decision makers on finances and policy? Determine if there will be any planned change of ownership within the term of the deal. Much of this information may not be located in a single package, neatly wrapped up for your review. However, by listening carefully and gathering secondary information through third-party services (e.g., Dun & Bradstreet, Standard & Poors), you can amass a fair amount of information.
5. Review accounting and sales order entry systems for accuracy, compatibility, and consistency. Are the expenses recognized similarly from year to year? Is the accounting system accurate? Does the inventory system reasonably track the value of the inventory without adjustments every year-end? Can the sales order entry system be aligned with your system with some means of internal controls? Are the two systems effectively compatible enough to support the deal and keep track of the required data? What is the measurement of the cost of quality, and is it consistent with your system? The thought here is that the two systems need to be compatible enough to measure performance of the venture to the satisfaction of both parties.

6. Investigate the cost reporting system and cost rollups (target versus actual). Investigate the specifics of the cost roll-up system so that product costs are kept accurate from year to year. This will ensure that there will be no disagreement on profits or margins in case the agreement calls for shared margin or percent of profit. If possible, it would also be of value to your company to assess the original and actual product costs historically within the partner company, because this will demonstrate their track record in achieving a design-to-cost regimen.

Level 4

Company fitability: Yours, theirs, and overall plans of both

A level 4 investigation is a more comprehensive analysis of the partner organization from an acquisition perspective. It is part and parcel of the total valuation of the company that will be purchased. It is the measurement of hard assets, know-how, and the ability to please the customer and widen its presence in the marketplace. However, a company is not a thing to be weighed and bought by the pound, so to speak. It is a living, dynamic organization composed of nonlinear, emotionally driven, and often unpredictable people whose diligence and effort contribute to the numbers being evaluated dispassionately. As such, people have hopes and dreams and goals. How are these transferred into the organization? What are the goals of the organization and how do they compare with your collection of contributors? Knowing this is essential to effecting a good marriage between the two companies. This effort must transcend the new product development portion of the agreement and permeate the relationship.

B. Valuation of company

The decision has been made to purchase the company, and now it becomes necessary to evaluate its worth. In many ways, worth is in the eye of the purchaser, and desired value on behalf of the seller is based on circumstance. In the simplest terms, there are a minimum of five ways to value a going concern: owner's investment value, a balance sheet analysis, a rate of return, income statement analysis, buyer investment valuation, a liquidation alternative, or a replacement value calculation. Each of these circumstantial ways may affect the perceived value of the following items; however, it is still necessary to have a starting point for each of these.

1. Assets, cash property, plants, and equipment

These hard assets comprise the visible portion of the balance sheet where a buyer can touch tangible things. Although these items are procured and depreciated over time to a very low value, they still represent a significant portion of the business's value at the time of acquisition. Watch out for overstated intangible assets in a deal.

2. Liabilities

These represent the obligations of the business. They are the bills and expenses that your company will incur after the transition.

3. Goodwill

This intangible represents the reputation of the organization in the marketplace (e.g., brand names and longevity of relationships with customers), and is often used to accumulate dollars beyond hard, identifiable assets in an acquisition. This is true where the purchase price exceeds hard asset value alone.

4. Orders

This is valuable from a study point of view to see where the orders are coming from and what the orders consist of. By examining the orders of an organization, several useful things can be learned. Elements such as number of orders, line items per order, purchasing authorization, and pricing and discount classifications all tell a story about the business in question.

5. Backlog

The backlog is useful in determining the inertia of the company. It provides the stabilizing influence during the transition. A large backlog gives the acquiring company some breathing room in making changes and absorbing the acquisition.

6. Receivables

This represents the monies owed to the company. A close examination of the accounts receivable records will indicate how the company does business. Are terms extended to customers as an inducement to offset product shortcomings? Are late payers overlooked as an appeasement for product issues? Are the receivable days out of step with the industry norm?

7. Inventory

By and large, the inventory is one of the most uncertain aspects of an acquisition. Of all of the tangible assets, the inventory is the only one that is recoverable out of operations. Inventory is of very low value on the street or to anyone other than the company. It must be called out by a clean set of product documentation and be useable as delivered without significant addition of labor to recover its cost. Aftermarket inventory support levels may be large due to obsolescence and growing aftermarket business. This is cause for further investigation.

8. Headcount

How does the headcount compare to industry norms? What is the dollar contribution for both direct and indirect employees? What are the trends in headcount, past and projected? What is the average longevity of the employees? A lot of information is available by researching this area thoroughly.

9. Trends analysis and pro forma projection

Finally, what does the trend analysis say about each of these measurements? Does it foretell a growing, vital organization or a stagnant one? What has the organization been through recently, and how are they positioned for change?

C. Structure the valuations with risk analysis

As a purchasing decision is being made, a summary of the valuations and options and their substantiating data needs to be reviewed. With each one, a risk analysis needs to be incorporated as part of the review. This risk analysis will shade the valuation with a bit of realism in pricing each alternative.

D. Comparative analysis with industry averages

Every company is different and must be evaluated based on its own merit and future contribution to your company. It is therefore only a measurement and not a valuation to compare industry averages between companies. In the acquisition arena, when the reason for the acquisition is to generate a new product, the comparative analysis and industry averages assessment can only be used as a guide. As mentioned earlier, there are five basic ways to value a business. They are as follows:

1. Set the price five ways

Consider the following:

1. Owner's investment value
2. Rate of return
3. Buyer investment value
4. Liquidation value
5. Replacement value

Each of these has a justification for the valuation and each can be used to modify a valuation; however, there is an additional consideration that must be given when acquiring a business—when the primary driving force is to obtain intellectual property, trade secrets, or know-how.

If additional new business is to be built on the existing business, then the complexion of the original will change. Investments may have to be made, and profit structures and cost structures may be different. In these cases, it becomes necessary to perform a more in-depth analysis of the business and the pro forma investment in it, not only from an acquisition perspective but also from the incremental investment perspective.

2. Set terms

Finally, set the terms of the deal. If the partner company must generate a new product as part of the deal, then the sales price and terms of inflow of cash will alter accordingly because the intellectual value does not exist until these programs are well on their way to completion. When constructing deals in these areas of new product development, linkages and demonstrable results take on new importance as monies change hands.

3. Matrix of analysis

Figure 2-18 represents a matrix of analysis for the different product partnership arrangements discussed in this chapter. It is a matrix of essentials for investigation in negotiating a brand label, joint venture, or acquisition arrangement. Although this list is not presented in its entirety, it does serve as a useful tool in determining scope of investigation in an arrangement.

4. Miscellaneous topics to agree on

There are several miscellaneous topics that should form the basis for negotiations as a deal is put together. Although each is not a deal-maker or -breaker, they should be addressed from the perspective of completeness in that resolving the issues will make the deal go that much more smoothly. Each of the topics, along with a short explanation, is included here for your reference and should serve as a check-off list in negotiating a deal.

Matrix of essentials in a brand label, joint venture, or acquisition arrangement			
	Brand label	**Joint venture**	**Acquisition**
Level of investigation			
Level 1	X	X	X
Level 2	X	X	X
Level 3	N/A	X	X
Level 4	N/A	N/A	X
Miscellaneous topics	X	X	X

Figure 2-18. Matrix of analysis.

- Exclusive rights
  - Domestic, North America, global, time frame
  - Payment for certifications
  - Product label—buyer, seller, both
- Guaranteed sales volumes
  - Unit sales per month, default volumes, escape clauses
  - Monthly forecasts—production, advance notification of product changes
- Salesforce (markets served)
  - Buyer's relationship to existing salesforce, channel management after the agreement
  - Buyer's channel profile
  - Seller's channel profile
- Authorizations
  - Engineering changes, product evolution
  - Chain of command for interactions, organization chart
  - Returns based on final specifications
- Payment responsibilities
  - Projects versus enhancements versus normal engineering support or applications engineering support
  - Individualized services
  - New products research and development
  - Field services, trouble calls, and authorizations for field returns
  - Travel costs
  - Warranty
  - Training—buyer's people, sales channel, field service
- Research and development
  - All development projects through seller? Or buyer? Or combination?
  - Employee utilization and priority requirements if seller is a small firm
  - Projects tied to contract

- Production—sellers
  - Quality control to be proved, meets minimum standards
  - First level of traceability
  - Gear-up time to manufacture
    - Training
    - Inventory levels
    - Product testing
  - Support needed from seller
    - Setup
    - Repair
    - Training
  - Lead times
    - Determine the lead time (in weeks) on 90-day projection
    - Schedule production with shipments starting on 91st day
  - Storage of units (warehousing, seller and buyer)
- Production—buyer
  - Tied to quality requirements
  - Tied to quality specifications
  - Return authorization requirements
- Default clauses
  - Sales
  - Payment
  - Quality
  - Manufacturing flow
  - Development
- Cancellation clause by either party
- Investment and stock exchange
  - Right of first refusal
  - Restrictions apply to transfer
  - Cap on controlling shares
  - Preemptive rights, month and year dilution of stock

5. *Checklist for brand-label arrangements*

The following is a checklist Figure 2-19, that can be used for setting up a brand-label arrangement. It contains the necessary topics for discussion and resolution between two partner companies.

- Diligence checklist for a joint venture
  The following Figure 2-20, represents a diligence checklist that can be used when crafting a joint venture or even as an initial acquisition survey.
- How to handle a forced arrangement
  There are times when the new product manager may not be part of a brand label, joint venture, or acquisition decision but may be required to manage a new product development as part of it.
  Often decisions are arrived at rather quickly without a lot of thought given to the details. If this is the case, there are several questions that you as the new product manager need to clarify as your company proceeds down this pathway.

Issues to resolve in a brand label arrangement				
Marketing				
	1	How was the marketing study done? Primary, secondary		
	2	What is the product cost structure?		
	3	Pricing, feature, and competitive comparison		
	4	Commercial feasibility		
	5	Fit with overall strategy		
Development				
	1	Profit and cost sensitivity		
	2	Functional testing		
	3	Quality track record		
	4	Technical feasibility/analysis		
	5	Standards testing and conformance		
	6	Design validation		
Channel Issues				
	1	Exclusive arrangements and noncompete		
	2	Reseller revenue structures		
	3	New market identification, development, and promotion		
	4	Intended route to market		
Production				
	1	Manufacturing plan		
	2	Manufacturing venue		
	3	Capital equipment		
	4	Manufacturing documentation		
	5	Change notices and support		
	6	Vendor linkages		
	7	Production testing		

Figure 2-19. Brand label checklist.

Quality					
	1	Quality system and reporting			
	2	Corrective action system			
	3	Track record for similar products			
	4	Warranty costs			
	5	Field feedback system for improvements			
	6	Quality event chart			

Figure 2-19. (Cont'd)

Diligence check list					
Item	**Description**	**Responsibility**	**Substantiation**	**Date 1**	**Date 2**
1	Product opportunity fits scope of corporate strategy				
2	Consistent organizational system for evaluation				
3	Consistent criteria				
4	Documented and followed procedures				
5	Linking department objectives				
6	Sound market planning				
7	Positioning				
8	Segmentation				
9	Capitalization				
10	Pricing				
11	Paced development to capture opportunity				
12	Product distinctiveness				
13	Supporting evidence of opportunity				
14	Sizable market with research and forecast				
15	Precise product definition				
16	Solid design criteria and review				
17	Technical feasibility in the company				
18	Achieving target factory cost objectives				
19	Projected competitive response				
20	Stability of the team/company				

Figure 2-20. Diligence checklist.

There are five major areas of concern when a deal is forced into the organization. As a manager, it is in your best interest to resolve the issues defined in the five areas as part of the overall plan. The five areas were previously outlined in Figure 2-19 and the diligence checklist is presented in Figure 2-20.

4. Keeping Balance Internally and Externally

Managing a joint development program, or even a brand-label program, is an exercise in pleasing many principals. The marketplace has requirements that need to be met, the partner company has needs, and your company has needs. At times the demands of each may conflict with internal pressures, and the manager needs to arbitrate issues. Keeping these requirements in balance is essential to the program's success. The interrelationship between the issues and the individual factions needs to be identified at the outset and managed throughout the term of the deal. If one area fails to make its own share of the sacrifice, the relationship eventually will be in jeopardy.

Since you're charged with the responsibility of acting in your company's best interest, and the partner company is a participant, your role is one of chief negotiator; each issue must be resolved to the company's and market's satisfaction in a timely and decisive manner. Don't waste time getting to market by focusing on arbitration and trying to please everyone. Execute the deal and bring the product to market in a form that will be accepted by the market.

5. Dealing with Entrepreneurs, Licensing, and Technology Transfer

Dealing with technology in a raw, unprocessed form (as in the case of entrepreneurs) can be difficult. The challenge occurs when taking raw technology and embedding it into a manufacturing organization. There are differences between the entrepreneur and the manufacturer that cloud the perception of completeness and manufacturability. Inventors solve a customer's problem by creating "designs to suit your needs." The manufacturer serves a cross section of a market, builds unchanging product in volume, and resists change by its desire for uniformity demanded by its structure.

Both the entrepreneur and the manufacturer have similarities that must be preserved, such as desire for quality designs, commitment to customer satisfaction, and commitment to the venture. There are differences, however, that must be resolved, such as designs that must be frozen and manufacturable and problems that must be jointly and swiftly addressed. Both parties must diligently work toward equitable participation in the venture.

A. The benefits of short-term and long-term trade-offs

1. Double-edged sword

When sculpting a new company by executing a long-term growth plan, there is a balance that must be struck between short-term and long-term trade-offs. The long-term view requires investment and time not recognizable in the short term. This means that the manager and the organization must exercise patience in achieving the long-term objective and be decisive in preserving the investment during those times when it is financially tempting to terminate the expenses.

Negotiating and executing a brand label or joint venture agreement will require a certain amount of street smarts in navigating both companies through the relationship. Each of the company's allegiances, pressure points, sensitivities, and goals desired out of the deal must be recognized. If you watch where the monies change hands, it will be easier to understand the dynamics of the relationship.

2. Don't get held hostage

Administering a venture between two companies can leave your company vulnerable in that a commitment has been made to see the venture to completion. Alternatives are no longer as beneficial or available after you have locked onto the partner company. Consequently, you run a risk of being held hostage with their technology and know-how. Being held hostage can have long-lasting negative effects that transcend a single obstacle or crisis.

The only way to counteract an attempt by an employee or a partner company to hold you hostage is to have a good alternative waiting in the background to take over. In fact, you may want to provide for this at the outset in the contract clauses. In any event, the chances of being held hostage must be mitigated by having alternative means to implement the program and demonstrating the resolve to exercise it.

It is said over and over that a good deal is the result of a win–win situation for both parties. It is fundamentally based on the fact that if both parties do not benefit from the venture, sooner or later the deal will fall apart. Neither party will knowingly continue a relationship indefinitely without tangible benefit.

3. Making it last

Given that win–win is the preferred result, the best way to preserve a deal or to continue a deal that has come to term is to structure additional benefits for both parties. It takes a significant effort to make a deal work in the first place and, if one can be extended, additional work can be saved. Leveraging the ventures now adds to the worth of the company by continuing to achieve results without significant additional effort. In addition, the individuals involved begin to know each other and overall communication improves.

4. Benefits: Quick access, experience, willing partners, and instant critical mass

To reiterate the benefits of a venture: your company gains quick access to a market and benefits from the partner's experience in the market or the technology, or both. With a willing and cooperative partner throughout the term of the venture, you can add critical mass to your company for a fraction of the investment cost.

5. Familiarity and market contention: Exposing your position

If the partner company is bringing a product technology to the deal, and if your company is providing the marketing and channel, remember the "camel in the tent" syndrome. This describes a competitive situation whereby your company exposes its customer base to the partner company, who has the means to take the customers away from you.

It is not unheard of to have the partner company secure customers *directly* after a time and effectively remove you from the transaction, thus allowing the "camel in your tent." It is most dangerous when your company is not the manufacturer.

Sales channel contention						
Channel	Partner cost	Your cost	With your markup	End user's price	Partner's margin	Your margin
Partner	100.00	200.00	285.71	285.71	100.00	85.71
Direct	100.00	Bypass	Bypass	200.00	100.00	0
			"Partner's"	−85.71		
			Advantage			

Figure 2-21. Channel contention.

In this case, the manufacturer has a built-in financial advantage based on the minimum margin your company would use in selling the product. By bypassing your channel, they can afford to sell at your cost and still secure the same margin as if they were selling to you for resale. Figure 2-21 illustrates the issue.

As can be seen in Figure 2-21, the partner can bypass your company to get to the customer and effectively deliver the same goods at an advantage over your company, which preserves their manufactured margin. When negotiating a venture with a partner company, legislate steps in the agreement to ensure they remain actual partners.

6. Plan for the future

A final note on the subject of brand labels and joint ventures: It is important to keep in mind that, with the exception of the acquisition case, a partnership deal doesn't last forever. Two companies will eventually go their separate ways after a venture because each will develop different strategies, needs, and goals. The two companies may have originally partnered as a result of expediency, and that expediency will not continue forever. Therefore, it is wise to plan for the future by absorbing the technology and manufacturing know-how (or whatever the partner company brought to the deal) to position yourself for after the term of the venture. Participate in the venture at 100% effort and make it work for both parties, then plan for the future. Remember the objective is to create long-term growth and value for *your* company, so it's incumbent on the venture manager to provide this continuity.

SUMMARY

This chapter's objective was to familiarize the reader with the basic concept of market opportunity. A market opportunity was related to the solving of a customer's problem and to making a profitable business out of the opportunity while satisfying the customer need.

Evaluation of ideas within the strategic framework of the business and competitive activities was presented. Tactics and strategy differences between large and small firms drive different results, and means for accomplishing those results through partnerships were presented as an alternative to executing a long-range plan. Finally, perspectives on partnering within the framework of new product development was discussed.

The reader should now have a basic, practical understanding of how a new product idea relates to a business and how it must relate to sustain the business in the long term. You should also have a working knowledge of partnership arrangements, their caveats, and benefits.

This serves as the basis for the next section, where the product opportunity and market opportunity will be refined into a product.

THE BUSINESS CONCEPT TO THE NEW PRODUCT

THE PATHWAY TO THE CUSTOMER NEEDS

Corporations often tend to make a mistake at the product definition step. In larger organizations there can be a disconnect between the corporation proper and the end-user. A market development team may start by doing some survey work and canvassing customers, and then digest the customer wants and needs into a set of benefits provided by features. Many times the information can be correlated incorrectly or certain bias is injected into the process. When the pathway to the customer is obscured, the product definition is at risk and subsequent success is in question. In this section we will discuss the pathway to the customer's needs.

Corporations have found that the longer the pathway to the customer (via sales channels), the less accuracy there is in the feedback information for product decision making. The end goal is to generate profitable sales by presenting the desired product to the customer, then *clear the pathway* to that customer so the proper product can be developed.

1. The Product Concept

A. Generating the idea

How does the new product idea get generated? There is a variety of ways this is accomplished. As will be seen, listening to the customer is a very important part of generating the idea, and screening is a very important part of taking action. Here are six generally accepted venues for new product idea generation:

1. Customer-defined needs and wants

These are the result of several actions:

- *Direct customer surveys*. Surveys are a valuable tool to engage a customer to solicit feedback and to obtain information in a directed manner.
- *Focused group discussions*. This method allows multiple-person input and prompted discussion to generate ideas.
- *Suggestion systems and communication from customers*. These are the results of unsolicited feedback from users and relatively anonymous input from stakeholders of the company.

- *Customer complaints.* This is an excellent feedback system for those who truly listen to the customer complaint for what it actually is and take the additional step to correct it through a product implementation.

2. Scientific research

This method generally occurs as a result of a scientific breakthrough of totally new technology or, more often, from applied research. Scientific research is usually responsible for breakthrough products or processes.

3. Competitors

Watch what your competitors are doing in the way of new product development and introduction. They are a yardstick measurement of marketplace activity and also an indication of how your company measures up in the marketplace. Previously, we discussed how spheres of influence of individual companies extend beyond their boundaries to employees, suppliers, and customers. In the same manner, you can monitor competitive product development activities.

4. Company dealers and representatives

These sales channels can feed back customer ideas and also generate ideas of their own. However, this medium must be evaluated in light of any individual agendas that may exist. The sales function is market-share driven, with a top-line orientation, and is driven by orders. This is not necessarily the best source of new product ideas because feedback may be slanted to make their sales job easier, rather than toward what may be in the best interest of the firm in total. Watch where the money changes hands and under what circumstances, and you won't get confused.

5. Top management

This can be a valuable source or a destructive influence in product development. Top management has the largest bat with which to drive a new product development idea through the organization. A program can be initiated and implemented in short order and with amazing results. However, less-than-constructive results can occur in cases in which supporting data, primary customer feedback, and due diligence are not practiced in favor of exercising executive privilege.

6. Miscellaneous sources

Finally, there is a wide variety of miscellaneous sources that can generate new product ideas. Remembering where the new product idea came from is not quite as important as evaluating and screening suitable ideas, locking onto them, and executing the best choices. Some of the miscellaneous sources are inventors, patent counsels, university laboratories, government licensing offices, industrial consultants, marketing research firms, and industry journals and periodicals.

There are several techniques for generating new product ideas, including the following:

Attribute listing: This method requires generating a detailed list of the major attributes of the existing product and envisioning modifying each attribute by maximizing, minimizing, substituting, or rearranging, and generating new combinations to improve the product or secure a new niche. As an example, the electric screwdriver was a result of studying a basic screwdriver and modifying the attribute of manual rotary motion to extend to motorized rotary motion. A next logical step would be to enhance the feel the traditional device user has through the hand and set torque limits in some numerical fashion.

Forced relationships: This method considers several products simultaneously, examines the interrelationship of these products with the others in the group, and generates alternatives that combine functionality into a single new product. For example, it could be said that the personal digital assistants (PDAs) on the market today are the result of combined functionality of a calendar, calculator, and address book.

Morphological analysis: This technique involves creative problem solving by examining a problem or objective, then singling out the most important dimensions of the problem and analyzing the interrelationships among them. For example:

Objective: Getting from point A to point B		
Type of vehicle	Medium of travel	Power source
Cart, cycle, wagon	Air, water, ground	Rail electric, magnetic

Now generate combinations that will be appealing, lower cost, more desirable, and/or novel.

Problem analysis: This method is different from the previously mentioned three methods in that it does not start at the company and work outward to the customer and marketplace; it starts with the customer and traverses back into the organization. The customer is interviewed to find out about problems with the usage of a specific product or category of products. The manufacturer then determines the product enhancement required to resolve the customer issue or complaint. Obviously, every issue cannot be addressed by adding enhancements, so the manufacturer needs to categorize the feedback into categories such as seriousness, incidence frequency, and cost to implement a remedy.

In effect, this is just a feedback method of engaging the customer to determine their satisfaction level. The theme repeatedly surfaces: Stay close to the customer to get an accurate assessment of product acceptance and business status.

7. Brainstorming

This method is oriented to generating ideas primarily in a free-flow fashion. It is specifically structured to generate ideas by removing the participants' mental paradigms. Generally limited to six people and one hour, a session consists of the leader defining the problem to be solved, as specifically as possible, and encouraging idea generation by observing the following rules: criticism is not allowed; freewheeling thinking is encouraged; a large quantity of ideas is desired; and combination and/or improvement will be done afterwards.

An iterative approach of the previous methods is where issues are not defined specifically at the beginning of the exercise. The problem is presented in the broadest frame of reference possible and, as the group exhausts the combinations and the creative suggestions, the moderator interjects facts that further define the issue, driving the group to a specific solution for a specific problem. For example, if the objective is to resolve a problem regarding an airtight body suit as a safety device, the moderator would spark responses by discussing the overall issue of protective coverings. As the group discusses different protective coverings, the moderator interjects more specificity into the discussion. This technique can be very helpful because it minimizes preconceived notions about how to solve a problem.

B. A new product concept must be tested to be refined

It is said that Rome wasn't built in a day; neither are new product ideas and their translation into new business. An idea doesn't become an overnight, long-lasting success without a significant amount of consideration and reexamination during its gestational period. Ideas become refined through the influence of three major factors: the passage of time, the venue where it is considered, and the frequency of review.

With respect to the passage of time, some concepts initially seem to be sure winners, but the excitement soon wanes. This is because the concept's authenticity and appeal are fleeting and cannot support a business model.

Venue plays an equally important role when evaluating ideas because ideas will be received differently. Consequently, testing an idea with a "traveling road show" has enormous value in generating responses representative of a cross section of people.

Frequency of review also has value in testing an idea. Every day, the mind receives new inputs and stimulation. Tomorrow's opinion of today's idea will be more informed than today's, so it's a good technique to revisit those concepts and ideas to see if they still have as much merit today as they did yesterday.

C. Third-party influence and feedback is key

The key to valued feedback is to get informed responses as distant as possible from the source of ideas generated. This will result in an unbiased assessment of the concept and also allow the new product manager to draft the promotional material and present it to see if, in effect, the story works. This third-party, detached opinion can be your worst critic or your best supporter. In either event, such feedback will drive you one way or the other. Either the feedback is good and you proceed, or it's negative. There can be no shame in negative feedback or rejection of the product concept at this point. In fact, it is essential to know early as possible because it will save time and money to *not* pursue something that will be unacceptable to the marketplace.

D. Primary customer contact has big payoffs

The use of primary customer feedback cannot be stressed enough. It serves two major purposes: it provides direct product feedback in terms of features, pricing, and configuration; and it provides feedback on how the customer perceives your firm in designing, manufacturing, and marketing this product. This positioning feedback is very important to determine customer and market acceptance of your firm's association with the product and business.

The use of primary customer influence ensures specificity in defining the product. Consider a case in which the product is *not positioned* acceptably in the marketplace because it has a broad definition. Subsequent to release, the manager will spend precious launch energy modifying the product, repricing it, modifying the feature set, and retargeting the market. While this is taking place, the competition is gaining momentum with its own product lines. As will be shown in the next section, placing these factors in a matrix is an easy way to keep track of the positioning of the product within the marketplace, and will aid in defining it.

E. Test the customer acceptance

Testing the customer acceptance as a means for positioning the product goes further than soliciting either a positive or negative response. It is desirable to establish a gradient of features, functionality, and customer cost. When launching a new program, it is important to refine the gradient because the product concept can serve a wide variety of people (however, not in the same product configuration and at the same price). By establishing the gradient, one can find the proper breakpoints and versions as illustrated in Figure 3-1.

In Figure 3-1, the matrix gradient of features and pricing is the result of primary customer contact, conducted to refine a product definition. As shown, the overall gradient is organized to have increasing pricing along the x-axis and increasing features and functionality along the y-axis. The individual squares indicate two things: the number of responses that indicated features and pricing in that section of the matrix, and the percent of total that it represents. The actual number is in the upper left-hand corner of the square, and the percent is in the lower right-hand corner of the square. As one moves along the x-axis to the right and the down the y-axis, on a diagonal, the matrix gradient indicates increasing price and increasing functionality. In this example, a total of 200 responses were received as part of primary customer interviews. The respondents fell into one of six market segments. The bulk of the responses centered on a medium-priced, medium-featured solution.

The customers wanting features desired in the lower range of this segment could be packaged with the features in the higher range in one product offering, perhaps. Together, they represent 128 (55 + 73) responses or 63% of the respondents. Conceivably, a single product platform could serve both segments without incurring a cost penalty in the design.

Conversely, 36 respondents (roughly 18% of the responses) were interested in the product concept having low functionality at a low price. At the high end of the spectrum, there was also interest. Roughly 21 respondents (11%) were interested in a highly functional, highly featured, premium-priced solution.

If the platform would allow it, a functionally separate product version could be created to serve this segment of the market. Since a premium price is acceptable, performance rather than platform cost would be the issue.

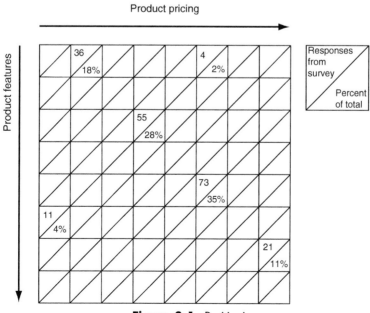

Figure 3-1. Positioning.

At the low end, cost is the key issue because the market segment wants low features and low price. In this case a different, a lower-cost platform may need to be used to serve that segment.

If this is the desired result from the customer contact, then the interviewing guide or questionnaire must be designed to produce this type of information. Consequently, the responses need to be directed into quantifiable criteria in order to be placed on the gradient with numerical weight. For example, the interview should solicit a response indicating, "This specific set of features are desirable at this price," and that can be entered into the matrix gradient.

To have these categories available to present to a respondent, it may be necessary initially to canvass several raw responses from the marketplace before designing the actual survey, to establish the categories first. Then the responses will fit the gradient more accurately because the survey will require a response to the query, "Would you buy the product under these circumstances with this feature set at this price level?"

2. The Route to Market

A. What is the route?

Every company needs to have an effective sales channel. Are they company-owned stores? Are they distributors, resellers, or agents? Each channel exists for a purpose and each one has benefits and drawbacks. If you think that new product success depends only on development, specifications, and features, then read on. The effectiveness of the channel will make or break your product introduction and sales effort.

Figure 3-2 diagrams the possible routes to market. It is by no means complete, but it does show the links possible in the pathway of a product from manufacturer to customer. More important, each of these possible links can be a strength or a weakness in your marketing effort.

In order to leverage off of their strengths and add them to your marketing effort, you must align with a formidable partner. Beware of the fallacy of partnering with weak but "hungry" partners. Hunger cannot make up for presence, staying power, and having a trained organization in place, able to meet the demands of new product introduction. "Weak" will sap your corporate energy and drain resources rather than accomplishing your objective of leveraging resources. Select your marketing partner to add strength and critical mass to the effort.

In many cases you will not have the luxury of selecting your route to market; however, go through the exercise of designing a route from scratch, and this will yield the requirements needed to improve the existing channel's effectiveness. Think of each of the links pictured as a question in the new product developer's eyes, asking the questions: Is this the best route to the customer? What does this link contribute to the overall transaction? Is it a leverage point or is it a hindrance? It is important to understand the effect because each link removes you farther from the customer. It makes it harder to communicate information from the factory through the channel, and it diffuses the feedback from the customer back to the manufacturer.

B. Where lie its strength and weaknesses?

If a sales route to market has been established, determine where the strengths and the weaknesses are. As the product is being developed, the sales channel must be improved and

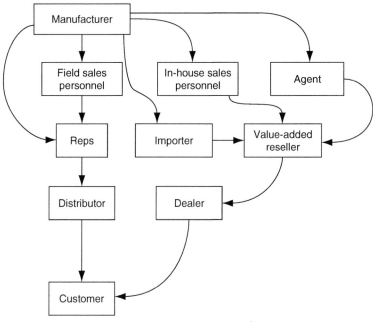

Figure 3-2. Route to market.

trained to successfully launch the new product. This means the strengths must be built upon and the weaknesses must be eliminated. As a starting point, let's examine some of the elements of the channel and their responsibilities in the sales transaction.

The *manufacturer* has the responsibility to develop, manufacture, and ship product. They assist in creating the user demand through media advertising and market awareness. They train the channel in the product. They generally have sales personnel—inside, outside, or both—to facilitate moving products through the channel.

The *representative* cultivates the demand among original equipment manufacturers (OEMs) and users. They may appoint distributors in cooperation with the manufacturer. In the case where there is a representative and a distributor, the product may be more engineered and need technical expertise to develop the market. The representative will report gathered market intelligence back to the factory as part of a feedback mechanism. They are independent business people who have other principals to generate income. They must balance effort among the lines that provide local expertise and product knowledge. They serve an established group of industry-similar or technology-similar customer bases. The representative does not have financial risk in the product because they do not take title to it; rather, they are compensated by a commission.

The *value-added reseller* is similar to the OEM in that they both add value to the product. However, the OEM incorporates your product *within* their own product, whereas the value-added reseller completes the product offering it in a way that the manufacturer cannot complete, or in some other specialized way the market accepts more readily. They take title to the goods and resell them.

The *distributor* provides local stocking, availability, and collateral products as part of one-stop shopping. They have inside sales personnel and field sales personnel. They provide feedback on products to the representative or directly to the manufacturer.

C. Offset these weaknesses with the new product embodiment

Assuming the sales channel is somewhat fixed and the product opportunity demands strengths in the channel not readily implementable, is there a way that the product can account for the shortfall in the channel? If the channel is technologically weak, can the product have a help screen, menu-driven diagnostics, or auto setup to make up for the shortfall in the channel? If the channel serves a wide variety of customer bases with different requirements, does the product have to be made modular and be configured at the point of sale or delivery? This is to satisfy the diversity in the marketplace, either by venue or type, which is not matched by corresponding diversity in the channel. An example would be personal care products in which the chemistry is formulated to be safe under mass-market sales campaigns sold in discount stores, not employing trained personnel in recommending application.

D. Motivation in the chain: Where does the money flow?

To truly understand the motivation in each step of the route to market, it is important to understand the role each member plays and where the pressure points in the transaction take place. Figure 3-3 outlines the relationship of each of the parties in the transaction, where the money and goods change hands, and where the value in the transaction takes place. Although all of the players in the chart may not necessarily be required in a transaction, they are listed for reference. The chart may be interpreted by looking at the manufacturer first, then selecting the players for your particular route to market.

This chart is by no means complete, since one chart cannot cover all of the financial arrangements that may exist between a manufacturer and the members of the channel. To further illustrate the point, a question mark was placed at the importer entry for Payment Trigger and the Financial Horizon. This is presented primarily for effect, since you may lack information as to their agenda in the transaction. It is best to evaluate where the value is added and if that value warrants the financial outlay for that part of the transaction.

As can be seen in Figure 3-3, the exchange of revenue happens at different points in the transaction, depending on where in the channel you are operating. It is also interesting to note the financial horizon for each and how they differ relative to each other. Along with the horizon can be seen the financial strength of each of the players.

These two factors contribute to the "commitment factor" in the overall relationship. This means, for example, that the relationship needs to satisfy the representative on a relatively

Summary of market participants							
	Manufacturer	Manufacturer Representative	Distributor	Agent	Importer	Dealer	User
What they sell	Goods and services	Time, knowledge	Availability	Connectability	International pathway	Local support	N/A
Payment trigger	Shipment/terms	Receipt of A/R	Shipment	Retainer/A/R	?	Delivery	Goods and terms
What they buy	Supplies/labor	Gasoline/lunch	Goods	Goods	Goods	Goods	Complete goods
Motivation	Volume	Volume/principle	Turnover	Manufacturer's shipment	Manufacturer's shipment	Customer delivery	Customer receipt
Financial horizon	Years	60–90 days	Monthly	30–60 days	?	30–60 days	N/A
Financial strength	High	Low	Medium	Low	Low	High	N/A

Figure 3-3. Summary of market channel.

short cycle. Their sales efforts must pay off in a cycle or two; otherwise, they will spend their earning power (time) with another principal.

The motivation for the rep is volume and principal. In other words, each day they ask themselves, "Which principal will contribute to my income 60 to 90 days in the future?" This is generally what they spend their time on. A distributor, by contrast, has inventory to sell and is interested in turns of that inventory. Their purpose is to ship product against local demand and assist customers in that regard. They are not motivated to develop a market for a manufacturer.

The overriding important difference is that, from the manufacturer's perspective, all of the players are independent businessmen. The only control your firm has over their time is to show them how they can make money with your product.

E. Keep close to the customer, while maintaining coverage

This is an important point. A repeated pattern, occurring as companies grow, can cause them to become farther and farther removed from the customer and the all-important dialogue that must occur between the two. Figure 3-4 shows that as the channel becomes more complex and has more and more players in it, the company can lose touch with the customer needs and feedback on product performance.

In the illustration, the manufacturer gets to the customer via several players in the channel. Each participant (CH-1, CH-2, CH-3, etc.) diffuses the information flow. This diffusion affects the outward flow from the manufacturer to the customer and feedback to the manufacturer. This can isolate the manufacturer from the market and potentially place them in a dangerous position of losing business. Furthermore, since members of the channel are often independent businessmen, the agenda and motivation of each will factor into the overall communications integrity.

F. Test the route

Don't wait until the product is ready to launch to test the market. It's almost too late at that point. Start testing the market channel at the outset of the development program. Decide what information can be communicated to the field without compromising a competitive position. Then start testing the channel's response and enthusiasm to the program.

As part of the development, you need to test the channel's commitment and resolve to execute tasks in promoting the product when it's released. There is no better measure than evaluating their behavior during the development. Select and assign tasks that you will need completed as part of the development. Draw them into the process and evaluate their performance. If they fail to perform during this stage, there is little chance of them performing to your expectations later on.

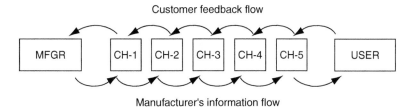

Customer feedback flow

Manufacturer's information flow

Figure 3-4. Channel isolation.

ENGAGING THE MARKET AND THE CUSTOMER

In this section we will examine the issues on engaging the market and the customer. In the business development process there are no absolutes and no one remains at rest. When your company launches a product or readjusts a price, the market reacts. The competition may take some action or the customer may take some action, and you must anticipate these actions.

To better position your firm to win at this, consider the following process:

- We accept that the customer buys the value in a product. Therefore, to best align with the customer's needs and position yourself, you must seek the *value-added solution* in your offering.
- The embodiment of this value can fall into one of three categories:
  - *Product embodiment*: Create customer value by solving a problem with the embodiment of the product.
  - *Process orientation:* Create customer value solving a customer problem by altering how the product is made, lower cost, faster lead time, venue.
  - *Supply chain staging:* Create customer value by solving a customer problem by altering the supply chain in delivering the product to the customer.

Each of these methods enhances the firm's ability to secure the business and provide a better situation for the customer.

- It starts with truly listening to the customer. However, we must listen in a completely different way than we have traditionally done. We must disconnect from our role in the corporation and place ourselves in the frame of reference of the customer and evaluate matters form a wider perspective. The customer has an issue or problem to resolve.
- The customer is not predisposed to your solution.
- The best evaluation of your competitive standing is to remove yourself from the sales arena and place yourself at the customer's perspective. When you create it, the value-engineered solution addresses direct and latent customer issues.
- You must use the process of discovery—a customer interaction is a dialogue and the interchange is a process of discovery where the seller probes for conditions, customer alternatives, and market situation.
- Discovery should produce the customers and hidden opportunities, the competitive situation, and threats to securing the business.
- The value-engineered solution is the preferentially compatible deal, which is a result of this discovery and planning.

The entire firm must understand and frame the customer's business, alternatives, decision breakpoints, and product pathways, and then use this understanding to synthesize the better situation for the customer.

Some additional steps are to:

- Scope out how the competitors will react in the proposal stage.
- What are the competitive alternatives?
- Capitalize on the weakness in the competitive arena.
- Set up the conditions for customer evaluation.
- Step into the customer's mind to evaluate your proposal.
- Evaluate the competition dynamic from the customer's perspective.
- The value-engineered solution positions the seller and benefits the buyer.

Your sales focus should then guide the transaction down the planned pathway and take action to accommodate changing competitive conditions. The value-engineered solution is the better deal for the customer.

1. Global Marketing

A. Will the new product be marketed globally?

It is difficult to conceive of a product that is developed and won't be marketed globally. The world is becoming more homogeneous and products developed for the world markets are becoming more prevalent. With countercyclical behavior in several markets, most manufacturers are making export an increasing part of their operational portfolio. Dependence on export relies on products defined and developed for the global market.

B. How is the specification affected?

If a product is to be globally focused, the specification must include the global requirement at the beginning of the development. Few products can be accepted globally without being developed as such. The requirements simply cannot be laid on top of an existing design. If you're going global, go early and go prepared.

C. How is the cost affected?

A product line capable of adherence to domestic as well as foreign standards may require a cost structure that may be slightly higher than that of the separate products. The product platform needs to be reevaluated to determine how best to address the global opportunity, factoring in all of the support and maintenance issues of product management.

D. Manufacturing venue

Another thought to keep in mind is the venue of manufacturing. Are all versions manufactured in one location for distribution throughout the world, or will each market be served by a separate manufacturing venue? Where will the parts come from? These issues need to be thought out initially, which may negate the need for product rationalization later.

E. Sensitivity to currency exchange rates

The issue of exchange rates was discussed in Chapter 1 and the issue of import duty can also affect the success of a product. Unfortunately, strides in cost reduction through engineering effort can be overridden by changes in tariffs or import duty, as can changes in the relative standing of the currency. A sensitivity analysis at this point may be in order as part of the overall plan.

F. Control of technological advantage

There is also another factor to consider when marketing products in foreign countries, the issue of proprietary technology. How do you maintain control of competitive advantage through technology when engineering and manufacturing may be overseas? The control of this advantage is critical to the long-term prospects of growing the business. There are several examples that could be cited here; however, it is sufficient to just be aware of the issue at this point.

G. Motivation for profit abroad

If there is a partner involved, again evaluate their motivation for profit and their tax structure. These elements will govern their primary actions. If taxes are high, the partner may be motivated to operate at no profit, content to only fund their growth. This can make repatriation of any funds difficult in a timely and worthwhile manner.

H. Export compliance regulation

This issue is not one to take lightly. Export compliance is one of the strictest, most unforgiving arms of the federal government. Violations go to the security and the foreign agenda of the federal government. If your company deals in goods that are prohibited from being shipped to certain foreign countries, you must adhere to the law regardless of knowledge of the law, changes in controlled products, and destination countries. For exporting, your company must set up a real-time evaluation system to maintain compliance.

2. Competitive Analysis, Advantage

A. Evaluate the competition prior to product definition

Many companies make the same mistake when it comes to competition and new product development. The best time to do a comprehensive competitive comparison chart is at the onset of a program. All too often this is done as part of the promotional package for the product launch, if at all. The comprehensive chart should be done at the project inception, and another one done at the time of product launch. The one done at inception will be required to define the product. The updated one at product launch will update your file and yield fresh information to the sales team, but it also will indicate the movement and nimbleness of the competitors. If they have made changes and introduced them in the time that your company has spent developing the product, it tells you something about their path, speed of development, incremental improvements, and market segmentation techniques.

A word about competitors in the marketplace: Competition is like a chess game, which is, in effect, patterned after war. If the war analogy is used, then most competitors fall into one of four major groups forming the complexion of the marketplace. The individual elements are as follows:

- Defensive orientation
- Direct pursuit orientation
- Oblique pursuit orientation
- Opportunistic pursuit orientation.

The *defensive orientation* posture is based on the market leader's position wherein the major market share is already controlled and the objective is to prevent abdication of the share to anyone else. Their focus of operation is that only the market leader can play a defensive game. They need to maintain market share by attacking themselves as others would attack them, only doing it first and establishing countermeasures to maintain share. Any strong competitive moves should be blocked immediately.

The *direct pursuit orientation* posture is based on competitors that are generally number 2 or 3 in the marketplace. They are interested in growing market share by taking it from the leader's share. Their main consideration is the strength of the leader's position. They are most successful when they find the inherent weakness in the leader's strength and attack at that point. For example, a low-cost, volume-oriented producer of goods with a formidable, established sales channel would be difficult to take market share away from. However, their weakness is the inertia of the entire system, leaving a market opportunity for a mass customization offering to compete directly for the business. The offensive player needs to launch attacks on as narrow a front as possible or diffusion of effort will occur, causing failure of an initiative.

The *oblique pursuit orientation* posture is an interesting one in that it is designed to seek out an uncontested area. Its major tactic is surprise. Since it is a surprise initiative, the pursuit of the business is as important as the attack itself. Many of these launches are manifested by developing functional alternatives to solving the problem that is currently solved by the defensive player.

The fourth posture is the *opportunistic pursuit orientation* posture. It is generally a small-company orientation and is manifested by ability to mobilize and serve market needs with originality and speed. The player needs to find a segment small enough to stake out and defend. They also must never forget that no matter how successful they are in a specific niche, they are to never act like a leader. If the playing field for that niche becomes hotly contested, they must be prepared to exit at a moment's notice.

These represent the four major types of competitive players in the market place and are generally easily recognizable. It is important to categorize the individual competitor's posture on the competitive comparison chart and understand their strategy. Know who you're competing against; it will clarify your actions.

B. List of competitors and attributes

The competitive comparison chart is an invaluable tool in positioning the new product in the marketplace. A properly prepared chart outlines the features, pricing, offering, business condition, and competitive standing.

With reference to its use as a sales tool; it has a tactical orientation, in which the product evolution flow chart is more strategic in nature. Its use as a sales tool is only for the benefit of the sales channel in training and communication of the current conditions. The document has its real value when used at the onset of a development program. Figure 3-5 is a format that can be used to complete a competitive comparison chart.

As shown in Figure 3-5, the vertical axis represents the product attributes. These are the tangible elements of the product line that have value to the customer. Some are simple listings of features and pricing that can be entered in hard data. Some can be entries such as target segment, strategies, effectiveness of the channel, relative standing, and aggregated rating, that may require additional research and tabulation of data. The target segment is a statement of where the product is aimed. Who will purchase it, and under what conditions? A simple sentence explanation here is sufficient. Perceived strategy is the market's assessment of the manufacturer's strategy and product positioning. Channel to market is fairly straightforward: simply chart the pathway from the manufacturer to the end-user. Comment on any weakness that may be exploited and comment on the overall effectiveness of the channel. Next, cite the relative standing of the manufacturer in terms of market share. As will be discussed next, these data will range from highest to lowest because of how we will organize the competitors.

Finally, based on the present assessment of market need, place an overall rating on each manufacturer with respect to their features, specification, pricing, and the other factors. A more detailed analysis may require reducing these data to numerical values to be tabulated and posted, or an overall grade or rating may be used on observed objective data.

Competitive comparison summary chart		Market players					You
		Market leader	No.2	No.3	No.4	All others	You
Product attributes							
Features							
	1						
	2						
	3						
	4						
	5						
	6						
Specification items							
	1						
	2						
	3						
	4						
	5						
	6						
Costs							
Pricing							
Target segment							
Perceived strategy							
Channel to market							
Effectivity of channel							
Relative standing							
Aggregated rating							

Figure 3-5. Competitive comparison chart.

This essentially completes the vertical axis. The horizontal axis is arranged by manufacturer, wherein the market leader is in the first slot. Because they have the lion's share of the market, the assumption is that their package of values brought to the market is relatively complete. The second-, third, and fourth-largest competitors are in the next slots.

Depending on the size of the market and the number of players, All Others may be lumped together or considered separately. Next, place your company and the proposed product against all of the others. Remember, lead the target in new product development, which means your target specification for the product should exceed all others by a wide margin. This is because the collection of competitors will not remain motionless during your product development cycle. Lead the target so that at product introduction your firm will still be in a favorable position.

C. Functional alternatives to solve the problem

An entrenched competitor is never invincible. The primary driver in the sales transaction is the satisfaction of the customer need. If there is an alternative way to solve the need, a new market niche may be created by those market constituents who accept the alternative solution mentioned previously in the discussion of flanking posture; this can allow participation in markets that otherwise would be inaccessible.

For example, in the 1980s and 1990s the American station wagon lost market share and virtual existence by the development and introduction of the minivan. Designed from the ground up as an alternative with more functionality and less design compromise than the station wagon, the minivan dominated the landscape of highly functional family vehicles. More recently, the emergence of sport utility vehicles has merged with the attributes of the minivan to create yet another category

D. Engineer the competitive advantage

The competitive advantage cannot be added on at the end of a program. Added features or other inducements to invoke market acceptance often add cost or remove advantage in other ways. To successfully beat the competition, you must keep a watchful eye on them during the development process, and strike a delicate balance between adding features to keep ahead and engaging in creeping functionalism, which will be discussed in more depth later.

The competitive advantage must be implicit and engineered into the product. For example, many products are electronics-based. As systems become more complex and needs vary, programming or setting of electronics becomes more complex. Internal menu structures, which prompt the user through the choices, are a means by which the advantage is engineered. Generally, these added features only consume memory space and add no cost other than development time, thus greatly enhancing the usability of the device for the customer and aiding acceptance.

Another example of competitive-advantage engineering is evident when product functionality begins to merge and formulate one new product. A good starting point could be handheld calculators in the early 1970s. Starting off as four-function devices, each competitor built upon the other and added features. Memory functions, trigonometric functions, and programmability all were engineered in competitive advantages. The calculator has begun to merge with the day schedulers, text editors, and worksheets to form the palm-sized personal computer. With each new introduction, additional features and operating systems to enhance usability were added. As the technology develops, these products are merging with cellular telephone technology and connecting to e-mail servers, which will take functionality and portability to an unprecedented level.

The key to market share is to satisfy the customer need better than the others in the arena. Engineering the competitive advantage has big payoffs at product introduction that cannot be achieved as well by using other means.

E. Anticipated competitive actions

As discussed previously, the world we live in and the marketplace we participate in not fixed entities; they are dynamic in nature. Actions that your firm takes will be countered by actions that your competitor will take to counteract your initiative. To win at the interchange, it is necessary to anticipate a reaction and plan for it. Then, when your action occurs and their reaction occurs, your position is that of a *net advantage*.

To get a feel for the competitor's reaction to a specific situation, a reference to the strategic posture of each one is in order. This posture defines how the competitor conducts their business. If they are an opportunistic player, they will react a certain way versus if they are a leader playing a defensive position. By understanding the opponent and their posture, it becomes easier to predict their reactions. What makes this complex is that reactions can be a result of one or more competitors working together to counteract your moves. Think of it as a chess game as you anticipate the market dynamics.

1. The nonhomogeneous consumer

Not every customer receives and acts on information the same way. Despite what we would like to believe, there is nonhomogeneity in the marketplace. If you are introducing brand-new technology that is the greatest available in the field, not every customer may warm up to it. The "early adopters" are most often the ones that will take on the new technology and begin to work with it. Demonstrating generally more patience and tolerance to initial failures and problems, the early adopter is a recognized saint of new product developers.

DEFINITION OF THE PRODUCT

1. The Configuration

A. Coalescing the product idea to specifications

This is one of the most important and difficult tasks in the new product development process. It is the essence of taking the loose collection of market data, wants, needs, preferences, and biases and evolving a specification that is achievable, cost-competitive, and able to be developed by the organization. One can neither master this activity overnight nor within one product development cycle. It takes several products, generated from concepts and developed and introduced to the marketplace and obtaining the feedback, to refines these skills. The mechanics may sound simple but the feel of the market and the accuracy developed from that feel are what takes the time and numbers to cultivate.

The outputs from this exercise are answers to questions such as number of versions, packaging options, bundling with other products, configuration of platforms, and configuration mapping to the target audience. Revisit this exercise often as the development progresses to verify positioning and targets.

B. Focusing on wants, needs, and musts

"The customer is always right" is how the saying goes. This means that you must listen to the customer's interaction with you and glean the wants and the needs. It is very important to segregate these in today's marketing climate. In years past, it was necessary to determine the basic option structure of products. For example, an automobile addresses the need of basic transportation as well as a want of climate comfort at a considerable extra expense. It is necessary to know which area of the customer base is interested in these options so as to design and structure an option system for the product line.

In today's computer age and flexible platforms, it is important to know the segmentation from the perspective of how to market and bundle the options to make the product more appealing to the specific segments. For instance, features can be added to a computer-based system at virtually zero cost by adding them in software. There is only the development cost. However, if all of the features researched and possible were to be included, it would make for a very cluttered function menu, or make the product too difficult for the average user. Consequently, many products are segmented by these functions and different versions are introduced, targeted to these specific customer segments.

C. Product timing is king, costing is queen; there isn't much else

The features and specifications are only part of the new product equation, as mentioned earlier. Timing is king in new product development. A product must be available to introduce in a specific time frame to capture the opportunity and to maintain advantage. And if timing is king, then factory cost is queen. It is the basis for financial success or failure. Many introductions are timed to specific annual events. Compromises in functionality and costs are sometimes sacrificed to meet the even timing.

D. Platforming the product

When setting up the initial platform of the product, it is best to understand the technology that will be employed. In fact, it is best to be very conversant on a technological level to be able to make mental trade-offs between cost-adders and no-cost-impact items. If items do not add incremental cost and can be easily be absorbed in the user's tolerance overhead (i.e., the ability to tolerate the overhead that accompanies functionality), then add the feature. If, however, the cost or ease of use is impacted, reexamine the target customer base to be sure you are not overspecifying the product.

It is a natural consequence of development of a product to add functionality as you progress in development, so you will have to keep that in check without completely missing the target when initially setting up the product platform.

With products evolving at ever-increasing rates, the issue of platform management becomes most important. In some firms, the focus is to get the team on the correct platform and spin product versions off of the platform. In this way several products can be used to test the marketplace. If they do not work out, a new version could be spun off and tried. However, in cases where the product platform does not allow this, the team is in trouble. A single product platform or a close-ended platform could cause product failure.

E. Defining the product platform pathways

Organizations need to think in terms of product pathways and product platforms. This is because the development effort needs to have some leverage. There is a very low probability for mass-market success for an individual product started from scratch. Instead, the company should develop platforms to house the product initiatives and launch several initiatives. One-product successes are rare. By using well-considered platforms, the pathways to success allow flexibility and are somewhat mistake-tolerant.

There are two types of platforms we will define here: the open-ended platform and the closed-ended platform. The open-ended platform allows seamless substitution of technologies, seamless integration of new technologies not yet defined, and seamless substitution of functionality.

The closed-ended platform has a defined and limited platform life, defined functionality, and presents obstacles to the cost reduction pathway. If used, they can result in an optimized product, but for a short period of time. They cannot easily morph into the next platform and can be somewhat limiting.

When deciding which route to go with, consider the cost trade-offs, development time trade-offs, and product complexity trade-offs. Then address the marketplace by doing the following:

- Determine product scope and versions
- Determine cost structures
- Determine longevity of platform
- Determine cost pathways
- Project product versions
- Fit versions to platform
- Project technologies and improvements.

F. Refer to the product evolution flow chart often, integrate into the plan

At this point it is a good idea to revisit the product evolution flow chart to determine whether the product concept has the same scope and basic configuration. Determine whether the target is the one on the chart or has crept to the follow-on version already. Next, integrate the identified product into the strategic plan and reassess whether the plan still makes sense. This will ensure that the plan is executed via the new product rather than falling by the wayside as a result of the development.

2. Is it a Product or a System?

Many times product planners, especially inexperienced ones, call an offering a product when it is actually a system. There are distinct differences between the two offerings. There are differences in what the customer expectations are, differences in the embodiment of the offering, and critical differences in how the firm needs to be structured to make and support the offering. Within the product category there are even differences along a gradient from brand-new impact products to more commodity products. Within the systems area there is also a gradient. Where you are on this gradient determines where you need to position the company in terms of offerings and services.

Consider the summary in Figure 3-6 and try to place your offering within the market segmentation chart.

A. Systems orientation

In the systems area we are defining three levels at which the company absorbs uncertainty for the customer.

Mega: These are projects, systems, and programs require highly specialized application knowledge from the manufacturer. The edges of the component performance envelope are understood and often require development personnel to be involved in the engineering for configuration. These types of systems require a high level of customer intimacy in terms of support, technical competence, and interface. The skills must be technically strong from the supplier's and the OEM's business perspective and approach to their marketplace. The manufacturer's representative must be well-versed in the industry and industry practices; a

Systems	Uncertainty profile			
	Mega	**Intermediate**	**Low**	Commodity
Products	Proprietary core knowledge	Proprietary development knowledge	Application knowledge	Highly evolved diffusion of knowledge

Figure 3-6. Uncertainty profile.

consultative approach is generally used. Project management interface is required in order to get the product specified and placed within the target organization. In-depth product knowledge is required. There is a high level of uncertainty in the project that must be absorbed by the supplier. The key elements looked for in a relationship from the supplier are integrity, honesty, tenacity, and commitment.

Intermediate: These projects, systems, and programs require seasoned application knowledge from inside support personnel from the manufacturer for system configuration. These systems require a medium level of customer interface in terms of support, technical competence, and interface. The skills must be technically strong from the supplier's and the OEM's business perspective and approach to their marketplace. The manufacturer's representative must be well-versed in the industry and industry practices; the projects are generally specified clearly. Capability and pricing are key issues in the negotiation. The level of uncertainty is medium.

Low: These projects, systems, and programs require application knowledge from field support personnel for configuration. The customer expects a traditional level of customer interface in terms of support, technical competence, and interface. The skills must be technically strong from the supplier's and the OEM's business perspective and approach to their marketplace. The manufacturer's representative must be well-versed in the industry and industry practices. Projects are always specified clearly, with industry de facto standards and methods of implementation. Pricing and delivery performance are key issues in the negotiation. The level of uncertainty is low.

B. Products orientation

In the product area we are defining four areas of uncertainty absorption for the customer.

Products Requiring Proprietary Core Knowledge: The sale of these products needs a high level of core technology product development knowledge from the manufacturer. The edges of the component performance envelope are stretched and customized solutions requiring new platforms will often require development personnel to be involved in the engineering. These types of products sales require high level of customer support, technical competence, and interface. The skills must be technically strong from both the supplier's and the OEM's business perspective and approach to their marketplace. The manufacturer's representative must be well-versed in the industry practices, trends in equipment, and implementation pathways. A consultant or, better yet, an individual subcontractor mentality in approach is generally used. Direct engineering interface is required in order to get the product specified and placed within the target organization. In-depth product design, product evolution, and product development experience is required.

Products Requiring Proprietary Development Knowledge: Products with highly specialized product development knowledge required from the manufacturer. The edges of the component performance envelope are stretched and customized solutions requiring modifications to existing platforms will often require development personnel to be involved in the engineering. These types of products sales require high level of customer support, technical competence, and interface. The skills must be technically strong from both the supplier's and OEM's business perspective and approach to their marketplace. The manufacturer's representative must be well-versed in the industry, industry practices, trends in equipment, and implementation pathways. Direct development engineering interface is required in order to get the product specified and placed within the target organization.

Products Requiring Application Knowledge: For sales of this type, product application knowledge is required from the manufacturer. The performance envelopes are adhered to with definite purpose analysis and planning for a specific account. These types of product sales require a medium level of customer support, technical competence, and interface. The business and technical skills must be strong and the manufacturer's representative must be well-versed in the industry and industry practices. Application engineering interface is required in order to get the product specified and placed within the target organization.

Products Requiring Highly Evolved Field Knowledge Base: These products require application knowledge from the field support personnel. The performance envelopes are adhered to with catalog guidelines and planning for a specific account. Traditional customer support skills must be technically oriented, from both the supplier's and OEM's business perspective and approach to their marketplace. The manufacturer's representative must be knowledgeable about the industry and industry practices; a "corporate flag-bearer" mentality in approach is generally used. Purchasing and some engineering interface are required in order to get the product placed within the target organization.

When coalescing the market opportunity and creating the offering, it is important to understand where you are on the grid. This is critical not only for positioning the product but also in ensuring that the organization is configured to take on the business.

PRODUCT PLANNING

1. Mass Customization from Generic Platforms

A. The key to successful exploitation of markets and segmentation

As the number of participants in a given market increases and the market grows, it becomes more and more difficult for a company to be successful in that market. One way of competing in this arena is to adopt a strategy of targeting specific customer groups with specific, customized offerings. There is a marketing caveat that you cannot be all things to all people and be effective. This strategy of mass customization allows your company to stretch the envelope a bit.

This strategy, however, cannot be executed in a vacuum of product development, since the concept of product platform must be the basis for design. By starting with platform design that is flexible to target different customer groups with customized feature sets, the company can meet increasing customer demands, expand market share, and increase revenue. This is accomplished by manufacturing a wide variety of products in small quantities while having a large overall unit volume.

B. Defining mass customization platforms

Mass customization is a term based on the concept of the product platform. A generic platform is designed and optimized for two somewhat diverse requirements. The first is the product range that will be offered and the second is the preferential acceptability of features that combine to make up a unique product offering.

For this concept to work, there must be a thorough understanding of the marketplace requirements. This defines the scope of the offering and the scope of the basic platform design. Although the concept affords flexibility down the road, the scope must be exact or you will lock yourself out of certain market opportunities. If the scope is too wide, the cost of the basic platform will be too high and the configured offering will be noncompetitive.

Second, the individual features must be designed to be somewhat portable and combinable to create a unique offering. The overall scope should allow for adding features to capture individual market opportunities. Figure 3-7 is an illustration of how the new product progresses from a wide variety of market requirements to distinct product versions.

As shown in Figure 3-7, the market is made up of diverse requirements, represented by the different shaded boxes. Each represents a distinct need from the product by the customer. Product version 1 is conceptualized to establish the feature listing and configuration. From this the basic platform is conceptualized. If the basic platform is too wide in scope, then the product will be too costly. If the manager undershoots by locking onto a too-narrow basic platform, the versions available for offering will be limited. Once the product platform is sized, all of the product versions can be generated, completing the product line.

C. How to compete with mass customization

Competing with mass customization at your disposal is a very powerful means to obtain and retain market share. It allows you to target and secure new customer bases and to widen

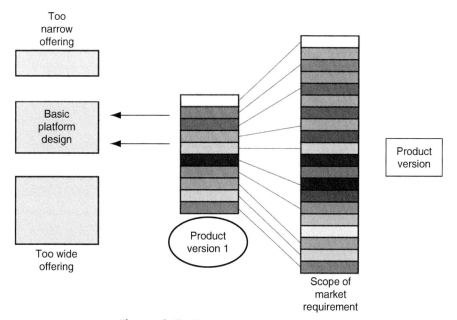

Figure 3-7. Mass customization.

your company's offering beyond original expectations. It allows entry into markets you may traditionally be locked out of. It yields a great amount of product flexibility, in that if a new customer base is discovered and an opportunity exists for the right configuration of features, the company can capture this customer expediently rather than having to try to develop a product from scratch.

D. Working the niches to your advantage

Now that you have the flexibility, it is necessary to work it to your advantage to grow the business. Select and target customers to secure and configure the products for that capture. It also affords you the opportunity to group feature sets to develop a tiered product line or an inclusive product line. The difference is in the way the features are grouped, as shown in Figure 3-8. In the following there are three distinct product offerings, namely A, B, and C.

In the top part of Figure 3-8, Products A, B, and C are generated from seven possible features. Product C has all of the features of Product B and, likewise, Product B has all the features of Product A. Product C is an all-inclusive product configuration in this example.

In the bottom part of the illustration, Product A is composed of two features, configured to target a narrow audience. It consists of Features 1 and 2. Product B consists of three features, namely Features 1, 2, and 3. Product C has no more functionality than Product B; however, it has a different target customer base, requiring that of Features 1, 2, and 4. It is separate and distinct from Product B and can be launched separately and priced separately.

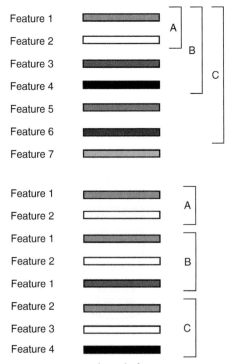

Figure 3-8. Product platform management.

E. The benefits

The benefits of this type of arrangement are numerous. Mass customization offers flexibility, speed, and direct targeting of market segments. It offers an expanded customer base on which to build a business.

F. Disadvantages

The benefits do come at a price, however. They require more effort on the part of product maintenance. This is manifested by having to lay in updates to the basic product line in each of the versions. If a correction needs to be made to the basic platform, each version of the product generated from this platform needs to be updated. This requires strict revision control and tracking through the product life cycle. Although this is suitable for some types of products, some companies cannot tolerate the product maintenance requirements.

G. Timing

Mass customization impacts timing of the development program. Designing to target cost, with the flexibility of the mass customization platform, takes longer than will a single-dimensional offering in terms of product definition and development. It offers the flexibility after the development; however, it will take more resources and maintenance to support. It should be used as a means for capturing diverse market share, not as an investigative means into a market or a correction for poor market intelligence.

H. Cost impact

Manufacturing cost also impacts mass customization. The mass customization platforms are not as optimized as the single-dimensional offerings and therefore will have a higher factory cost to accommodate the requirements for flexibility. The company may need to lose a little margin to capture market share. It is the price to be paid for flexibility. If the product line is a tiered offering with a low- to high-price gradient, the overall performance of the product line will be good, whereas just the lower tier may generate a little less margin.

2. Design to Cost

A. A must in any market

Design to cost is a new product development philosophy that makes the factory cost a marketing requirement. It differs from traditional historical approaches in which a product was conceived, designed, and priced based on cost-plus basis, in which added value added cost. The design to cost philosophy is based on the principle that the customer has established a perceived value for goods and services, and the manufacturer must develop, manufacture, and market that product at a price not to exceed that perceived value. The difference in design philosophies is illustrated in Figure 3-9. As shown, and emphasized by the direction of the arrows to and from the marketplace, there is a vast difference between the two approaches.

The design to cost approach conducts accurate marketing research to determine customer needs and wants and attaches a value to these elements. It then works backward from the customer, through the points of customer interface, back to the company, and eventually the development lab. By first determining the acceptable market price, the various added value

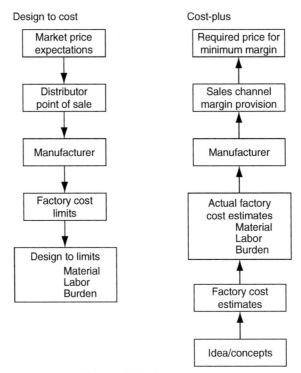

Figure 3-9. Design to cost.

elements are factored into the calculated maximum factory cost. The product produced by the organization must meet the market requirements.

In contrast, the cost-plus approach differs in direction and starting point. It starts out with an idea of the customer wants and accumulates rationalization along the pathway to product introduction. The development team starts with the concept and determines the approximate material cost to implement all the functionality. The labor is estimated based on the material content and burden is further estimated based on historical averages. The estimated factory cost is then rolled-up, and later in the development process the actual factory cost is determined. There are instances in which the actual cost varies as much as 1.75 times as much as the original estimate.

The minimum acceptable gross margin is calculated and the minimum price at which the product will be shipped to the sales channel is determined. The pathway to market is then determined and its associated costs are factored in to generate the minimum required price the product must sell for in the marketplace. The numbers involved in the example notwithstanding, it is the basic philosophy and direction of analysis that is flawed in the cost-plus approach. It simply views the market as the end stage in the process, when it should be the driving force and focal point. By viewing it as the end stage, it places the effect of all of the unbridled variability and uncertainty at the customer's door, when it should be factored and managed in the development lab.

The design to cost example in Figure 3-10 starts with a product that has an acceptable market price of $200. The various value-added activities are calculated in determining the maximum factory cost allowable at $8. The material labor and burden are then planned. The numbers at the left of the descriptions indicate the order in which the activities are executed. In design to cost, the specification is confirmed in Step 10.

Design to cost vs cost-plus implementation				
	Design to cost methodology	Financial $	Cost-plus methodology	Financial $
1	Market price expectations	200	9 Minimum saleable price	261.36
2	Sales channel profit 20%	40	8 Sales channel profit 20%	52.27
3	Distributor point of sale	160	7 Distributor point of sale	209.09
4	Manufacturers profit margin 45%	72	6 Manufacturer's minimum required margin (45%)	94.09
5	Manufacturer's factory cost	88	5 Manufacturer's estimated factory cost	115
6	Manufacturer's factory cost limit	88	No stated limit	
7	Material	61.6	2 Material	75
8	Labor	13.2	3 Labor	15
9	Burden	13.2	4 Burden	25
10	Specification and performance requirements	Satisfy	1 Specification and performance requirements	Idea

Figure 3-10. Design to cost versus cost-plus.

The cost-plus example starts at the idea and is often a loosely defined specification. This prompts the conceptualization of the product to be wider than needed in an attempt to accommodate the uncertainty due to the lack of accurate market information. An initial material estimate and factory cost estimate are made. The value-added activities are then factored in at traditional values and the minimum market price is determined. In this example, there is a significant difference between the two. Some customers can tolerate the difference; however, most will not.

B. Sooner or later, every product must deal with the cost issue

Virtually no product is immune from the eventual market pressure associated with the design to cost issue. There is a short time in an emerging market fed by healthy demand in which a participating company may be temporarily shielded from the cost pressure issue; however, as these markets mature and more competitors come into the market, the same characteristic forces come into play. Eventually the companies participating need to change over from cost-plus to design to cost.

There is another proprietary reason to stay away from cost-plus: in cases where the market price is established and the design and purchasing team effect cost reductions in material content. With a cost-plus mentality, the savings in material will be immediately passed through to the customer without any benefit to the company and, if the company management only evaluates percentages, actual margin dollars can be lost.

C. Learning curve is proactive

The *learning curve cost reduction* is an often misunderstood term, and for good reason. Its use lulls management into a false sense of security by sending a message that high costs today will be better tomorrow by virtue of the learning curve cost reduction. The fallacy of this assumption is that the discipline required to hit the cost target initially is the same discipline required to effect these cost reductions. If the team cannot hit the cost target initially, there will be no learning curve, thus exacerbating the eventual price pressure.

The learning curve requires consistent development work to effect a reduced factory cost. If there is no work, there will be no net reduction. In addition, the fallacy is that the purchasing function is the primary driver of the learning curve phenomenon, whereas in actuality it is a group effort of purchasing, manufacturing, and development that is needed to coordinate the reduction in cost.

D. Sensitivity of design to cost and cost-plus

The lack of design to cost control of a product line left unchallenged, either initially or through normal increases (i.e., vendor substitutions, increases in labor rates), can place pressure on the profit margin. Figure 3-11 illustrates the point that the manufacturer can be placed in a squeeze play in which they cannot respond.

With a fixed market price and increasing costs, the margin will deteriorate as shown in the first chart in Figure 3-11. The margin problem is exacerbated when the market pricing levels are dynamic and also deteriorating. This is evidenced by the drastic change in slope of the margin with both the price and cost affected as shown in the second chart in Figure 3-11.

In any competitive situation, a company needs a certain amount of flexibility in price negotiations. If there were no controls on cost or maintenance of costs, the amount of flexibility is reduced. In addition, the profit is impacted significantly depending on several factors.

As shown in Figure 3-12, there is a range of products proposed in a bidding situation. The products have an initial baseline cost and gross margin based on a market price. For this example, the volume of units sold will be inelastic to price. The profit is generated from the difference between the original cost and the selling price. This is shown in the upper section of the spreadsheet. If the factory cost goes up, the profit erodes, as shown. If the pricing level deteriorates, the profit level will erode further. As shown with the last three products in the list, the margin goes negative because of the increase in cost and the erosion of the pricing level. This puts the bid into a squeeze play whereby the manufacturer is locked out of participating.

Because a bidding situation is dynamic and the market seems to follow the latest bidding price level, prices can deteriorate with time. If the cost creeps up, you may be in a situation where you cannot afford to take the business. Furthermore, if the customer is seeking a bundled solution of equipment, you may be prevented from even selling the profitable items. Consequently, the design to cost initiative is essential because your company needs the latitude to navigate in the competitive arena.

E. Accuracy at product definition: Checks and balances

As evidenced by the various examples, it is essential to gain control of factory cost early in the development. Failure to do this cannot be recovered later and prevents your ability to effectively compete. Since all subsequent development activity is generated from the initial definition of a product, it must be accurate so as not to send the development team in the wrong direction.

Impact of cost on margin								
	Cost 1	Cost 2	Cost 3	Cost 4	Cost 5	Cost 6	Cost 7	Cost 8
Fixed market price (dollars)	50	50	50	50	50	50	50	50
Factory cost (dollars)	18	20	22	24	26	28	30	32
Gross margin (dollars)	32	30	28	26	24	22	20	18
Declining market price (dollars)	50	49	48	47	46	45	44	43
Factory cost (dollars)	18	20	22	24	26	28	30	32
Impacted margin (dollars)	32	29	26	23	20	17	14	11

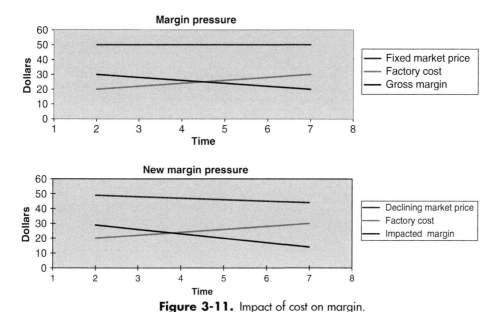

Figure 3-11. Impact of cost on margin.

Design to cost is a good methodology for product development but it cannot tolerate midcourse corrections or changes in the product concept. Generally, product development programs are not affected as much by wild inaccuracy as they are when sufficient definition is lacking in the initial stages. The problem occurs when companies ignore the uncertainty by not resolving it early, proceed to start the development, and add clarifications along the pathway of development. This changes direction, reduces momentum, and stalls the program.

F. Designing for low manufacturing cost

As stated in the previous section, the definition is critical to designing for a low manufacturing cost. The specification is the basis on which all the alternatives will be generated; as such, if the specification is floating then the alternatives generated are not comparable. In addition, there is no pathway for the evolution of the product from an implementation standpoint. This makes it a one-time program with no subsequent leverage. The optimum means for product implementation is to start with a solid specification and determine the lowest-cost, most flexible *platform* to implement the design.

			Cost (dollars)	Base sell price (dollars)	Base gm	Base margin (dollars)	Base volume	Profit (dollars)	Dollars volume ($)
					Profit impact of cost pressure				
Item	Description		Cost (dollars)	Base sell price (dollars)	Base gm	Base margin (dollars)	Base volume	Profit (dollars)	Dollars volume ($)
1	Unit 1		40	68	0.4118	28.00	195	5,460	13,260
2	Unit 2		38	65	0.4154	27.00	150	4,050	9,750
3	Unit 3		26	78.5	0.6688	52.50	199	10,448	15,621.50
4	Unit 4		50	80	0.3750	30.00	125	3,750	10,000
5	Unit 5		23	65	0.6462	42.00	185	7,770	12,025
6	Unit 6		55	65	0.1538	10.00	175	1,750	11,375
7	Unit 7		78	99	0.2121	21.00	95	1,995	9,405
8	Unit 8		67	68	0.0147	1.00	200	200	13,600
9	Unit 9		89	92	0.0326	3.00	200	600	18,400
10	Unit 10		76	78	0.0256	2.00	200	400	15,600
								36,423	12,9036.50

Item	Description	% Add	Impacted cost (dollars)	Impacted sell price (dollars)	Impacted gm	Impacted margin (dollars)	Constant volume	Impacted profit (dollars)	
1	Unit 1	5.0	42.0	63.92	0.343	21.9	195	4,274.4	12,464.4
2	Unit 2	5.0	39.9	61.1	0.347	21.2	150	3,180	9,165
3	Unit 3	5.0	27.3	73.79	0.630	46.5	199	9,251.51	14,684.21
4	Unit 4	5.0	52.5	75.2	0.302	22.7	125	2837.5	9,400
5	Unit 5	5.0	24.2	61.1	0.605	37.0	185	6,835.75	11,303.5
6	Unit 6	5.0	57.8	61.1	0.055	3.3	175	586.25	10,692.5
7	Unit 7	5.0	81.9	93.06	0.120	11.2	95	1,060.2	8840.7
8	Unit 8	5.0	70.4	63.92	-0.101	-6.4	200	-1286	12,784
9	Unit 9	5.0	93.5	86.48	-0.081	-7.0	200	-1394	1,7296
10	Unit 10	5.0	79.8	73.32	-0.088	-6.5	200	-1296	14,664
								24,049.61	12,129,4.31
							Difference ($)	12,373	7,742.19
							Difference (%)	-33.97	

Figure 3-12. Cumulative impact of cost on margin.

G. Costing methodologies

There are several different costing methodologies in use today. The important analysis to include is the relationship of the product's cost components from one product to another. If, for example, labor is accounted for and is a larger percent of sales in the new product than in the existing product, then there may be cause for concern. At the same time, if the material is drastically reduced, the product cost may be able to tolerate the additional labor. More often then not, however, the trend in the industry is to absorb additional material cost to reduce labor.

H. The triad of performance, cost, and design speed

The triad used in product development graphically describes the interrelationship of product factory cost, time to develop, and performance of the product. It is shown in Figure 3-13. It is fundamentally based on the premise that once a product is established and a specification locked onto, the triad shows compromises and trade-offs in the process.

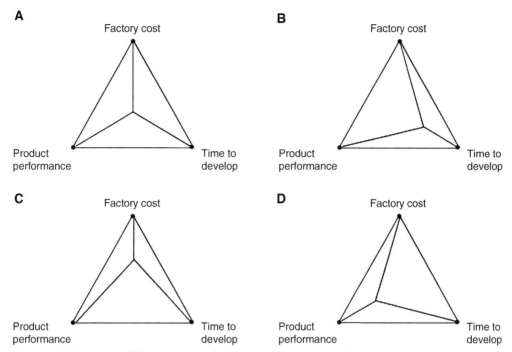

Figure 3-13. (A) Equilibrium; (B) Development time compromised; (C) Factory cost compromised; (D) Product performance compromised.

In Figure 3-13(A), the factory cost development time and product performance are in equilibrium. If any of these three parameters is more optimized, it will come at the expense of the others. The concept is that you can push any two of the parameters to optimization, but not all three. Figure 3-13(B) is an example of factory cost and product performance being pushed at the expense of time to develop. Figure 3-13(C) shows time to develop and product performance being pushed at the expense of product factory cost, and in Figure 3-13(D) the factory cost and the time to develop are pushed to the expense of product performance. The best way is to accurately assess the product performance requirements, the factory cost requirements, and the time to develop the product and operate the program in equilibrium.

3. Development Engineers' Influence on Factory Cost

There are several components that make up the factory cost of a product. This section discusses these components.

A. Bill of material cost

If design to cost is that critical to a new product success, then where does the majority of the cost occur and how is it levied on the product? Development engineers are capable of influencing the majority of the product factory cost, up to 85% in some cases. They select the materials, designate the platforms, match components to features, and specify processes. There are also a differences between prototype, initial volume, and production costs that need to be factored into the analysis.

B. Production labor—direct

The direct production labor can make or break a new product's cost structure. It is a measure of the development engineer's ability to apply materials and processes to embody an idea. The production labor is a measure of the effectiveness of these processes. If the processes are not accurate, effective, and deterministic, the labor will be excessive.

C. Production labor—indirect

The indirect production labor is a measure of the after-labor processes. It draws on the elements of the organization that are not specifically associated with the direct labor content but are required for support of their labor. If a high involvement is required here, then it may be a sign that the processes are poorly defined and need remediation to be effective.

D. Test

Test, being a part of the production, is a good example of this. If test is required in interim checking of the product during manufacture, and production labor is not performing this function, there may be a problem with the process being poorly defined or not in control. If the test function is indiscriminately applied to solve production problems and anomalies, you are not in production; you're still in the development stage. This is a serious problem for quality and consistency.

E. Cost drivers for product development

New product development touches each of the areas of production labor (either direct or indirect labor), the processes, and the manufacturing issues. If there are structural problems in the manufacturing organization, this will point them out. Often this is why new product development requires so much time and energy. The organization must be redesigned on several fronts to accommodate the product, the assumptions, and the planning that was done initially.

The management statement, "That's OK, we will just have to handle it when we get to that point" tells the story. If that was the prevailing attitude initially and no changes were made as part of the program, management can expect longer-than-expected delays while the organization changes to accommodate the new product.

F. Comparative cost structures

One of the best ways to highlight the cost issues being discussed is to do a comparative cost analysis. Figure 3-14 clearly indicates the difference between actual design and production capability and assumed capability for the new product and the need for improvement.

As illustrated in Figure 3-14, the various components of the cost structure can indicate organizational differences or required changes. For the analysis to be truly comparative, the y-axis must be normalized and be presented at a percentage of net sales revenue. To facilitate this example, the existing product is on the right-hand side and the new product is on the left; refer then, from right to left.

The change in material content from the existing product to the proposed is an indication of platform management by the development engineers. In this case, higher material cost for the new product is offset by the decrease in labor to assemble it. This can also be an indication of investment in tooling. The change in labor goes to the issue of improved

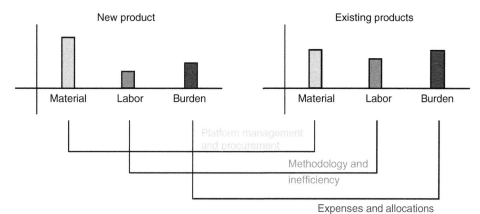

Figure 3-14. Comparative cost structure.

methodology and efficiency in the labor force. Will the new product demand personnel being directed in a more efficient manner with processes and procedures that remove variability?

The third area is the burden and manufacturing expenses. This is often a result of allocations and extra expenses associated with the manufacturing organization.

G. Value equation at product's inception

A value engineering exercise needs to be completed at the beginning of a program. It can be the means by which value is attached to the various features of the product. It is a structured way to differentiate the needs from the wants and to determine the degree to which the customer will pay money for them.

As shown in Figure 3-15, adding priced features to the basic product will affect volume and profit. Looking at Feature 4 reveals that improved gross margin, based solely on cost and price without the volume, will not drive profit. However, an incremental price adder with incremental cost can drive over expected volume to generate more profit.

This type of analysis and planning is useful when developing the requirement's specification for the development team.

Value engineering								
Item		Sell price (dollars)	Added cost (dollars)	Total cost (dollars)	Gm	Margin (dollars)	Volume	Profit (s)
	Basic platform price	87.5	0	45	0.49	42.5	200	8,500
1	Add feature 1	91	0.25	45.25	0.51	45.75	200	9,150
2	Add feature 2	93	0.37	45.37	0.52	47.63	200	9,526
3	Add feature 3	94	3	48	0.52	46	189	8,694
4	Add feature 4	104	4	49	0.57	55	125	6,875
5	Add feature 5	89	0.56	45.56	0.49	43.44	225	9,774
6	Add feature 6	95	0.35	45.35	0.53	49.65	175	8,689

Figure 3-15. Value engineering.

DEVELOPMENT MANAGEMENT

1. Creeping Functionalism

A. Powerless to the lure of added functionality!

Creeping functionalism is like extra bricks in your backpack. It is added weight and distraction that can take a program off course from the original intended focus. The creep starts quite innocently and is generated by both the marketing and development personnel throughout the program.

The marketing personnel, constantly watching developments in the marketplace, will have a tendency to add onto the specification to keep pace with the competitive offering. The other way this occurs is if the marketing people failed to accurately define the product initially. This is one of the most dangerous lacks of diligence that can occur because it dispatches personnel and expense to chase a wandering dream.

Another way creeping functionalism occurs is when the development personnel discover interesting additions and features that add value to the product, and have a tendency to unilaterally add them in. If the manager is not diligent in screening these ideas, creeping functionalism prevails and costs increase while the program drags on.

If the development team is talented and motivated properly, and the marketing team did their job properly, the manager's focus will be to screen and slot development creep features into the program where it makes sense, does not add cost, and doesn't result in changing the customer acceptance of the product. The objective is to encourage the creativity and harness the output for the benefit of the overall program, without missing the window of opportunity.

B. Creeping adds cost and complexity, and wastes time

Given the above, you may infer that creeping functionalism should be rejected at all cost. Contrary to this, the objective is to evaluate the various ideas that come up and decide if it makes sense to add these items into the requirement to be competitive in the marketplace at product introduction. There needs to be an objective means for evaluating these items for inclusion or deferral to the next generation of device. Figure 3-16(A) illustrates the evaluation and planning method for feature inclusion into the product line.

The table in Figure 3-16(A) lists the features along with the basic cost for the product. Each feature is analyzed in terms of its factory cost impact and its acceptability to the customer base. This is indicated by a plus or minus in the column next to the feature. The factory cost impact and percent impact are also noted.

Next, the selling price is established. This was established previously by market research and product planning. Beside each feature is the assessment of the market price adder or detractor for each feature. The significance in the negative entries is that some features could be depleted and would require a price reduction in the marketplace.

The new projected margin, both in percent and dollars, would be calculated based on cost impact and market pricing evaluation. Next, evaluate and post to the analysis the expected volume impact from the base assessment. In some cases for example, Features 2, 5, and 6 resulted in less overall profit, even with a cost and price reduction, because of the volume reduction. Also, added features such as Features 4 and 7, resulted in less-than-expected profit because of increased factory cost and less volume because of the higher price.

Item		Cost dollars		Total cost (dollars)	%Impact	Sell Price (dollars)	Gm	Margin (dollars)	Volume	Profit (s)
	Basic platform cost	38.58	< +/–>			65	0.41	26.42	200	5,284
	Feature creep	Cost impact								
1	Description	0.25	+	38.83	0.6	68	0.43	29.17	195	5,688
2	Description	0.37	–	38.21	–1.0	65	0.41	26.79	150	4,019
3	Description	3	+	41.58	7.8	78.5	0.47	36.92	199	7,347
4	Description	4	+	42.58	10.4	80	0.47	37.42	125	4,678
5	Description	1.24	–	37.34	–3.2	65	0.43	27.66	185	5,117
6	Description	2.35	–	36.23	–6.1	65	0.44	28.77	175	5,035
7	Description	10	+	48.58	25.9	99	0.51	50.42	95	4,790
8	Description	0.35	0	38.93	0.9	66	0.41	27.07	200	5,414
9	Description	0.56	+	39.14	1.5	67.5	0.42	28.36	200	5,672
10	Description	0.87	+	39.45	2.3	69.5	0.43	30.05	200	6,010

A

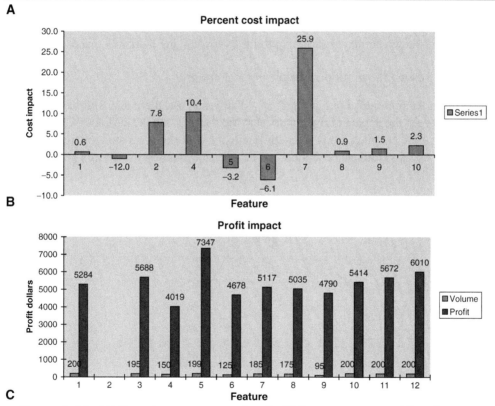

B

C

Figure 3-16. (A) Creep impact on cost and profit; (B) Percent cost impact; (C) Profit impact.

The one to seriously evaluate is Feature 3. Although it adds cost, it has negligible impact on volume because the marketplace has recognized the added value and can support the price increase. It retains the volume and greatly enhances the profit, even with the added cost.

As with any objective evaluation, the input data are most important to the output conclusion. The challenge is to be accurate in the assessment of the individual features that will be posted to the analysis.

Another way to look at the impact of the feature creep is the graph in Figure 3-16(B), which shows the cost impact relative to each other. Feature 7 shows a dramatic impact to the factory cost. As previously discussed, this represents only part of the overall decision. The other part is the profit impact, which is shown graphically in the graph in Figure 3-16(C). In this analysis, Feature 3 generates the most profit. This means of assessment can also be used in initial product feature selection.

C. Strike a balance between the need for stretch and expediency

There is a delicate balance that must be struck between the need for freshness in the product offering and the need for expediency in completing the development. The manager needs to keep creep in check and keep a watchful eye on the marketplace. This balance is best illustrated in Figure 3-17.

There is a baseline feature set that corresponds to the baseline market price tolerance and a baseline factory cost. The added features may tend to creep in at a rate in excess of the market desire. As the development team adds features, factory cost is affected and program schedule limits how you may measure up to the ever-changing market demands.

D. The unforeseen harm: undo complexity to the user

Even with features added to a product at virtually no cost, there may still be impacts to consider. There is the impact of time to develop the features and the customer's willingness to accept the more complex product. In our high-octane world, users are looking for immediate satisfaction and immediate solutions. They are not looking for research projects with

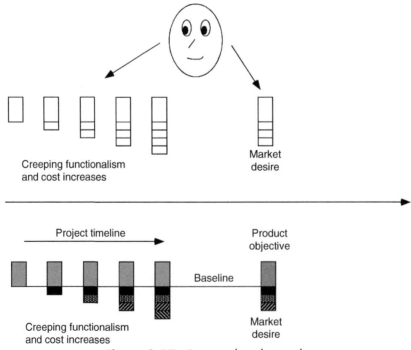

Figure 3-17. Creep and market needs.

each new product they use; consequently, manufacturers are continuously striving to simplify the complexity. This is manifested in application-specific products in which the feature set is customized to a specific application for the ease of the user.

E. The creeping functionalism trap

There is a characteristic curve that describes creeping functionalism in a program. It can best be described as a chase in uncertainty. The curve is presented in Figure 3-18, where the product definition undergoes scrutiny and bears uncertainty a couple of times during the development. There is an original level of product understanding upon which the program is launched. When it is in development, features get added to the product and the team experiences uncertainty because of this creep of features. This is the first danger point.

As the project progresses, the manufacturing element needs to lock onto a design that can be tooled and set up for volume manufacturing. Whether the manufacturing-driven understanding bears close resemblance to the market requirements at this point is yet to be determined; nevertheless, the product is locked onto for manufacturing's sake.

At market introduction, the company obtains it first feedback on the product. Lackluster initial sales or selective rejection in certain markets may cause the team to second-guess the product definition, and the second danger point occurs. Hopefully, this introspection is unwarranted and the company can enjoy growth and volume and can position itself for the next development.

F. Cost drivers not limited to only the product

The cost drivers may not only be product- and hardware-oriented. They could be software-oriented, causing development time problems. In addition, corporate infrastructure can greatly be affected by operation cost increases to hire application and customer support engineers to assist customers in using the new product that has been compromised by complexity. Where a simpler version would have served the application well, a new, more complex version is offered, requiring more warmware (customer support personnel) to get the product in use properly.

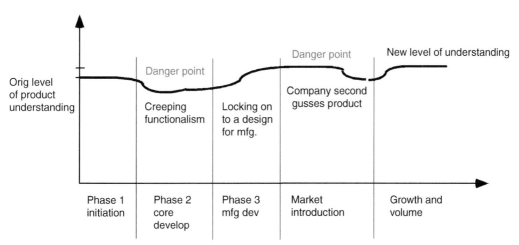

Figure 3-18. Creep characteristics.

INTERDISCIPLINARY CONSIDERATIONS

1. The Prevention of New Product Failure

A. We know the don'ts. What are the do's?

When discussing product and business development it is easy to get into a negative focus in citing things that we shouldn't do. More difficult is to focus on those things that we should do and to hold these items inviolate in our prosecution of a program. A checklist that the manager can review periodically to ensure diligence in the program is presented in Figure 3-19.

B. The checklist

Start with the first listed item and assess the organization's performance in ensuring that the new product opportunity fits the scope of the strategic plan and company capabilities. There is no honor in trying to develop something that is outside of the firm's capabilities. Discuss the idea with several people at the firm to get an idea of the company's ability to carry out the development.

Next, determine if the present criteria for evaluation are consistent with the overall plans of the company. If the long-term product plan is forecasting one way and the product opportunity does not support it, the plan will not be met and the development will be eventually undermined. Make sure there are documented procedures and that they are followed. Finally, ensure that operational plans have department-linking objectives so that individual departments are not at odds with each other. Have you ensured that the company is capable of developing the product?

Diligence checklist					
Item	Description	Responsible	Substantiation	Date 1	Date 2
1	Product opportunity fits scope of corporate strategy				
2	Consistant organizational system for evaluation				
3	Consistent criteria				
4	Documented and followed procedures				
5	Linking department objectives				
6	Sound market planning				
7	Positioning				
8	Segmentation				
9	Capitalization				
10	Pricing				
11	Paced development to capture opportunity				
12	Product distinctiveness				
13	Supporting evidence of opportunity				
14	Sizable market with research and forecast				
15	Precise product definition				
16	Solid design criteria and review				
17	Technical feasibility in the company				
18	Achieving target factory cost objectives				
19	Projected competitive response				
20	Stability of the team/company				

Figure 3-19. Diligence checklist.

Market planning is somewhat of a misnomer in that markets aren't generally planned. However, it is necessary to assess what the market direction is and what your firm's role will be in it. Has the product been positioned properly, such that your firm's offering is plausible in the perception of the marketplace? Has the market been segmented such that the individual product embodiments represent the proper configurations?

Has your firm demonstrated serious commitment to the development by way of development monies and capitalization for manufacturing? Is the pricing right? Are you positioned to charge a market-level price and not suffer margin pressure?

Make sure that the pace of development is sufficient to capture the opportunity in the timetable that is actually being charted.

The product originally conceived was distinctive. As the development unfolds and compromises are made, reassess that distinctiveness. Has there been control of creeping functionalism during the development?

Have you diligently collected the required supporting evidence to justify the development? Has anything changed? Are the forecast numbers for volume still there?

What, if any, compromises did you make in the development? Did marginal results of tests go uncorrected? Are the proper tests, design qualifications, and production in place and does the product pass these tests?

Has the technical feasibility initially assumed proved to be correct, or have compromises in the specification been made to overcome technical shortfalls within the company?

Manage the design team to meet the target factory cost. This is as important as any other specification item. The market price is given; the margin, profit, and retained earnings from the program are governed by the factory cost. Pay close attention to the cost issue and keep on track all the way through.

Try to anticipate the competitive response to your introduction and strategize on your next move. Competitors can be formidable opponents, so it is best to anticipate their reactions.

Finally, preserve stability in the team, the corporation, and the funds flow for the program. Interruptions sap energy and add time to the project, reducing your chances for success.

C. Relate to an organization (internal structural problems)

Most (if not all) of these issues relate to the organization internally (i.e., they are under your control). This being the case, exercise control for the betterment of the program. Do not gloss over these issues because they will come back to haunt the program. Correct any internal structural problems that can directly or indirectly affect the program.

D. Relate to external problems

Those issues that are external to the organization can only be addressed by market intelligence and knowledge of what is happening. Make it your mission to fully understand the market, the opportunity, and the areas of uncertainty.

2. Operational Planning (Getting Support)

A. How the new product will be presented to the marketplace

The product is identified, competition is analyzed, and the product positioning is accurate, but how will you prepare your organization to tell the new product story to the marketplace? What makes your firm think it is the firm of choice to develop and introduce this to the marketplace? The answers to these questions need to be woven into a story that the marketplace will accept. If it cannot be *told and sold* as a believable story, then that marketplace will have difficulty in accepting your company as being committed to the product and the industry segment. This makes it an uphill battle for recognition and leverage because the same perception must be overcome with each new target customer.

B. Prepare the organization for the new venture

If the market may need to change its perception of your firm, then perhaps your firm needs training in paradigm shifting to make way for the new product! Old organizations and complacency do not successfully generate new, innovative products and businesses. They simply do not possess the skills to break the bonds of uncertainty to develop and platform new ideas into new products. To successfully generate new business through new product development, create an atmosphere in the organization to meet the demands of the new product. Don't be compromised by corporate complacency and preservation of a comfort zone.

C. Prerequisites need to be implemented

Consequently, what are the specific changes that the organization must undergo? At the stage where the program is, you are visualizing the future with the product, so it is easy to visualize the corporate shortfalls and how they can negatively impact the program. List any additional prerequisites for the program and institute the changes in the organization.

Not all of the issues are related to internal organizational changes. If the development program involves an external player in the form of a brand-label partner or a joint venture, there are additional elements to consider. For example, if your corporate strategy has you collaborating with new players, external players, or both, review the section in Chapter 2 on partnering. Take the time to develop an understanding of the players, their personal and professional agendas, and their goals.

Also develop an understanding of the degree of vulnerability your organization has in embarking on this program. An example of this comes from a personal experience. We disclosed a brochure for a product we were going to brand label to the supplier company as a courtesy. Once they saw that the brochure was printed and 25,000 to 30,000 copies were in stock, they felt quite comfortable in raising the price slightly. We had little choice but to acquiesce because the announcement was made to the sales organization and the design was proprietary enough that it would have taken too much time to elevate our company's negotiation position by seeking an alternative. The lesson: Always understand the partner's motivation and the internal personnel agendas.

If it is beginning to sound to you as if the business of new product development effects change in all areas of the organization, congratulations. You are now undergoing the necessary paradigm shift to make the product a success.

D. What tactics will be used to introduce and shift loyalties?

There are built-in loyalties in the marketplace that must be dislodged to obtain market share with a new product. Human nature being what it is, there is hesitation to change from the known to the unknown. As part of the new product development, the loyalties must be changed by your organization's actions. What will these tactics be—price, quality, service, knowledge, availability? During the development of the program, determine what these tactics are and attack at that point to shift purchasing patterns. Bring something exciting, special, and of higher value to the customer base. Launch with this and keep the pace difficult for the competition to match.

E. Getting due attention in the marketplace

You have taken the necessary steps to ensure that the new product is not a "me, too" offering. It is unique and of high value. How do you get the marketplace's attention and focus on your firm's solution as opposed to the entrenched competition? Short of reciting a litany of media blitzes, promotional hype, and spin control, suffice it to say that the launch is your time in the sun. Make the most of it and use it to create a base to build on.

If a success story needs to be created, start planning it now at program initiation. If you need some other means for promotion, develop them now and start laying the groundwork early.

3. Manufacturing Considerations

A. Manufacturing is integral

Manufacturing is an integral part of the new product development system. The term *system* is wholly descriptive because new product development is not an isolated series of tasks performed in a vacuum. It is an integrated, interdependent program requiring the cooperation and commitment of the organization. This is why it is critical to involve the various functional areas at the onset of a program. This subject will be covered in more detail in Chapter 8.

One of the most critical and often-overlooked functional areas *is* the manufacturing element. They are overlooked until the project reaches a certain point where they are materially impacted. Usually, by then it is too late to benefit from their input. The manufacturing element can affect the factory cost, the production quantity, and the quality of the goods produced. The manufacturing element, whether it is internal or external to the organization, is a partner in the new product. There must be give and take in the interplay between development and the manufacturing disciplines.

B. Do not underestimate manufacturing's impact or value

The term *over-the-wall syndrome* is used in the new product development arena to characterize a fractured development team in which the development engineers design a product without input from manufacturing. When the bills of material and drawings are complete, the engineers turn them over to manufacturing for their first look. Manufacturing then must develop manufacturing strategies to build the product. There are no trade-offs at this point, so what you get is what you get.

C. The key to momentum

The key to momentum is to get very proficient at moving a product concept through development and manufacturing. Rather than this being a sequential process, it must be more integral—both disciplines must facilitate the program through to completion.

As shown in Figure 3-20, the integrated approach will execute a program in a more complete manner than the sequential approach. In addition, the sequential method always results in a longer time line than was originally anticipated because it is fraught with delays, restarts, cost overruns, lack of optimized designs for manufacture, and tooling compromises. This is due to surprises that may occur at the development-to-manufacturing interface.

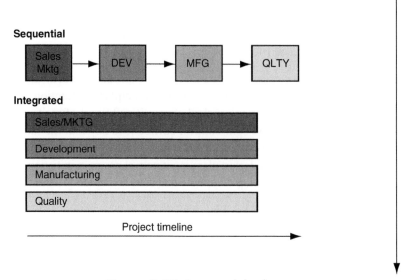

Figure 3-20. Integrated development.

D. Purchasing and procurement

Development engineers design products based on forecast volumes. The purchasing and procurement function aligns all of their activities based on these numbers. The procurement function affects the factory cost of the products produced. In addition, the quality of the product is affected by purchasing by virtue of configuration control of purchased parts and substitutions. In the long term, they affect quality by managing the conversion of obsolete parts in continuing to build product.

E. Design control

Engineering change control processes are also an integral means of affecting the quality and cost of a product. They need to be managed to retain cost structure and performance of the product.

F. Product configuration

A large portion of the problems encountered in manufacturing can be designed out at the product definition stage. There are methods for product configuration that can facilitate the product line. These fall into following three basic areas:

- Raw parts count
- Portability of assemblies over the product line
- Forward compatibility

At this point in configuring the product, always try to minimize the raw parts count. Undue complexity involving human interaction through assembly generates a host of training, retraining, and variability issues. This causes difficulty when trying to trace down root causes of problems.

Try to leverage subassemblies so they can be used over the product line. This will increase volume in custom assemblies, thus lowering the cost, as well as assist in balancing the production line. Design the product line such that actions taken today do not lock you out of supporting the product in the future. This means that original decisions need to be forward-compatible with subsequent product support decisions. The three factors to consider here are duration of the life of the product line, degree of supportability, and cost.

G. Reporting, feedback, analysis, corrective actions

Once a product is designed, manufactured, produced, and sent out to the field, there needs to be a feedback mechanism to allow customer comments (clear of interpretation from outside sources) to be channeled back to the development people. The field feedback serves the following two major purposes:

- First, it feeds the necessary information about product weaknesses, performance, and ease of use back to the source.
- Second (and in a longer-term sense, more important), it plugs the development group into the customer. A specification can articulate needs and wants, and internal people can try to influence development; however, nothing has as much effect as a customer visit or direct feedback to the development personnel. For improvement, nothing can influence the urgency of a corrective action faster or more effectively than direct customer feedback to development.

H. Forecasting and scheduling

These are two basic requirements for the manufacturing sector that directly contribute to the success of a product line. The manufacturing sector depends on an accurate forecast to drive materials and manpower allocation. This is one of the most difficult areas in which to get resolution in some companies. The difficulty is encountered when the initial market development work is not done properly. At the time of product introduction, there is no accurate information on the volume or mix of the line. If the proper work was completed, it is a simple matter to update the forecast and issue it to manufacturing.

The forecast also directly affects the company's financial performance through the balance sheet item known as inventory. If the forecast is inaccurate, the incorrect materials will be driven and there will be no consumption to offset the procurement. Also, if the forecast is incorrect the purchasing function does not purchase in the patterns originally planned on which the factory cost was based. Consequently, material variances show up due to expedited purchases, wrong parts, changing quantities, and missed shipments.

PATHWAY THROUGH THE ORGANIZATION

1. Background, Format, and Risk

A. Track record of the business

1. What has the company traditionally been good at?

The issue of traditional success was reviewed previously, as it related to the overall strategic direction of the company; however, at this point you may want to conduct a more in-depth review of products that the company has been traditionally successful with. Is the new product you are ready to embark on developing consistent with these previous winners? How do the products differ? What are the significant differences that need to be understood and addressed?

2. *What type of programs have historically failed, and why?*

Conversely, what types of products have traditionally failed to meet expectations and how do they relate to the product under consideration? What were the root causes of failure for these incidences, and have they been corrected?

3. *Are the same presumptions still true today?*

Are there organizational attitudes that must be changed to allow success? Are there remnants of the root causes of failure that need to be eliminated? Is the organization ready to execute a program?

4. *Trace successful programs and cite their contributions to the success*

Develop an in-depth understanding of the historical successes and the failures in the organization. More important, what were the contributions to success and can they be built upon? This series of questions is for your consideration as a double check of your company direction, wants, needs, and tenacity. You are near the juncture where significant funds will be committed to develop a product and a business, so it is important to gain the understanding at this time.

B. Preparing an operational plan for the new product

Facilitating a new product into a company, with all the associated changes that come into play, is a formidable task. It is itself a project that needs management. It is generally a good idea to formulate a plan to take care of all of the needed organizational items. These are not developmental items but, rather, are organizational changes and modifications that are needed to ensure success. It is best to take care of these changes as swiftly and as early as possible because energy must be devoted to the development tasks at hand, not to organizational tasks.

C. A discussion of formats

Set up a critical pathway with a list of critical changes and dates. This will ensure that the organization changes to meet the needs of the new product. There are as many formats for program management as there are third-party software suppliers. Almost any format will do as long as the evaluation, measurement, and follow-up are consistent. The difficult part is not just keeping track of the changes; it is effecting the change through people and making it stick. An organization is people, and people have to make the organization work. Failure to decisively make a change and integrate it will cause problems all along the pathway to introduction.

D. Integrate tactics to execute a plan

What are the push-buttons of the organization? How do the necessary changes get made and enforced? Do you work through executive decree, by influence, or by other means? Whatever the method, remember that you are sculpting the organization to accept the new product development program. Not everyone will share your vision and enthusiasm!

E. Determine risk

Each new product development carries with it a measure of risk to the corporation as an ongoing enterprise. There is the financial risk of the investment, the risk of lost opportunity, and the risk of any liability associated with the product and its performance. The measurement of risk falls into three areas that need to be evaluated:

1. The scope of the program (i.e., what portion of the total asset base of the company is the investment?). Some companies place all their hopes for the future in the development area and virtually lose control of the corporation when the development drags on with no return to offset the investment.
2. How achievable is the program? How likely is the chance of success?
3. What is the measure of acceptance of the program within the corporation? Who are the negative participants and what is their influence base?

Figure 3-21 is an example of risk assessment. Even if you do not do a rigorous assessment, this should serve as a reminder checklist. By assigning a numerical value to each item and comparing it to the total possible numerical value, one can get a measure of risk in each category by expressing it as percentage. A simple average of the percents will give a measure of the risk.

F. Managing risk

The other part of dealing with risk is to manage the risk to an acceptable level. This requires constant attention to keep the program on track. The mechanics of managing the risk of a program are as follows:

1. Be attentive to negative trends, issues, and problems that start out small and grow. Act early and decisively to contain them. You cannot be shy when containing risk.
2. Adopt a working philosophy that is central to the organization, such as, "Manage today and look ahead to tomorrow." Look for possible issues that have their origins in today's activities and could fester into tomorrow's problems.
3. Act quickly to mitigate any damage that may have already occurred.

2. Agreement of Principals

A. Agreement is often misassumed

This is more of a problem in smaller companies and closely held companies where strategy and direction are given by a select few people. An enthusiastic new product champion may go through motions to solicit agreement among the principals; however, in reality, there may not be agreement. Companies are composed of people, and people have human biases and fears of the unknown. They tend to protect their area of comfort in actions taken while verbalizing the need for dynamic change. It is therefore the responsibility of the manager playing the advocate role of champion to precipitate agreement among the principals and reinforce it throughout the program. All too often, this agreement is tacit only and lacks commitment behind it. Many times the agreement comes quickly and is shallow, only to be second-guessed and overruled later. Companies demonstrating this type of corporate behavior exhibit weak management and rarely complete a program through to success.

B. Commitment versus agreement

There is a distinct difference between commitment and agreement. Saying the words to agree on a program is relatively easy and risk-free. Committing corporate funds to back up the agreement is more difficult. It takes faith in the program, faith in the team, and faith in the new product manager. That is a lot of faith; a more comfortable position is to initiate the development and take a "see how it goes" approach—"If it progresses, we will fund more money to the program."

Development, however, is a game of conquering uncertainty. Progress is measured by tasks (surrounded by uncertainty) being completed in a timely fashion. When uncertainty arises, that is the time to have the faith and the stamina to see it through. Unfortunately, the "see how it goes" approach generally has management viewing the program as going bad

Risk factors in new product development				
Summary	**Category**	**Rating**	**Maximum**	**Percent**
	Product size	49	70	70.0
	Business impact	45	100	45.0
	Customer characteristics	49	70	70.0
	Process definition	72	80	90.0
	Technology employed	100	100	100.0
	Staff size and experience	32	80	40.0
	Cumulative total assessment	347	500	69.4

Product size		
Item	**Description**	**Value (1–10)**
1	Estimated size of the product in development effort	10
2	Estimated size of product in number of programs, files, versions	9
3	Percentage deviation in size of product from average for previous products	8
4	Size of database created or used by the product	7
5	Number of users of the product	6
6	Number of projected changes to the requirements for product? Before delivery? After delivery?	5
7	Amount of reused software/hardware/packaging	4

Business impact		
Item		
1	Effect of this product on company revenue	1
2	Visibility of this product by senior management	2
3	Resonableness of delivery deadline	3
4	Number of customers who will use this product	4
5	Interoperability constraints	5

Figure 3-21. Risk analysis.

6	Sophistication of end-users	6
7	Amount and quality of product documentation that must be produced and delivered to the customer	7
8	Governmental constraints	8
9	Costs associated with late delivery	9
10	Costs associated with a defective product	0
Customer characteristics		
Item		
1	Have you worked with the customer in the past?	7
2	Does the customer have a solid idea of requirements?	7
3	Has the customer agreed to spend time with you?	7
4	Is the customer willing to participate in reviews?	7
5	Is the customer technically sophisticated?	7
6	Is the customer willing to let your people do their job (i.e., will the customer resist looking over your shoulder during technically detailed work)?	7
7	Does the customer understand the software engineering process?	7
Process definition		
Item		
1	Have you established a common process framework?	9
2	Is it followed by project teams?	9
3	Do you have management support for software engineering?	9
4	Do you have a proactive approach to software quality assurance?	9
5	Do you conduct formal technical reviews?	9
6	What tools are used for analysis, design, and testing?	9
7	Are the tools integrated with one another?	9
8	Have document formats been established?	9

Figure 3-21 (Cont'd)

and seeking out other opportunities to fund. The result is that only easy programs are completed, with loss of *impact* products to the organization.

C. Testing the commitment

The issue of commitment is so important that it must be tested for. Remember that you are leading the development group to accomplish a program. It is a corporate trust wherein

Technology employed		
Item		
1	Is the technology new to your organization?	10
2	Are new algorithms, Input/Output technology required?	10
3	Is new or unproven hardware involved?	10
4	Does the new software interface with the application?	10
5	Is a specialized user interface required?	10
6	Is the application radically different?	10
7	Are you using new software engineering methods?	10
8	Are you using unconventional development methods?	10
9	Are there significant performance constraints?	10
10	Is there a doubt that the functionality requested is doable?	10
Staff size and experience		
Item		
1	Are the best people available?	4
2	Does the staff have the right skills?	4
3	Are enough people available?	4
4	Are staff committed for the entire duration?	4
5	Will all staff members be full-time?	4
6	Do staff have the right expectations?	4
7	Have the staff received necessary training?	4
8	Will turnover among staff be low?	4

Figure 3-21 (Cont'd)

you cannot waste technical talent or corporate funds. If you are leading the charge, the organization must back you up to the extent they agreed to.

It is wise to test commitment with something that is noncritical but is important enough to be telling. Depending on the circumstances, you may want to test the organization as early as possible. The earlier you know, the better.

The other dangerous game that can be played is that two or more projects may be initiated with funds to complete only one. You have two basic choices: either wage a street fight for the balance of the funds or demand that, in the interest of a successful development and fiscal responsibility, only one program be generated based on its own merit. Do not allow the organization to put off the new product decision or strategic choice by starting two developments and then trimming back. It's better to start one and drive it with the energy of two, and complete it.

D. Effecting staying power

How do you effect staying power in the overall development? Constantly update management and the financiers of the progress and problems encountered and tie them into the strategic plan. Use resources of the organization to resolve problems and make management and the organization part of the resolution. Bring them into the development to become part of it. If this is done effectively, management will have less of a desire to discontinue a program. It will also bring more power to bear on to the problem than if you go it alone.

3. The Assignment
A. Refine the fit

Initial plans are made from a macro outlook. Now it is time to refine with detail the prospect for new products coming into the organization. Where within the organization will the new product be placed? Decide which group is best-suited to champion the program and carry it to completion. This depends on their interest level, their training and understanding in the specific market segment, and degree of leverage achieved in the use of that particular group versus any other group.

Keep in mind that initial selection of a group within the organization is telling. If there is no sponsorship or acceptance of the program by the initially selected group, the program may become branded and avoided. This must be avoided at all cost. Do not undertake a program without the entire organization's support. You will be left out on a limb with no recourse and, more important, the organization has not bought into the program. Obstacles will take on another dimension of difficulty with this arrangement.

B. Become detailed on actions, responsibility centers, completion dates

The concept of establishing responsibility centers is a good one. It presupposes agreement and commitment and names personnel and departments as responsible for certain critical outcomes. There are accountability and consequences for failure. It is a workable, practical means for assignment and follow-up because it focuses on results rather than on behavior and tasks. A motivated group charged with responsibility and given the tools can accomplish the objective and conquer obstacles without excuses.

The responsibility center needs to identify the responsible parties, what they are to accomplish, when it is to be complete, a general direction of how it is to be accomplished, and where the activity is to take place. Be sure to include all of these elements, and be sure to amass a motivated group.

PRODUCT DEFINITION DOCUMENTATION

1. Requirement Specification
A. The pathway to understanding the customers' expectations

The requirement specification is a vehicle to organize and document a list of various customers' wants and needs into a specification that defines the new product to be designed and built. It is not just the sum total of all of the suggestions, ideas, and new product nuances; it is, rather, the compilation and selection of those features and requirements that are most appropriate to the success of the product. The pathway to the marketing requirement specification is illustrated in Figure 3-22.

As shown, the new product is an idea that may originate anywhere, from the most internal parts of the organization to the customer. This idea must then be coalesced in a new product opportunity. This opportunity can then be qualified in terms of secondary market research and then primary market research. The combination of these two studies defines the company's perceived value of the product to the marketplace.

Next is the definition of the value to the individual customers. This occurs in the iterative process of product positioning, cost evaluation, and determining customer value. This is one of the most important steps because the product definition must be accurate and a measure of what the customer will purchase. This generates and finalizes the marketing requirement specification, which is now ready to issue to the development team.

B. Basic elements of a requirement specification

There are several basic elements of a requirement specification that can be reviewed. The basic outline of these elements is as follows:

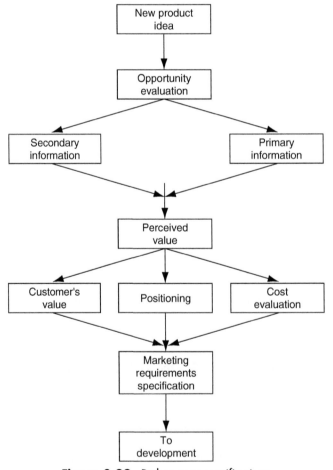

Figure 3-22. Pathways to specifications.

1. Background

The background section of the specification should identify the general market and customer base that the new product will be involved in. It serves to direct the audience to the specific area for consideration. A historical perspective is in order here, as well as a general projection of what will be in the future. This will help the reader focus on the specific product under investigation and serve to better evaluate the rest of the specification.

2. Industry trends

The industry trends section allows the manager to discuss the general trends in the industry under investigation. Specific references to products and technologies should be included here. This area should be specific enough so the reader can understand the basic industry, learn the trends, and picture the opportunity. There are several items to be included here, namely, the general application of technology, the speed of transition, the industry's acceptance of change and improvement, and tolerance for learning.

3. Market opportunity

This section outlines the selected opportunity from all of the background information. It discusses how the development of this new product will capture market share. Make sure to be specific in the identified opportunity rather than dabbling in generalizations. If the specification is to have meaning and be useful, the identification of the opportunity should be accurate and specific enough to be able to be referred back to and redirect any company or development efforts. If the team gets lost. they should be able to find their way back via this area of the specification.

4. Tie to strategic plan

The tie to the strategic plan is also critical to the specification's value. It lends legitimacy to the development effort. It also will be invaluable to reiterate the program's importance and positioning after the program starts and second thoughts may be occurring. Again, the information needs to be brief, tangible, and specific.

5. Scope of product line

The opportunity also needs to be framed in scope. Identify the product's positioning, the markets served, the breadth of the product line, and the number of versions. Outline the size and rating performance requirements in terms of market segments, and identify the scope of how the product will satisfy the market demand. Take care to be specific in identifying how selected versions address specific segments.

6. Component parts and product configuration

This section discusses how the product line needs to be configured to address the market. It outlines the component parts and how they will fit together to comprise the line. It needs to be as accurate as possible because this part of the document will be used to generate the manufacturing scope and plans and any other downstream activities.

7. Functional sequence of operation

This is a detailed description of the operational characteristics of the product. It should be written from the user's perspective and be as complete as possible. In addition to describing the function of the product from a positive aspect, it should describe the function of the product under misuse and wrongful use (i.e., what happens when the wrong key sequence is entered?). This aspect is often overlooked but is useful to the development engineers to perform their function in a more complete manner.

8. Performance requirements

The performance requirements are the complement to the functional specification. They specify the operating envelope for the product. It is the responsibility of the specifying engineer to take care in this part of the specification, since overconservative specifications and superlatives here can cause undue complexity at the development level. In addition, the degree of overdesign is specified here. The degree of overdesign is that amount of additional capacity beyond normal advertised use. It correlates to several items, such as technology employed, clarity of the channel, the relationship between the manufacturer and the customer, standards, and manufacturing variability.

9. The operating envelope

The following outlines the operating envelope for the product.

a. Electrical
  i. Performance: describe how the product should perform electrically.
  ii. Environmental: describe how the product should endure in an electrical environment (e.g., spikes, noise) and perform to specifications.
b. Mechanical
  i. Performance: describe how the product should perform mechanically.
  ii. Environmental: describe how the product should endure in a mechanical environment and perform to specification for the product's expected life.
c. Temperature
  i. Performance: describe how the product should perform over a temperature range and still meet specifications.
  ii. Environmental: describe how the product should perform when subjected to temperature extremes.
d. Humidity
  i. Performance: describe how the product should perform over a humidity range and still meet specifications.
  ii. Environmental: describe how the product should perform when subjected to humidity extremes.

10. Standards

What are the relevant specifications that the product must adhere to in the markets to be served? This critical element requires accuracy because designing to the wrong specifications can waste development time and create a situation in which compliance is either not possible or practical after the design is complete.

Unfortunately, compliance to standards cannot be overlaid onto a product line to achieve compliance. It must be integrated as part of the original design. If not, the trade-offs made during a design process by an individual engineer or an engineering team will now have to be executed through engineering change control. In addition, compliance to several standards after the design fact may not be possible given an existing design, parts availability, and cost constraints. It is critical to get this accurate at the outset.

11. Cost target

Product opportunities are only financial opportunities if the factory cost is consistent with all of the other assumptions. To that end, the factory cost target is as important a specification item as any other. Failure to achieve this can ruin a program in the short and long term. Be sure to factor in the trends of cost items and forecast the factory cost target in the time frame when the product will be produced. This is especially true of electronics-based products in which the cost decreases rapidly.

12. Timing to introduction

If cost is critical, so is meeting the window of opportunity. A product introduced late cannot be offset by improvement of the other elements. You simply must deliver a design on time and at cost to be effective in the marketplace. In fact, time is usually the most critical and least recoverable element in the new product development arena.

13. Human factors engineering

If humans use the product, then factor in their needs and wants for ease of use, accuracy in use, and overall acceptance. For example, complex electronics-based products with complex readouts and menu structures are simply not good choices for elderly people. Consequently, ignoring these human factors in engineering the product will result in non-acceptance. Conversely, designing a product for ease of use by the elderly will prove to be a success in the marketplace.

14. Safety

Aside from the altruistic reasons for engineering a product for safety, there are financial reasons and, in some countries, personal liability reasons for engineering safety into the product line. The marketing requirement's specification should outline the safety reviews and their expected measure of effectiveness. In addition, when documenting a program during development, these safety issues and resolutions and safety reviews should be recorded for possible future use in a legal defense.

15. Longevity and product life

Every product has a useful life that is finite. It performs its function for some time and then it fails. If it is a repairable product, it can then be repaired and have continued life. That is the meaning of the following terms:

1. Mean Time Between Failures (MTBF).
2. Mean Time To Repair (MTTR).

The requirement's specification should outline these issues and prescribe the expected life of the product.

16. Service plan

If the product is repairable, then there needs to be a plan to integrate the service of the product. Is it designed so a third-party repair facility can repair the unit or does it have to be repaired at the factory? Where do repair parts come from? How is the design controlled? What about parts substitutions? Is there a warranty on the repair? All of these issues need to be defined, and the design needs to reflect the plan.

17. Field replacement parts

Following on with the service plan, are there going to be field replacement parts or subassemblies? If so, these products need to be designed as part of the product and be factored into the manufacturing strategy. What is the control plan for these parts and subassemblies?

18. Corporate standards

Finally, identify what specific corporate standards must be followed in the development of the product. Resolve any conflicts between the market requirements for standards and internal requirements. For example, if the market standards allow for lower quality and price pressure causes little or no margin or derating, and your company's standards are higher, ask yourself whether this is the market segment in which your company can be successful. It might represents a divergence from traditional positioning of products by your company, which goes to the overall strategy and product planning.

2. The Interrelationship of the Specifications

During the course of development of a product line, there are several specifications that the development team needs to concern themselves with. Each basically traces and documents the pathway of the product though the development and manufacturing process.

A. Marketing requirements specification

As discussed earlier in this section, the marketing requirement specification identifies the customer's needs and wants and coalesces these requirements with internal corporate and external market requirements and conditions. It is the basic foundation of the program and must be accurate to be effective and achievable. It is issued to the development team for analysis and planning purposes. Most important, it initiates a discussion between the marketing, development, and manufacturing elements of the company.

B. Design specification

The design specification is the development team's response to the marketing requirement specification. If the marketing requirement specification is the summary of what is wanted, the design specification is a summary of what can be done. The challenge is to lead the target by time-phasing the two to take advantage of changes in fast-moving technologies.

C. SPAF: Design verification, qualification, and production testing

SPAF stands for the *significant performance and application features* document. This is the performance document produced after the development is complete and the product is released to manufacturing. It outlines the specifications, from an operating perspective, for what the product actually does. Consider an example where the market requirements specified "Needs to tolerate a temperature of 65°C," the design specification outlined an upper tolerance of 70°C, and the SPAF included test data that the actual unit can tolerate 68°C. In this example, the manufacturer achieved the advertised ratings but the actual performance fell short of the design specification; however, the unit was still ahead of the marketing requirement specification.

D. Typical specification

The typical specification is a sales tool that describes the operating envelope of the product. It is generally used as a means for third-party project engineers to specify your product in projects. It is generally written feature-specific without specific reference to names of manufacturers. This means that it is written such that the degree to which the typical specification is accepted by the specifying project engineer is the degree to which your company can bid the project and not have competition without that competition having to take exception to the specification. A cleverly prepared specification can lock out competition to a great extent.

SUMMARY

This chapter reviewed the activities necessary to take a new product concept from identification of an opportunity, through all of the marketing, design, and manufacturing aspects, to position it as a revenue- and profit-generating business for the company. It began with how to generate the new product idea and coalesce it into a business prospect for the company. The all-important route to market for the product was reviewed and outlined, and the competitive assessment and market conditions were covered. The product configuration (including mass customization) was discussed, leading to the hard features and functions of the product. The concept of design to cost was reviewed in detail in order to enhance the chances of the new product's success in the marketplace. The issues of creeping functionalism and its impact were reviewed, along with tools to mitigate its effect. The product value equation was discussed and a tool for evaluation was presented.

Also discussed were how the new product will be integrated into the company, and all of the caveats involved. Risk assessment and an objective way to evaluate it were also presented. The important issue of *commitment* was discussed, including its importance in implementing the new product development. Finally, the basic elements in producing a specification were presented, as well as a discussion of the interrelationship of the various specifications encountered in a development program.

The reader should now have a good understanding of how to determine the market opportunity and coalesce it into a workable specification from which the company can develop and introduce a new product, and generate planned revenue. This provides the basis for Chapter 4, The Product and Business Plan, which addresses creating the operational plan—the road map to the company's future!

THE PRODUCT AND BUSINESS PLAN

BUSINESS PLAN—BLUEPRINT TO SUCCESS

1. Perspectives

A. Formula for success

A business plan is a culmination of all of the data, conclusions, projections, and creativity of the business development team's vision for the new product. No one format for a business plan will generate instant or guaranteed success. Each business and product opportunity will require different treatment, analysis, and planning. If there is a formula for success in preparing a business plan, it is in factoring the objectives, the opportunities, the caveats, and the contingencies into a cohesive, executable plan and sticking to it through completion.

The business plan is even somewhat of a misnomer. It is a well-considered guideline that outlines the assumptions and requirements; however, it is not a fixed and rigid pathway. It cannot be, because the arena of marketing battle is dynamic. Following a strict business plan in a fast-changing market and expecting success is like a gladiator walking into the coliseum with an instruction book! Therefore, at the risk of sounding trite, the business plan is a *work in progress*. When executed and modified, it is really the story of how the success took place!

There must be a distinction drawn here, unfortunately, because many companies and business planning departments use the terms *work in progress* and *evolving document* as an excuse for not doing the diligence and work in a complete manner. These platitudes underscore failure in business development.

B. Include all of the elements

There are temptations to gloss over certain parts of a plan. Depending on the circumstances, however, this is never advisable. It is these loose assumptions that will haunt the program later. For example, in a larger company with an established sales channel it may be tempting to gloss over the route to market. This may prove to be fatal for the program if the product does not fit into the existing channel or the existing channel is too expensive to secure the market. The lesson here is to include all of the basic elements of the plan so as to double-check all of the assumptions.

C. Pervasive and complete

The business plan should touch all areas of business initially. This is to ensure integration. Failure to address all the aspects of the business may incur risk that the new product development will not be absorbed by the business

If there is such a thing as trial ballooning a business plan, then it should effectively trigger the various organizational immune systems in the corporation. The desire here is to see where the organizational boundaries lie and what corrective actions need to be taken to secure commitment to the endeavor. Make sure that all areas of the plan address all functional areas of the business. This is an effective way to garner support for the program. Failure to address this early will cause problems later because the business will not take ownership for the product, as it should. The program will be best served by determining the organizational issues. The business plan is an easy way to surface these issues and gather support.

2. Mechanics

A. Process of development

The mechanics of producing a business plan are not as simple as filling in some forms or making spreadsheet projections. In fact, the business plan is the summary of all of the new business and product development activity. Accordingly, the business plan should be developed over a long period of time and reviewed in one sitting periodically to assess its relevance to the present climate. Many times the business plan becomes its own goal and, once completed, there are no updates nor assessment of relevance to the present market.

B. Contingency development

The business plan needs to have contingencies. Often this is a glossed-over consideration. In order to effectively generate a contingency action for a business, one must think through the scenarios and create a pathway for each possible scenario. This is a difficult task because we are projecting possible scenarios and projecting our actions to them. As frustrating as it may seem, this is a valuable exercise to perform. It checks our thought processes and also provides tactical practice for the future.

C. Loss mitigation

Part of the contingency development process for a business plan is to factor in loss mitigation. The objective is to launch the new product or business; however, at some point there must be triggers established to effect loss mitigation. In these cases, the plan must have some reference to these triggers. This is because at the time the plan was developed and assumptions for the circumstances for success were known, there also was knowledge of when to call it quits. These should be identified and recorded at the time of plan development. It prevents ratchet error in risk assessment.

3. Format of the Plan

A. Elements of a business plan

The business plan for new product development can take several forms, depending on the circumstances and the type of product and businesses. They do, however, share many common traits and elements, and it is these elements that are mandatory for its value. The business plan is a means of obtaining funds, in addition to serving as a road map for the development of the business. Within this context, the business plan needs to be written such that it can initiate the appropriation of sufficient funding to execute a program. This is true for a group within a larger company as well as for a start-up company seeking venture capital. The story told in the business plan should be effective and address considerations and contingencies in a thorough enough manner to secure financing by whatever means.

B. Review of the plan elements

The following represents a typical example of the basic elements of a business plan. The purpose is to include all of the information required to assuage the concerns of the investor. As mentioned earlier, this is important for acceptance whether it is in seeking funds from a venture capitalist, mezzanine financing, or from the next layer of management in a corporation.

1. Executive summary

The executive summary is the quick overview of the entire plan. It is designed to allow quick review of several plans in order for the manager or investor to narrow down a list of opportunities to more closely scrutinize. It consists of the following four elements:

- *Business theme statement:* The business theme or mission statement provides the reader with several critical pieces of information. It positions the business in terms of its products and services and how they relate to the customers and marketplace. It also states the general objectives of the firm and how it perceives itself in the cadre of competition. It also defines the scope of what the company's operations are about and will serve as a check against the proposed opportunity. Finally, it provides specific focus for the company to discourage wild deviations to tempting but off-strategy opportunities.
- *General market information:* The next part of the summary is a legitimization of the market conditions upon which the plan is predicated. General market data supporting the case for the new product development should be documented here.
- *Brief statement about competition:* Since no marketing action is without reaction and competition is always at the doorstep, a brief statement about the competitive climate is in order. The objective here is to convince the reader that the idea, business, and plan will overpower any competitive threat, whether direct or a functional replacement.
- *Critical factors for success:* Every program has critical factors for success. These are usually company-specific and must be reconciled as part of the initiative. Cite these factors so the reader can weigh the financial risk.

2. State of the company

This section of the business plan establishes the state of the company—its financial health, the people, and the direction.

Status of the business

The first element to include is the current status of the business. It must be reviewed from two perspectives, internal and external. The format should consist of two basic elements:

- External review perspective: The external review is a summary of how the market and competitors view the business and the company's participation in it.

- Internal review perspective: The internal review is an introspective assessment of how the company and its managers view themselves with respect to the marketplace and its customer base.

Corporate objectives

The plan also needs to have the company objectives in it. Be sure to include both the near-term and the long-term objectives. This will indicate a certain amount of resolve on the management team's part, as well as formulate an opinion in the investor's mind. This will be important if the program should founder a bit. In addition, state the long-term ownership objectives of the company. Is it growth for near-term sale or for long-term measured growth and stability?

Management team description

This should not be a listing of résumés, since that will be included as detail information in an appendix. Rather, this should be a description of how the management team has accomplished something together. Again, this will give reassurance to the investor that the people in place can actually carry off the program. As part of this section, be sure to include the management team's objectives and how they perceive themselves as part of the enterprise.

3. The market and its need

The plan now needs to focus on the segment that will eventually generate the revenue though the new product development program. Define the market need in terms of specifics. What set of customers will buy what, when, under what conditions, and at what price level? The following are the parts to this section, which characterize the market:

- *Introduction:* The introduction is simply a means for the reader to become acquainted with the general area of the market under scrutiny.
- *Industry trends:* The industry trends should give the reader an idea that the industry is a vital market to participate in, as opposed to a declining market that may be over-crowded with players.
- *Market description:* This section describes the general market in terms of wants and needs. It should outline what the customers buy and why they buy those products. It is designed to familiarize the reader with the patterns of market behavior.
- *Market trends:* The market trends section needs to accomplish two basic things. The first is to cite and document specific trends in the marketplace from a historical perspective. The second is to project what the future trends will be based on past behavior and new technology available.
- *Market numbers and segments:* This is where to be specific and accurate. It is, in fact, the basis for the investor's due diligence in making the investment. The numbers presented here will be corroborated by a third-party source. Failure to be accurate and specific here will lead to rejection of the plan.
- *Market growth and stability:* The investor needs assurances that the program they invest in will have a certain amount of growth stability. The market numbers and segments analysis can substantiate the conclusions in this section; however, it will be helpful to prepare a narrative explaining the company's flexibility in creating revenue in this market with the new product.

- *Competition analysis:* Review what is happening with the competition in this area. Describe the competitive trends and how they may affect the new product plans. Include an explanation on the following:
  1. Domestic versus global: Are there global players who may displace domestic development activities? Are there any threats to consider?
  2. Functional displacement: Can the new product be functionally displaced by a flanking market move?
  3. Sensitivity to external partnerships.

Are you at risk by two or more competitors teaming up against your new product?

4. The product(s)

In this section of the business plan you need to clearly articulate the scope and configuration of the products. In addition, the plan must contain a description of how you will develop these products on time and on budget, and how you go from an opportunity to profitable products. What are the product descriptions? Define the scope, boundary, and use of the products in this description. What is the cost structure of the products and will they generate profit? Explain why the products are a better long-term solution than what is currently available.

Reemphasize the difference between research and development. Reassure the investor that the team understands the difference. Also, show that the product needs development, not research. This will go a long way to assuage their fears about funding endless activity without return. Finally, develop and present a detailed product development schedule and timeline showing expected progress and measurable milestones. The development plan should include the following:

- Schedule with dates and time line.
- Funds expenditures with time line.
- Critical factors for success.
- Technology basis for the products.

5. Manufacturing

In this section the business plan needs to describe the manufacturing system that will be used to manufacture the product. Hopefully, it will be consistent with the company's present manufacturing system; otherwise it could be perceived as an additional development in these systems. If the technology needs to be competitive on the world market, then the manufacturing system also needs to be competitive.

- Describe who will manufacturer the product.
- Describe what manufacturing systems will be used.
- Describe how the labor force will be trained.
- Describe how the certification of the labor force will be maintained.

6. Marketing and sales

In this section describe how the products will be taken to market. Cite the route to market and its cost, as well as the strategy used in that route. Show how the customers will be targeted and include a typical expected time line for a sale, from the introduction to the acceptance of the product. Describe who will market and sell the products. Factor in how you will capture your company's share of the channel participant's time and have them devote it to your product.

7. Product launch

Describe how and when the product launch will take place. Cite expected initial sales and forecasted demand. Project how the competition will react to the product launch. Develop and present countermeasure plans to sustain momentum. Describe the expectations of the channel participants and their expected performance for the company. Review the competitor's route to market and cite any differential advantages you may have. If there is an international element to the program, factor this into your presentation.

8. Program financial specifics

The format for the business plan has largely been composed of a narrative up to this point. In this section, specific numbers are expected to be presented for the investor's evaluation. The following are some of the essential elements:

- Development budget requirements
- Manufacturing budget requirements
- Marketing budget requirements
- Sales forecast [unit sales (not dollars), growth patterns in units, and timing]
- Revenue projections
- Profit projections
- Balance sheet impact
- Returns on investment and other measurables
- Risk analysis

9. Contingencies

Develop and cite contingency plans for each critical aspect of the plan. This will demonstrate that the plan was well-considered, that the management team who authored the plan understands that things can and do go awry, and that recovery needs to be in the new product vocabulary. Make sure to cover the contingencies that are affected by people, finances, and technical issues.

10. Aftermarket plans

If the nature of the business has aftermarket requirements, these also need to be included in the plan. If there are incremental additions of manpower required for the product line, they need to be factored in at this time. Include elements such as repair, service, and customer support.

11. Appendices

Product(s) specifications: Include a detailed product specification for the investor's due diligence.

Results of market assessment: Include the summarized results of any questionnaires, market assessments, or general feedback to the product concept. Use primary and secondary results.

PRODUCT–PLAN INTEGRATION

1. Planning with a Corporate Partner

There are acute differences in product planning with and without a corporate partner. When you are planning as a single entity, there is a set of company-wide assumptions about the business, long-term commitment, and approach to the marketplace.

When dealing with a partner in the equation, there may not be uniform, homogeneous agreement on these issues. In fact, the process of planning with a corporate partner will expose their intentions in several areas. In some cases, if the partner is an opportunistic player and you are using their technology to hold onto a market, immediate infusion of their technology into your organization is needed to protect against the possibility of them bugging out or losing interest. This is especially an issue when your company is slow to generate sales or other means to the contributing company's compensation. Whether it is a brand-label arrangement, a joint venture, or an acquisition, each partner company needs to plan the product and business plan with clear goals, contingencies, and dissolution protocol in the event one partner needs to leave the relationship.

When preparing the business plan, it is advisable to also elevate these issues to an actual protocol for discussion purposes. It also serves as a reminder checklist. An example is presented in Figure 4-1.

2. Product Mix, Offering

A. How will the products be broken out?

The determination of how the product will be configured and what will comprise the product line will help determine the financial risk. A product line consists of several parts,

Business planning guideline				
Category	Your firm	Partner company		
		Brand label	Joint venture	Acquisition
Expected market volume				
Expected investment amount				
Expected return amount				
Means for return				
Trigger points				
Cutoff points				
Protocol for dissolution				

Figure 4-1. Business planning guideline with a partner.

although it has a basic product as its core. These line extensions and peripherals add to the risk and cost of development. How the follow-on products will evolve and how the pieces will be added are decisions that must be made at some point in the development.

Figure 4-2 illustrates the risk and management of uncertainty in configuring the product line. As shown, there are four line completion products (numbered 2, 3, 4, and 5) generated from product 1. The increasing line indicates the investment increasing with each product segment. The top (increasing) line indicates the ability to manage the uncertainty. The more invested and the further through the development of the product line, the greater the ability to manage the uncertainty. The sawtooth (decreasing) line indicates the uncertainty itself. It decreases through the development cycle and rises to a new level with each start of a successive development. Overall, the uncertainty reduces through the development period but spikes with new subprojects introduced.

From the development standpoint, it is desirable to design all the pieces concurrently with all of the requisite resources. In this way the lowest-cost structure can be selected and implemented. Unfortunately, it is rare that all of the pieces can be developed concurrently. Rather, they are developed sequentially with planning for future models and forward compatibility. In other words, forecast the added models and the so-called hooks it will need to operate with the flagship unit and standardize on those elements in the flagship unit. Care must be taken to ensure the standardized platform will be a cost-effective and competitive one throughout the term of the subsequent developments and the life of the product run.

B. Margin planning and enhancement

When configuring a product line, there should be some effort to plan the margin initially and a plan to enhance the margin after initial launch. This margin enhancement can generally be effected through a cost reduction rather than a price increase. Depending on the commercial aspects of the development and the product marketing strategy, there may be price pressure immediately after launch, so the cost reduction is generally in the best long-term interests of the company. Cost reduction does not take place passively. It requires active involvement to accomplish.

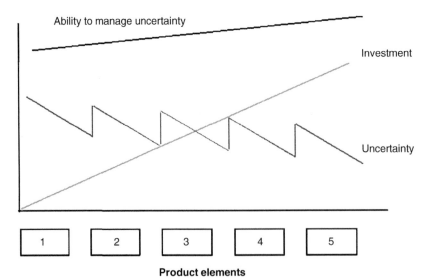

Product elements

Figure 4-2. Uncertainty management.

C. Configuring the mix for future advantage—manufacturing issues

In Chapter 3 we discussed the configuration of the product. When organizing the business plan, it is essential to know how the product and business will be configured for reporting and evaluation purposes. One of the chief contributors to low cost and manufacturing success comes from orchestrating the design to leverage off of commonality between versions of the product. By reducing costs of the product and uncertainty of future inventory, future success in manufacturing can be gained. If the design team fails to implement this effectively, inventory will swell and lead times for the product will extend because product versions and forecasts will rarely match. This contributes to always having shortages or the wrong subassembly in inventory to satisfy an order requirement. If the market absolutely demands a certain configuration that will result in new and higher inventory levels, the business plan must factor that in.

D. Discuss the vulnerability of the mix

The danger of missed implementation is positioning the company in a vulnerable position because of the product mix. Consider the example in Figure 4-3.

The market price for the product is linear with the size or rating of the product. The cost structure is as shown with decreasing cost for decreasing size or rating, down to a floor cost. The quantity is high at the low end of the rating with the fixed cost (at a low market price level). Profits here are low to possibly negative. Don't get caught being a supplier to the product 1 scenario, in which the higher margin, lower quantity segments like product 2 and product 3 may not offset the losses generated by the lower end of the market (product 1).

3. Pricing Policy

A. Background of pricing

"So what's it worth to you?" This old adage, which has almost folklore status, clearly articulates the basic concept behind pricing of a product and service. As will be presented subsequently, there are numerous strategies and approaches to pricing, depending on the circumstances of the new product. One tenet should ring through: Pricing should reflect the *perceived value* in the *eyes of the consumer*, or what it is worth to the consumer! The pricing is a fundamental component to the successful execution of the business plan.

All too often, product marketing personnel make the mistake of setting pricing without clearly understanding the dynamics of the buying decision, the buyer's alternatives, and the value the buyer places on the goods and services at the time of purchase. In addition, the market conditions may cause values to change or shift, resulting in a price that is less reflective of the consumer's original value placed on the product. The key element in setting pricing is to understand the customer's value system and structure your package of values (the product) accordingly.

B. External versus internal orientation

There are two facets to consider when setting pricing—the internal requirements for the organization and the external market price level tolerated. These two are opposing because the internal corporate need for profit is placed in check by the external pricing limits the market has tolerance for (as shown in Figure 4-4).

Vulnerability of the mix				
Description	Cost ($)	Price ($)	Quantity	Gross margin ($)
Product 1	110	80	1,000	−30,000
Product 2	110	160	500	25,000
Product 3	150	225	250	18,750
Product 4	200	325	125	15,625
Product 5	250	400	65	9,750

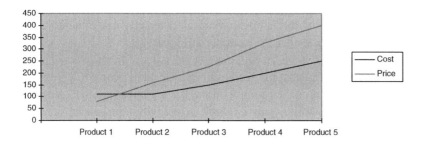

Quantity

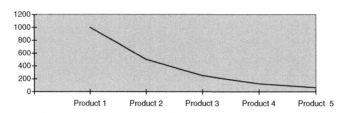

Gross margin

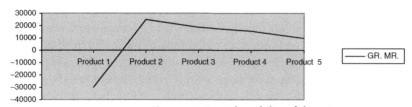

Figure 4-3. Vulnerability of the mix.

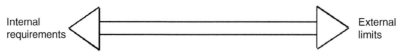

Figure 4-4. Internal versus external orientation.

A good starting point for a pricing exercise is to take the market price and determine the costs and overhead, then see how the company can make money with the product at the particular market price level. For example, the pricing breakdown in Figure 4-5 for a manufactured product shows a $100 price. The profit before provision for tax is at 9% of the sales price. Can the ownership structure's requirements and profit motive be satisfied with 9% of sales? This is the internal orientation of pricing when the market sets the level.

There are rare circumstances in which the manufacturer can dictate the price level in the vacuum of market pricing. This occurs where the manufacturer or service provider is a market leader; there are no tangible customer alternatives and time, space, and venue circumstances work to the goods and services provider's advantage. In this case (albeit rare), the world is in fact beating a pathway to your door to buy that better mousetrap!

There is also the issue of the morality of pricing. Pricing should be an equitable and even exchange between the buyer and the seller. Profits realized in the transaction are generated from the firm's efficiency, know-how, skills, and ability to market. Price gouging and taking advantage of a buyer's no-alternative buying decision will generate bad will and eventually work against you.

There are three basic, but differing, objectives in setting pricing. They are as follows:

1. *Sales objective:* The sales objective focuses on topline bookings and shipment figures. It may factor in profit requirements or competitive pressures; however, its main focus is to generate volume through attractive pricing.
2. *Profit objective:* The profit objective's main focus is the profit for the corporation. It may factor in other elements; however, maximizing profits is the main goal here. The profit objective, if left unchecked, can wreak havoc on the reputation of the organization.
3. *Competitive objective:* The competitive objective's main focus is beating the competition in the marketplace by offering the most attractive price for the goods and services. When taken solely, this (as well as the sales objective) can have devastating effects on the organization.

The pricing objectives are supported by a cadre of price strategies. As a starting point, it is helpful to think of pricing in a triad. The three vertices of the triad are costs, demand, and competition, as illustrated in Figure 4-6. This triad defines the basic operating arena for pricing strategies and market pressures. From this, each of the strategies will be defined and reviewed.

Price breakdown	
Price (net)	$100.00
Material	$40.00
Labor	$9.00
Manufacturing overhead	$16.00
Operating expenses	$26.00
Profit	$9.00

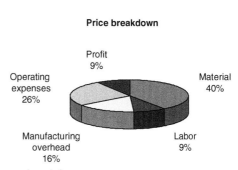

Figure 4-5. Pricing breakdown.

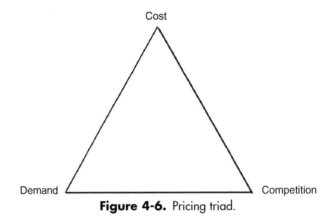

Figure 4-6. Pricing triad.

C. Relating pricing to value

The process of relating pricing to the value the customer places on the product is similar to the value engineering process when configuring the product. It is based on the ability of the company to understand the mind-set of the customer and to attach value pricing to each feature and to the entire package of values embodied in the product. As a starting point, let's examine the process in simplistic terms.

Figure 4-7 is a chart to assist in evaluating the features of the product and relating them to a price. Each feature is listed on the left. The estimated price relating to each feature is established. The product is then considered in total and a total price is established. Then each competitor is evaluated in a similar manner, this time factoring in the features which may or may not be included (as denoted by the inserted Xs). Next, the positioning of the product, with respect to the competition, must be decided and established. The overall complexion of the marketplace is now in place so that the pricing may be set.

This planning method can be used to configure your pricing.

D. Price versus quantity

The concept of a price–volume relationship, in which the volume increases if the price is lowered, has often been misapplied. In certain markets and under certain circumstances, this methodology does play out. However, it is *not* a universal concept and indiscriminate use of it will only sacrifice gross margin on each unit sold without generating additional sales. If you are attempting to use this concept, it is imperative to first understand the sales transaction thoroughly. Understand where and why the motivation for the exchange of funds for product takes place; otherwise, you may draw the incorrect conclusions from someone else's history.

The other caveat to remember when being tempted to drive volume by cutting price is that price is only one component of the transaction. It cannot be the sole driver of a buying decision considered in a vacuum of other factors. Since the buying decision is not a single-dimensional issue, the single dimension of price reduction cannot necessarily drive huge volume increases.

E. Cost-based strategies

Cost-based strategies consist of several approaches, as listed. They all ignore the demand side of the marketing equation and focus solely on internal needs.

Relating value to pricing						
Feature	Price estimate	Competitor A	Competitor B	Competitor C	Positi- oning	Price level set
1		X		X		
2		X	X			
3			X	X		
4		X	X			
5				X		
6		X				
7				X		
8		X	X	X		
As a package						

Figure 4-7. Value pricing.

1. Formula pricing

This is a strategy in which the cost elements or subcomponents are manipulated in a mathematical arrangement to generate a price large enough to absorb all other costs and transaction uncertainty, and provide for profit. It has pure internal focus and neglects market acceptance criteria. For example, if the product cost is $50 and the formula is a 3× multiple, the price is $150. Fifty dollars is the cost and the balance of $100 is for other costs and profit.

2. Cost-plus

The cost-plus strategy is like a snowplow in that it simply adds the required expense numbers and the required profit to the base cost to generate the price. This absorbs the uncertainty and provides for profit by decree. The premise is that the general level of consumption is stable, known, and understood. It also assumes that the market for the product is less sensitive to price.

3. Cost-plus by intermediaries

This can be characterized by accumulating price as the product progresses through the sales channel. Each intermediary will get his or her cut on the way to the customer. The dynamics are channel dependent and involve customary markups. They can be different for different classes of product within the same sales channel.

4. Targeted return

The targeted return approach seeks a specific return on the transaction, either in percentage or raw dollars. It is based on an assumed volume and quantity. The actual versus planned volume and quantity is a critical issue in securing the return. Overhead and other costs are absorbed across the volume (quantity) of products sold and are a key factor in the calculation of the return.

5. Breakeven

Breakeven is a strategy, usually short-term for healthy programs, in which the total revenue minus the total costs is zero, leaving no profit for the corporation. The caveat here is to thoroughly understand your entire costs so as not to have one product selling at the expense of another. There are numerous methods to calculate the breakeven base on absorbed costs; however, the basic philosophy is the same: move product, generate volume, and secure zero profit.

6. Experience curve

The experience curve strategy is based on two cost-related phenomena: economies of scale and the experience curve of the organization. Simply stated, the greater the volume of product manufactured by an organization, the greater the economies of scale. This is the application of the fixed costs over a larger volume base. In addition, other economies of scale can also be gained in procurement of goods, manufacturing setup, and cost reductions. This experience cost basis can then be applied to pricing in several ways, depending on the competitive circumstances.

The other component is the experience curve the organization takes with the increased volume. As the organization's rate of learning increases, the pathway to lower cost gets shorter. It is important to note that pure quantity of produced product does not in itself reduce costs; rather, it is the ability of the organization to absorb and learn from the volume that reduces cost.

7. Marginal-cost price

The marginal-cost pricing strategy is based on a price that allows an organization to recover primary or direct costs as a minimum. It is not the total cost recovery and therefore should be used temporarily and with care.

As an example, the manufacturer may want to lead the experience curve with this type of pricing; however, they need to ensure that the company does in fact progress down the learning curve.

F. Demand-based strategies

The demand-based strategies consist of several approaches. These strategies focus on a different perspective, that being of demand-side pricing.

1. Prestige pricing strategy

The prestige strategy is based on the premise that some product categories that can command higher prices can bring higher sales volumes. Particularly popular with safety items and purchases based on quality (e.g., food), the prestige strategy indicates to the buyer that increased price indicates increased value; in those categories, increased volumes are the result.

2. Price elasticity strategy

This strategy applies to products in which the price directly affects the volume of the product. There are three types of price elasticity strategies: elastic demand, inelastic demand, and unitary demand, as shown in Figure 4-8.

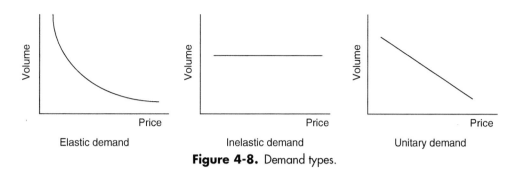

Figure 4-8. Demand types.

The *elastic* demand behaves in a manner in which a small decrease in price generates a large increase in volume.

The *inelastic* demand behaves in a manner in which changes in price, either up or down, will not materially affect the volume.

The *unitary* demand is a more idealistic behavior in which a percentage change in price generates an equal percentage change in volume.

3. Price range strategy

The price range strategy can also be thought of as a class of product. For example, the Mercedes-Benz automobile is in the high-end, high-price category and is in demand by a few select consumers. It, however, has a price range for the consumer. The lower limit of the price range still sets the perceived quality level as high.

G. Competitive strategy

The competition-based focus is an active pricing strategy that must be nimble and able to react to the rapidly changing marketplace. As the competition changes their price, your company must alter your price or change the package of values. Sometimes the competition does not initiate a pricing action; they initiate a maneuver to gain market share. Your company needs to respond to maintain a presence in the game.

Figure 4-9 illustrates some of the possible dynamics of a typical situation. There are three competitors other than your company in the marketplace. Competitor 1 initiates an offering with X as the price. Competitor 2 responds with a revised price (X'), and you respond to the marketplace with (X''). Competitor 3 is nonreactive in keeping price (X) during the

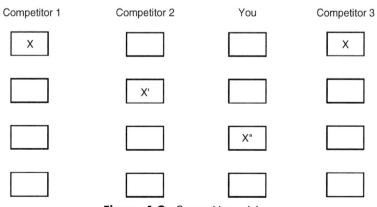

Figure 4-9. Competition pricing.

interchange. In this way the market is dynamic and competitors are participating in pricing exercises based on competitive price levels.

H. Other strategies: Situational and value pricing

There are other pricing strategies that are situation–specific and are based on value, which are summarized as follows:

1. New product introductions
2. Pricing when intangibles are important
3. Pricing in oligopolies
4. Pricing when buying is habitual
5. Pricing to reflect buyer behavior attitudes

These are basically offshoots or temporary applications of the previously discussed examples in order to gain advantage in certain situations. They, along with pricing over the product life cycle, will be covered in more detail in Chapter 11.

Whatever the strategy or for whatever the goal, pricing must reflect value to the customer and allow for profit to the corporation. It is therefore a key component to the business plan and must be well-considered.

I. Bundled versus unbundled pricing strategies

The other consideration in pricing is the use of bundled and unbundled pricing. In this case, the package of values can be adjusted to suit the market by including collateral goods and services with the product to make the entire package more appealing to the customer. These can be very popular and very successful as consumers begin to examine total installed costs in which bundling can lower their outlay for the entire package of goods and services.

J. Market feedback—what is working and what is not

One final note about pricing: Each assumption and each plan may appear flawless on paper; however, it must be verified in the field with actual transactions under the planned or assumed scenarios. This adds credibility to the assumptions and prevents marching down a path that will not yield results. Get early feedback as a point of reference and monitor periodically for any required changes.

SELLING AND FUNDING THE PROGRAM

1. The Importance of the Accounting Function

A. Engineers' broadened perspective

Development engineering is a highly specialized field of endeavor. As such, not all engineers broaden their scope of corporate understanding to include a solid background in accounting. It is, however, a good idea to round out the engineers with this discipline. The accounting function is the scorekeeper and the checks and balances within the corporation. A development team should never underestimate the power of the accounting function in an organization. All the plans and great product ideas can vanish in the wake of an accountant's fiscal recommendation. They are the keepers of the funds used for projects. They can turn off the money faucet as quickly as it is turned on. This is not meant to be alarmist in any way. The accounting function is not to be feared in an organization; rather, the accounting team is to be respected for the fiscal job they have to do.

B. Understanding the accounting function

To help make the people on the development team be more productive in terms of their understanding of the entire scope of the program, it is key to integrate the accounting function into the process. Let team members log hours and post costs to the project. Have them chart progress and keep track of lost opportunity costs. This will go further to imbue a sense of urgency within the team, rather than using the program's manager to preach the urgency.

A thorough knowledge of accounting is also required to design a successful business plan. It has been made abundantly clear to most development people that the yardstick in the measurement of success in the business field is money! It is what initiates the program, what keeps it going through the gestational period, and what is used to measure the return. In many cases it is what is used to reward the contributing members of the team.

All too often, development engineers see the accounting function as an adversary rather than as the partner they should be. The accounting function can be your ally, both in corporate good times and in bad times. They can sway the cutbacks away from a program or toward it. In preparing the business plan, bring the accounting function into the process of the plan development. They can steer you toward positive input and away from negative input. In the presentation they will be your advocates if they were part of the plan development. This can assist you in selling the plan.

Like it or not, accounting is one of the few objective, dispassionate, and universally accepted methods for evaluation. This being the case, the project team members should become conversant with the dynamics and the language of accounting.

C. Cost tracking

The accounting function is basically concerned with three things: the expenses, the return, and the timing. Consequently, you can expect that when funds are released for development the accounting function will be watching the amount of funding, the speed at which funds are consumed, and when the return will start coming in. They simply will be reviewing the business plan you prepared and will expect performance to it. The operative point here is to design the business plan as close to what you believe the actual pathway for the program will be. Most planners do not overstate the revenue or understate the expenses; however, they do misread the timing and, in doing so, risk getting caught in a situation of not meeting the business plan.

D. Get close to the money

If there is one conclusion or idea that you need to draw on from this section, it is that the manager or development team leader needs to understand the money flow in the organization. Watch where the money comes in and where it flows out. This will round out the manager to be more effective in his or her role. The best idea for a new product cannot sell itself to management without the accounting function as part of the justification, and the accounting function cannot solely generate cash and profit for the company. Consequently, it is the combination of both that can generate and initiate a program with management support.

E. Competitive programs

Many times the venue for an appropriation is a budgeting meeting. In those cases many programs are being evaluated and are, in effect, competing for a fixed amount of corporate resources. Thus, there is a tendency to overstate the expected results of a program in order to ensure funding. This is a dangerous practice and actually sets the team up for eventual failure. The program should be stated exactly how it is expected to come out. It will be difficult enough to get that result, let alone an overstated one. One has to compromise between forward statements and the ability to eventually deliver.

2. Selling the Plan Proper

A. Getting the audience

A fair amount of preparation is needed to prepare a fundable business plan. Once the plan it completed, it must be presented to senior management for review and approval. The process is less about the employee–management relationship than it is about investor and entrepreneur. At this point it's strictly business, and your task is to secure funds for development that will generate future returns for the investor.

However, it can be a fun part of the process. The first step is to identify and secure the audience. Who is the audience? How is the funding controlled? Under what conditions are the funds released? What are the expected returns? Gather all this information as you are preparing the presentation. You need to control the meeting and presentation and its outcome, so the more information, the better.

Find out the hot buttons and biases of the members of the decision-making group. Prepare the presentation in such a way as to appeal to the members. You also need to assess the competition. There are other groups usually seeking funding for their programs at the same time you are seeking funds for yours. What are *their* strategies for obtaining funds and how does *your* program stack up against theirs? Position yourself to prevail against your competition.

B. Having the plan precede you

An individual has an individual's energy. The best that one person can do is to transfer positive energy or enthusiasm about a program on a one-on-one basis or in front of a group. On a one-on-one basis, the energy transfer is arithmetic (i.e., one for one). To get three people enthused, however, you need to expend three times the energy of getting one enthused. In front of a group, you can get more people enthused; however, the energy transfer is only for the length of time of the presentation and is subject to any scrutiny after the presentation.

Therefore, to generate enthusiasm or positive energy about a program you have to do it exponentially. This means influencing several people who can influence others. In this manner consensus can be built for your program. You convince one person, who convinces two people, who convince four people, and so on.

Generating enthusiasm exponentially multiplies the effect for you. This needs to be done before the pivotal presentation so that at presentation the audience is ready and eager to hear your message. The presentation then becomes a confirmation of what everyone else agrees with. Having your reputation precede you is even better as you have a demonstrated track record of success.

There is also the concept of consistency of inputs. When most senior managers make their decisions, they gain knowledge indirectly from a variety of sources. Make sure that the sources feeding the information about your program are consistent and positive to the senior manager. In this way the presentation becomes a confirmation of known data, with no surprises.

C. Trial ballooning

Trial ballooning is a technique whereby a concept can be tested in the audience to gain understanding or trigger an initial reaction. It is useful when the presenter may not have any prior knowledge of how the audience will react to an idea and wants to trial balloon it to see the results. It also can be an effective tool to build consensus by releasing information a little at a time, and it allows for modification at subsequent releases. These techniques are used when the presenter is uncertain about the idea.

The caveat in the trial ballooning technique is that audience reaction may not always be accurate. The conditions under which the information is transferred may not support the desired conclusion, or the lack of a complete, cohesive story may cause an adverse reaction. In addition, it is more difficult to recover from a negative reaction once the information is out and conclusions are formulated.

D. The presentation

The presentation needs to follow a basic format, regardless of how much embellishment you may want to include. The following is the basic format to be followed for the presentation. It should also be noted that each of the basic elements needs to have detail to back it up.

1. Overview

This is the first section of the presentation. It needs to provide focus to the audience to allow their scope of examination to go from a macro view of many opportunities and markets to the business view on which the product and business plan are based. Bring them into focus fast and clearly. Do not confuse the audience by outlining many choices or options for the same opportunity. Pick your story and articulate it!

2. Opportunity

In this section, clearly articulate the business opportunity and the business condition. Now that you have focused the audience to a specific area, show them how you intend to make money for them with your business opportunity. Define it in terms of narrative and numbers. Use just enough detail to communicate the opportunity and to substantiate it. The audience is interested in your vision of how you intend to evolve this new business for them. They are not interested in grading you on all of your details.

3. Product description

If it is your intention to create the new business by embodying the opportunity in a product, here is the opportunity to articulate what the product is and how it will serve the business condition. Be as complete as possible in outlining the product and relate the features to the market needs and requirements. Explain the cost structure and what gross margins are expected. Explain how the company will implement the technology and what the technology for the product is. Create a perception in the mind of the decision maker that this program is doable and should be done without delay.

4. Strategy and implementation

You must have a clear and concise plan of how to go about developing the new product opportunity. In this section of the presentation, present the plan you have developed to make the new business a reality and show how you intend to do it. Times, dates, places, responsibility centers, linking objectives, funding, management support, and commitment are all the items that must be integrally linked in this part of the presentation. The issue of linking objectives is a very important one and not to be underestimated. The product needs to be presented in what can be referred to as a five-for-four format. Cite five benefits the organization can obtain in pursuing this program that will only cost four cost units. In other words, show how this program will help others in the company. It will create the desire to fund this program because the management can get more for their money in this one as opposed to others.

5. Call to action

All the preparation work is done and presented in a positive light to the management. Now is the time to wrap up the presentation and ask for funding. This now becomes a bidirectional interface between management and the project team seeking funds. As the leader, do not leave this discussion without getting a firm answer.

You have done all the homework required and deserve an answer one way or another. Management can request and you can provide clarity of detail, additional detail, or supporting facts; however, it is decision time for management and they should give you an answer.

Call for action and drive it to a conclusion—yes, no, or get a follow-up review date.

E. Securing the commitment

Nothing speaks louder than money! The objective of the presentation of the business plan is to obtain funding. This being the case, there is no statement, reassurance, or conversation that can occur between management and the product team that can substitute for money. If management cannot fund the program, they won't be able to support it, either. Also, do not be lulled into a false sense of security or management buy-in with the suggestion that the program be partially funded. Statements and compromises like these are an indication they are unsure and concerned. As discussed earlier, management may choose to partially fund several programs and select the best one at a later date. This only starves good programs and inappropriately funds poor ones while wasting precious time. Funding is what you came for; don't leave without it!

F. Assignment of initial funds

Generally, an entire program will not be funded from start through to completion. You will, however, want to obtain funds to initiate and complete a stage of development. This serves several purposes. First, it allows for the accurate conclusions to be drawn from the work done. Second, it allows for an appropriate termination point for work completed. This is a very important point for technical continuity. All too often, companies stop a program without appropriate time to close it out for future revitalization, making it difficult for later start-up or transfer of technology to another group. Therefore, it is important to obtain the funding and to post it to a specific project with specific deliverables in a specific time frame. Otherwise, it can have a tendency to become a catch-all project to gather expenses for nonapproved programs.

G. Follow-on funding

On successful funding of an initial stage of a program, immediately start campaigning for subsequent stages. In this way momentum can be established for the entire program. Do this for each stage of the program.

CORPORATE OPERATIONS USING THE PLAN

1. Facilitating Change in the Business to Execute the Plan

A. Not all companies are structured to execute the plan

As was stated in previous chapters, the marketplace is a fast-changing place with interwoven dynamics. A company positioned for a specific type of business today may not be well-positioned tomorrow. If the business and product plan and the company's present direction diverge, then the company must be brought in line to execute the plan. The key is to recognize this at the time of the preparation of the business and product plan, and to integrate the necessary changes into the company as the development progresses. It also makes the corporate changes prerequisite to the plan, thereby mitigating the pressure on the team manager to effect all the necessary changes by himself or herself.

B. Slight modifications to major overhauls

The type of changes can vary in scope and by functional area. They do, however, fall into one or more of the following categories:

1. People
2. Systems
3. Equipment

Is the current workforce equipped from a tools and training standpoint to execute the new development and the new business? Are the manufacturing, management information, and quality systems that are in place functioning and sophisticated enough to support the new business? And is the company's equipment state-of-the-art enough to support operations? These questions must be addressed within the plan and linking action plans must be assigned.

The business plan should outline the scope and extent of the changes required. There should be a cautionary note, however. If there are widespread changes required in each and every area of the business across all of the functional areas, then the business plan may be too much of a stretch for the organization. Initiating change in all areas of the company at the same time and making it stick is a very difficult proposition that requires a level of management commitment far beyond the basic commitment for a new product development. If the business plan calls for sweeping changes, rethink your objectives; the company may not be able to cope with them.

C. Linking department objectives

The key point and recurring theme in new product development is that it cannot be initiated and successfully implemented without all parties being in lock-step in executing the plan. A resource-consuming new product development is so important to the company that you must have all areas working toward the same goal. Do not let noncooperation and noncontribution go uncorrected. It demoralizes the contributing members of the team.

2. Management Focus

A. Senior management focus is not a given in new product development

Senior management focus is that single-minded prioritization of program initiatives that supports the goal. Unfortunately, not every senior manager can have this single-minded focus. Their days are filled with new opportunities for products and businesses. Each may appear initially better than the current program. However, it is just as important to identify a program and see it through as it is to select the best program. Focus is important because you cannot have restarts over and over again and complete a project. The issue of focus is a management problem, not an engineering problem.

The subject of management commitment and focus is one of the most fleeting concepts in new product development. It sounds easy, straightforward, and permanent but it is, in actuality, difficult to garner and establish a consistent effort. Multiple agendas, changing priorities, and the nature of short-term profit goals make it difficult for the senior manager to support programs.

B. The difference between managers and vocational engineers

The role of the manager differs from that of the engineer. The manager generally has a wider scope of interests for the business and does not have nearly the depth in any one of the interests. Management traditionally focuses on multiple priorities, juggling of resources, and managing the day-to-day affairs of the business. The development personnel (including engineers) need to focus on the development tasks at hand and have a tremendous amount of depth in these areas. They need to have a single-minded focus to complete the program on time and at cost. The engineer-manager must balance both the day-to-day and long-term needs as well as provide technical guidance to overcome obstacles.

C. Product champions

The fact that senior management has a hard time supporting programs has prompted the emergence of product champions within the business to provide the necessary focus and sense of urgency to the product development team. Such a person is the spark plug to help keep momentum during the project. The product champion has an almost sacred trust with management to deliver the program on time, at budget, and under any prevailing conditions. The terms *zealot* and *crusader* apply to these personality types. Care should be taken in identifying the product champion in your organization to ensure the individual's personality type is consistent with the program. There are numerous stories about skunk-works programs, which are successful programs in the face of adversity and the like. The important point to remember is that product champions have the focus to complete programs. They are not distracted by other elements of the business and can devote 100% of their energies to the program execution.

D. The manager's actions

For the project to go through the various obstacles of program justification, product development, launch, and product management, consider the manager that is selected to lead the program. An experienced manager who has completed several programs is a good choice. To get a sense of how this manager will execute the program, evaluate how the initial pitfalls are handled. This will give a sense of the approach to the handling of major problems. If you are the manager, keep this in mind because *your* management is evaluating *your* actions.

E. The well-considered approach

Although it sounds clichéd, a competent new product development manager should anticipate the problems and have generated alternative approaches ready to resolve them. The energy expended personally and collectively by the product development team in anticipating problems and accounting for them will be less than the energy expended by a totally overreactive manager. These people burn corporate energy at every turn and do not provide the cohesive momentum required for a program. Do not let them into your plans.

3. Confirming the Technological Fit—Internally
A. Are we equipped to handle the technology?

Previously, we discussed the ability of the organization to absorb the new business, its technology, and its needs and requirements. This is especially true when new technology is brought into the organization from outside rather than grown within the organization. This goes beyond the human resource issue to include capital equipment procurement required to support the new product. If capital, talent to operate the new equipment, and training for the personnel are required, they need to be factored into the business plan.

How will the technology base be supported? Think in terms of pathways. This question needs to be thought out in detail. Not every company can afford the expense of leaping into a new state-of-the-art technology. Many times there needs to be a pathway to the new technology. This pathway allows the affordability as well as the time to assimilate it into the organization and allow the learning required to implement it effectively.

B. Are you in the right generation of technology?

The optimum time for evaluating this question is at the beginning of the project. As discussed earlier, the technological base cannot be changed mid-project and still meet timing requirements. The planning of the technology must be well-considered because you want to be operating on a platform that has longevity and will be in demand at the time of introduction.

An extreme example of failure to do this is in the case of instant-developing photographic film, and extending it to movie film. The extension was a natural conclusion for the chemical-oriented product planners. What they didn't count on was the simultaneous development of the magnetic tape medium for movies and its embodiment in video cameras for the mass market. Coupled with the VCR's popularity gains from its ability to play movies, the video camera was the end for wet chemistry in movie film—instant developing or not.

The lesson: Be sure the technology used in the product is the one that has market acceptance at the time of introduction and thereafter.

C. Technology evolution flow chart

In preparing the product and business plan, be sure to forecast the technology evolution for the product line. There needs to be a balance struck between the time to develop and capture the market, the degree of difficulty in implementing a technology, the market dynamics, the acceptance of new technology, and the longevity of the technology employed. Its good to include the forecasting of technology in a business plan, even to the point of breaking it out separately, since the overall technology needs of the corporation can be evaluated and similar requirements can be funded as part of a corporate research initiative to support the businesses.

D. Technology on the cost-reduction curve?

As a final check, make sure the technology employed will be on a cost-reduction or learning curve that will be consistent with the market expectations for pricing. For example, many products have changed from pure mechanical form to a combination of mechanical and electronics.

This has been driven by the market demand for long life, added functionality, and lower cost. As the transition to electronics hardware was taking place, there was also a trend to embed microprocessor technology and implement functionality in the use of software. An example is shown in Figure 4-10.

Taking it a step farther, the hardware electronics will progress down an integration pathway whereby more and more complex circuitry will be integrated onto smaller and smaller pieces of silicon substrate. The software will have its own pathway whereby structured techniques for software development will eventually give way to adaptive learning in systems, in which the operational characteristics will readjust automatically.

Each of these leaps into the newer technology eventually drive product costs down while increasing functionality. In many cases this increase in functionality opens up new market and customer bases for increased sales. The market will communicate the need for these leaps into the new technology, sometimes in an indirect way. The challenge is to be able to sense the need and react without leading the technology too quickly. Doing so can lead to product cost problems, reliability problems with immaturely developed technology, and general market acceptance issues.

DYNAMICS OF THE PLAN

1. Confirming Initial Market Assumptions

The action of *continuous market assessment* is critical to a good plan. In fact, when they refer to a plan is as a work in progress, a key contributor is the continuous feedback from the market. Every product has its roots in an initial market assessment. At the time of the preparation of the business plan, these initial assessments must be reconfirmed. Real-time market feedback can be an invaluable means for decision making.

2. Plan Modifications

In the same manner, the business plan needs to be slightly modified based on the real-time data streaming in. One thing to keep in mind is that the basic core of the original assumptions and the original plan should remain in tact. Slight modifications based on the real-time input can be made. If sweeping changes are made based on the input, the business may fail. If the original assessment is nowhere near what the real-time data are saying, a market shift has occurred or the original assessment was incorrect.

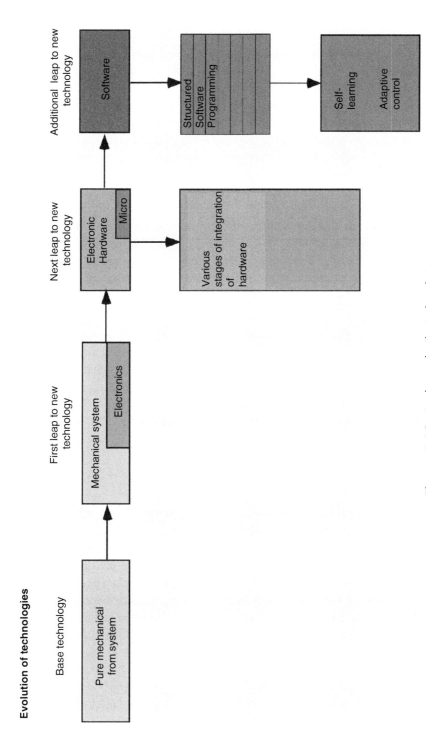

Figure 4-10. Product technological evolution.

3. Timing

A. Timing is everything

If timing is everything for the development and introduction of new products, the same is true for obtaining funds to execute a program. Funding is not always readily available for good opportunities. Many good ideas go unnoticed and unfunded because the timing was inappropriate or the presentation was ineffective.

The concept of timing needs to be viewed as a window of opportunity to align with, rather than as a blind sense of urgency. The manager needs to seek funding from the sources where and when they are available. They also must be consistent with the investor's parameters for investing. If the investor is a short-term player who is used to producing immediate results, it is foolhardy to approach him or her with a long-term program that will require massive funding.

A common misconception (and a source of the misunderstanding about funding) is that small funding is easier and more readily available than large funding. One needs to remember that the parameters for the investor's funding are based on timing, amount of returns, and risk—not necessarily the absolute amount of funds. If the parameters are in line, the specific amount of cash invested may not be that relevant to the speed of securing the funds.

B. Continuity of programs

Funding of a business plan may require several phases or, at least, the funds will be released in discrete steps. It is wise to align with an investor who can fund an entire program and allow you to finish what you start. One of the worst things that a manager can do is secure funding for only a part of a program and fail to provide the team with the necessary continuity to complete it.

The problem manifests itself where different investors will have different requirements and agendas. They may be at odds with the initial fund provider or, worse yet, the program will have to be compromised to secure the funding.

C. Hit the opportunity window

The best way to secure funding is to hit the investor's opportunity window with a program requirement that is consistent with their investment parameters. Keep in mind that, as odd as it may sound, the investor is under pressure to invest funds in viable programs. Your job in preparing and presenting the business plan is to help them in selecting and funding your plan by demonstrating that it is the best choice. From a product planning perspective, the plan must sell the investor into a full program, not a halfhearted attempt at one. If it is not funded fully, it may compromise the development, as depicted in Figure 4-11.

In this example (Figure 4-11), the first window of opportunity is available for product A'. This is the result of a development in iterating from A to A'. If the program is delayed, as shown in the bottom section of the illustration, the available window of opportunity for A' ceases to exist. Unfortunately, many companies see the second window of opportunity as a continuation of the first and assume the market will be there for a long time.

What is really happening is that a second window of opportunity has been made available for product B'. This product can address more of the market than A' can, so A' lost the window of opportunity and, more important, will not generate the revenue it needs. From a

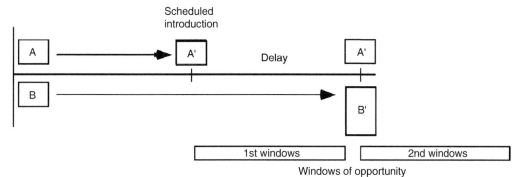

Figure 4-11. The opportunity window.

pure business perspective, the business plan failed to secure adequate financing, which slowed development and execution; this in turn caused delaying product introduction, thus losing the window and resulting in the reduced revenue.

D. The statistics of trying to catch up

The importance of timeliness to the marketplace cannot be overstated, especially when the funding stage is involved. The entire enterprise needs to create financial and development progress—a momentum that cannot be substituted by other means. Doubling efforts, reorganizing the program, or throwing additional personnel at the problems cannot be a substitute for steady progress. It simply takes nine months for the gestational period of a human being. It cannot be foreshortened by impregnating nine women to have a child in one month!

E. The financial impact

To get a feel for the financial impact of slipped schedules caused by underfunding, refer to Figure 4-12. Two equivalent programs are depicted. The top line in the graph shows the on-time program bringing in the revenue as planned. It initiates at t1 and contributes up to the t6 time period. The bottom line in the graph shows the same program, but delayed. It doesn't start until t3 and stops contributions at t6. The time base is fixed because this represents the window of opportunity; the area under the curves is the totalized revenue. It shows a significant revenue loss because of the delay, which was a result of poor funding or program management. Keep this graph in mind when securing funds to make a case for adequate funding, and also as a reminder to keep programs on track.

4. Trading Time Saved for Technology

A. Time is of the essence

As discussed earlier, time is a precious commodity in new product development. It is in ever-shorter supply, and the speed at which competitors can assimilate information to capture new opportunities is increasing. Therefore, the implementation of technology needs to be on a predictable schedule. There is little tolerance for the development of raw technology and its application during product development. The technology should ideally be well-understood and mastered before its application in the product.

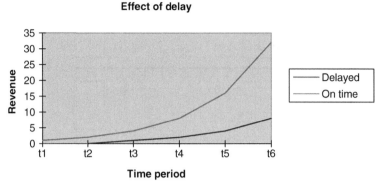

Figure 4-12. Effects of delay.

B. Technology doesn't sell; function, price, and performance do sell

Many times the lure of new technology can take over the focus of a product development. When this happens, the development of the technology, not the product, becomes the driving force. This is a prescription for failure because customers do not buy technology per se; they buy products with features that create value and benefit as a result of their use.

This is an important point because it is a vocational hazard for the business and product development team to become so enamored with technology that they lose sight of the program's objective. If the technology is not mastered, it is manifested by development engineer's trying to work around issues by placing constraints and qualifiers on the application and use of the product. In its worst form, the marketing and sales part of the corporation end up "holding the product's hand through life" while engineering tries to understand the technology. This is a certain prescription for product and new business failure.

Figure 4-13 illustrates the concept of targeting a technology for a product that is (or soon will be) mastered by the corporation and implementable in the new product. As time progresses, the technologies evolve from the Technology Base 1 through to the Technology Base 4. The product development manager then targets Technology 3. This ensures that, at the point in the future when the product is introduced, the corporation will have mastered the technology, and that the product is still in demand by the market. In this example, Technology 4 is too much of a stretch because it may be too new and not understood enough

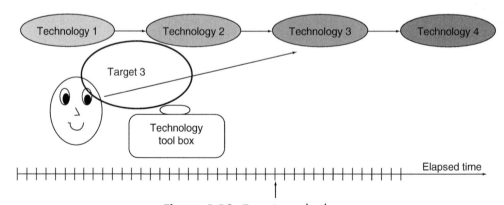

Figure 4-13. Targeting technology.

for product platforming. Technologies 1 and 2 will be less than state-of-the-art or minimally acceptable by the customer base.

C. Don't chase the Holy Grail of technology

This goes without saying, but will be repeated here for emphasis: Do not chase new raw technology as part of product development. Rather, you need to implement new technologies that are understood by the corporation, as part of the product development, with the primary focus being product development. Consequently, you need to be practical in its selection and application. The new product development manager and the team should strike a balance between implementing the latest state of the art in technology and project timing.

Many companies have made the fatal error of betting on the implementation of new technology in product development before the technology is internally mastered within the corporation. The degree to which companies have bet their future in this manner is the degree to which many companies have had to endure financial losses, changes in ownership structure, and product recalls.

The primary function of development is to service the customer in the best way possible. The team needs to assess the market's needs and desires. For some markets, the customers simply don't care about the technology. They are interested only in the functionality and price. Take this into account.

CUSTOMER AND MARKET INPUT TO THE PLAN

1. Testing the Market

A. Assess your new product position

The process of planning and executing a new product development effort is rarely initiated and completed using the same information, assumptions, and data. More realistically, the team needs to reassess the product positioning, the competition, and the cost and pricing. As we stated earlier, the business of new product development is a fast-changing, dynamic environment where fixed plans give way to navigation. The new product must be led through the competitive maze and the dynamics of the market, to introduction.

Failure to adequately assess the market during the development phase may cause a misintroduction of the product, overlooked or missed opportunities, or incorrect positioning, depending on what the competition may be doing.

B. Retest the marketability, pricing, and timing

Specific items to be reevaluated during the process are the general marketability of the product, the demand for it, the all-important quantity of produced units, the pricing of the product, and the timing of the program and introduction. As a matter of course, it is a good idea to keep a running list of these items, since it is easier to identify trends and potential pressure points.

C. Update competitive analysis and project changes

A properly executed program shouldn't involve changing the product during the development phase. The development should be fast enough to capitalize on an opportunity.

Alternatively, the program's objective should lead the market enough so that a sufficiently unique product satisfying the demand still exists at the end of the longer development cycle.

The product stays stable; however, the business plan may change to suit changing market scenarios. If price, availability, or usage pattern shift, the plan, the financial profile, or the distribution arrangement may be modified to suit the new conditions. Therefore, it is crucial to understand what is happening and to project the effects into the future when you will introduce the new product. Monitor the market and accommodate material changes, in order to position yourself.

2. Confirming the Technological Fit—Externally

A. Is the technology right for the marketplace?

At the same time, it is critical to assess whether the technology embodied in the new product is a good fit with the marketplace. For example, it is a mistake to use a technology in a product serving a market that may have an aversion to it. It simply becomes an uphill battle for market acceptance, and energy will be spent on selling the embodied technology rather than on the solution the product brings to the customer. Does the technology render the products too difficult to use?

B. Which platform are you on?

The three basic technology platforms traditionally used are industrial, commercial, and military (in order of technological advancement). Traditionally, the pathway for technology has been from the most advanced to the least advanced. Certainly barring security requirements, elements in the commercial sector have found their way into the industrial sector after hardening. The important point is that is you are working in one realm, so consider the source of technology and the pathway for the technology. Make sure you have some runway with the technology in order to create your own product platform scenarios.

The raw technology should define the product platforms. The product platforms should define the product versions. When creating the plan for the business, ensure the technology procurement and ramp-up times and costs are factored in. If the product line must jump from one technology to another or if the platform must be adjusted, make financial arrangements for the time and funds needed to do this.

3. The Customer as Part of the Plan

A. Involve customers early in the process

There is no substitute for direct feedback in understanding the need for the product being developed and hone your plan to capture the market. To the extent that you can disclose information without affecting security, the direct customer disclosure and feedback method is better than any secondary measurement. It serves several purposes: it keeps good, loyal customers informed; it fosters trust; and it allows an honest engagement with the customer to solicit unbiased feedback on the product in its current state.

Feedback also permits the customer to buy into the product and the development program. It also allows the company to test drive their marketing story on a real prospect.

Finally, it allows the manufacturer to double-check their strategy and begin to weave the customer's strategy with their own.

There are several caveats in this engagement that must be noted: You need to guard against information leaks, protect your intellectual property by confidentiality agreements, and double-check the potential customer's competitive affiliation. The use of the direct engagement technique is an excellent way to add credibility to the business plan. It provides incontrovertible evidence of a market for the new product, which can be used to nullify any doubts present in the audience.

B. A good customer will tell you things you can't learn elsewhere

One of the most desirable aspects of the direct engagement technique is that you can obtain information from the customer that is not available elsewhere. For example, let's say you are a manufacturer of cement mixers. Your secondary marketing information has shown that there is a worldwide market for 1000 cement mixers each year. By direct interview, you may find out that only 350 cement mixers are sold each year in the free world. The balance is sold in areas where there are intermediaries, third-party payoffs, or other methods different from the traditional methods you use. Consequently, these 650 mixers are *unavailable* to your company. If your business plan depended on a breakeven of 400 cement mixers, the "available market" is not sufficient to support your plans. This may not be apparent in the secondary market survey. It is only knowledge passed directly from customer to vendor. By obtaining this critical information, you can redesign your business plan to suit the business condition.

C. Put your team on the firing line

Finally, one of the most important reasons to use the direct engagement technique is that it puts your team directly into a customer's firing line. All the requesting and "specsmanship" you may bring to the marketing requirement's specification cannot outweigh the benefit of exposing the development team to the customer firsthand. Once they hear it from the customer, you will no longer need to repeat yourself.

PLAN CORRECTIVE ACTION

A plan is just that—a guideline for a team. The real world generally does not go according to plan. The is the job of management—to correct the activities and get the team back on track and salvage the program. However, should the program be salvaged?

1. When to Save a Program, When to Kill it

A. Your fiduciary responsibility

As a new product manager, your first priority is your fiduciary responsibility to the company. Management is entrusting you with the funds to prosecute a program and, as such, you need to monitor performance of the team and effect appropriate action. As discussed earlier, the development expense burns the resource candle at both ends. It consumes financial resources and has lost opportunity costs associated with it. Therefore, expending it requires diligence.

B. Cost patterns and profiles

The business of project management has two basic items that must be controlled: expenditures and time. The expenses, as related to time, can be outlined for each stage of a program and should be part of the business and product plan. The profile generated can therefore be placed on a graph and the actual expenses per time period can be compared to planned expenses. The comparison can thus give an indication of program progress.

Each stage of a project has an appropriation request. This is the amount of funds appropriated to complete that stage, and is the spending limit for that stage. There are certain milestones that must be completed during the stage, and each requires a certain amount of funds. The progress should track the milestones and the expenditures.

Some programs are front-end loaded. In this case a large amount of cash outlay occurs initially for equipment, and the balance is for labor to experiment with the equipment, for example. Others are back-end loaded, where the expenses are at the end of the stage. Still others are linear in which the expenses basically track the milestones. The linear approach is generally taken when pure engineering talent is used throughout the stage. Figure 4-14 demonstrates the three types of loading programs and their characteristics.

C. Some programs shouldn't be saved

This is not a popular position to take, but the statement is true. Some programs may be ill-conceived or they go sour and need to be redefined in a major way, or they must be terminated. There are some signs that the program may be in trouble based on the expense profile. Figure 4-15 is an example profile, indicating trouble spots.

The three horizontal lines indicate the different appropriations limits for the stages of the project. The lower profile is the planned profile and the upper profile is the actual profile. As shown, the first sign of trouble comes in Stage 1, where expenses are clearly out of line with the front-end-loaded projection. Corrective action is taken to mitigate the damage, as indicated by the change in slope, but the program is already off to a poor start. It is worse yet if the project milestones also slip. This would be indicated by a normalized x-axis for both the actual and planned expenses/milestones curves. In the second and third appropriations, the rapid increase in cash outlay would indicate possible premature capital expenditures.

D. Don't get saddled with a dud; fix it or kill it

From a pure career management perspective (and also from a corporate employee perspective), do not get saddled with a dud. It is bad for your career; it wastes corporate assets and resources; and demoralizes an otherwise productive development team. Initiate a fix, re-lay out the program, or kill it. Do not let it founder. It can tear down your career by undermining any future decisions or recommendations you may have to make to management. Remember: management is looking to you to watch out for their interests. Do not disappoint them.

E. Terminating a program, setting the stage for future funding

It then follows that if you demonstrate diligence for the company in your decisiveness in terminating a program when it is warranted, you will earn senior management's respect for your judgment. This can set the stage for future requests and grants of funds for programs. One of the most difficult management aspects of new product development is the need to

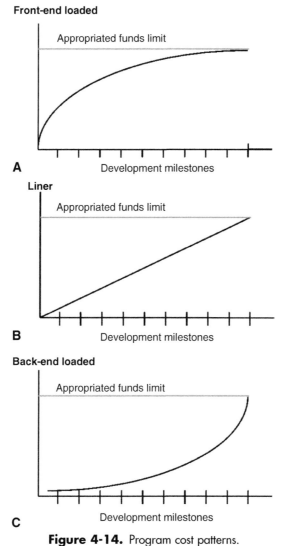

Figure 4-14. Program cost patterns.

request and secure additional funds beyond the budget. If you demonstrate decisive action in protecting company assets and resources, management will more readily trust your assessment of a situation and respect your wishes.

2. Continuous Assessment

The best way to keep on track of the business plan is to continually assess the conditions and the progress of the company toward the business development goal. Accordingly, the manager should implement the following process shown in Figure 4-16. Start with an assessment of the market condition and how the company is progressing toward capturing the market. Next, establish the validity of the incoming information. This is to ensure there are no wild swings in interpretation. Then, assess the impact to the program of any new information and determine its materiality. Corrective action (either internally in the case of

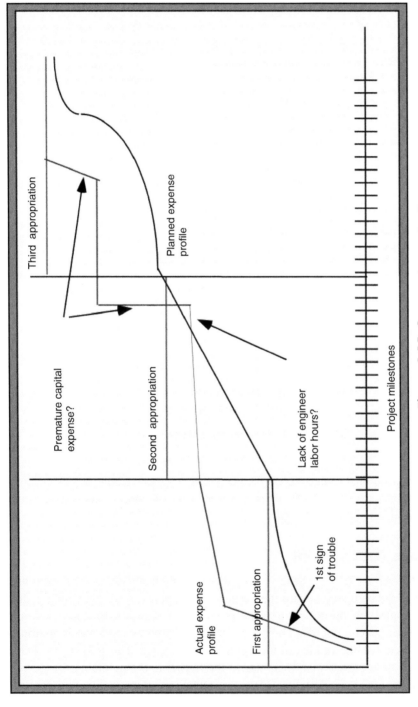

Figure 4-15. Program stage cost patterns.

Figure 4-16. Continuous program assessment.

lack of progress, or externally in the case of a modification to meet an external change) must be taken. Finally, implement the action and provide feedback to top management, where another assessment is made some time later.

SUMMARY

In this chapter we discussed the organization of a business plan: the *blueprint to success.* A format for the business plan was presented and discussed. The importance of timeliness and integration was stressed, and the relative standing of the company (in terms of where the market is headed) was also presented. How to sell the program to management was covered, as was how to modify the operations to accommodate the plan. Internal issues such as management focus and organizational changes were included as part of the business plan.

A section on pricing and the philosophy and justification for pricing was presented, all for the purpose of constructing a workable business plan. A section on the importance of the accounting function was included to serve as a background in preparing the business plan. Market testing and technological fit were also included to provide continuity to the business and product plan, as well as incorporating the customer as part of the plan. The reader should now have a good idea of how to structure and execute his or her own business plan. In the next chapter we will look at the program from the accountant's viewpoint and provide specific tools to ensure that the accounting function embraces the project.

JUSTIFYING A PROGRAM—THE ACCOUNTING VIEWPOINT

BACKGROUND

1. Accounting Background

A. The financial model

The financial model for a new product is different from that of a development engineer or team member. The finance model is characterized by funds flow, direction, exposure, and timing. Finance generally has limited understanding of the technical, marketing, or sales aspects of the transaction. Accountants' views of a new product development program can be represented in terms best accepted by accounting, as shown in Figure 5-1.

The basic model for financial analysis is that there is an initial investment in development that will return future dollars at an increasing rate in subsequent time periods. In addition, between time periods 3 and 4, for example, an additional investment is required to update the product or to respond to some competitive action. The initial and subsequent allocation of funds are offset by the returned revenue stream. The value of the revenue, the time to revenue generation, and the initial investment all factor into the quality of the investment from the financial perspective.

B. The business story in terms of narrative and numbers

When presenting the program to the financial team, be sure to use both numbers and narrative to ensure understanding in the accounting functional area. The numerical data will serve the analytical requirements of accounting, while the narrative will assist the team in understanding strategy, product planning, and the company's overall marketing direction. The combination of data and narrative will best serve the information needs of all concerned.

2. Financial and Economic Analysis

A. Background and general model for product development and sale

The mechanics of financial analysis for a new product are straightforward. They involve investment, revenue, and profit within a specified time frame. The most difficult task is to

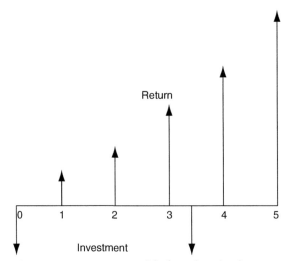

Figure 5-1. Finance model of product development.

determine the company's expectations and minimum acceptable performance for a program. Referring to the section on planning and assessment of operations in Chapter 1, the company has an established track record for projects and an expectation of the minimum acceptable return. These should serve as gauges for what would be acceptable for the new business opportunity under consideration.

Figure 5-2 represents the general model for a product sale that will serve as a basis for discussion in this chapter. This model deviates from a traditional income statement and balance sheet format for illustrative purposes only. The data presented are a designed example to show the investment and return and some of the expenses along the way toward growth. This model recognizes a booking concurrently with a shipment, and collection is immediate. Materials are also procured concurrently with bookings. In actuality, these dates can differ based on several internal and external factors as well as on requested ship dates and component lead times.

As the figure shows, the model begins with a marketing product planning expense. This is the expenditure to determine the new product opportunity. The next step is to expend funds in development to develop the new product to be sold. This occurs during the first time period with no revenue to offset it.

The sales expenses are based on a fixed amount of $4,000 and 11% of booking value. This means that as the booking value increases, the sales expenses also increase. Orders or bookings are a hoped-for result of the sales expenses. The company then responds by ordering material and adding labor to assemble it into the product.

Burden is a calculated value based on 1.75 times the labor expenses. The product is shipped and a collection is made. Administrative expenses are deducted and a profit is realized.

As shown in the graph in Figure 5-2(B), the allure of large sales numbers in the forecast quickly descends to modest profit figures. These profits are the foundation on which the

	Financial model for a product sale				Price each 5000 dollars					
Item	Description		Qty.	Qty.	Qty.	Qty.	Qty.	Qty.	Qty.
			2	8	16	35	65	105	135
1	Maketing/planning Expense (dollars)	−75,000	0	0	0	0	0	0	0
2	Development investment (dollars)	−200,000	0	0	0	0	0	0	0
3	Sales expenses (dollars)	0	−5,100	−8,400	−12,800	−23,250	−39,750	−61,750	−78,250
4	Booking (dollars)	0	10,000	40,000	80,000	175,000	325,000	525,000	675,000
5	Material (dollars)	0	−4,250	−17,000	−34,000	−74,375	−138,125	−223,125	−286,875
6	Labor (dollars)	0	−900	−3,600	−7,200	−15,750	−29,250	−47,250	−60,750
7	Burden (dollars)	0	−1,575	−6,300	−12,600	−27,563	−51,188	−82,688	−106,313
8	Sales (dollars)	0	10,000	40,000	80,000	175,000	325,000	525,000	675,000
9	Administrative expenses (dollars)	0	−500	−2,000	−4,000	−8,750	−16,250	−26,250	−33,750
10	Profit (dollars)	0	−2,325	2,700	9,400	25,313	50,438	83,938	109,063

A

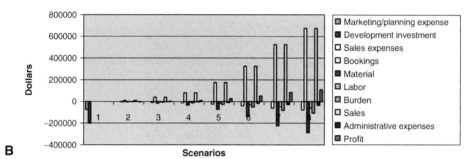

B

Figure 5-2. (A) Financial model for a product sale. (B) Graphical representation of product sale model

enterprise must operate and with which it must grow. Another factor to consider is that the company is spending money on an ongoing basis—purchasing materials, buying labor, and paying other expenses. In addition, all of the profit added up throughout the run of the product barely offsets the development expenses and the marketing expenses. This calculated profit does not even factor into the temporal value of funds, which we will discuss later.

B. Other types of transactions

The other types of transactions may not be so straightforward. Certain cooperative arrangements, such as joint ventures, are structured contractually where the component product is built by one company (e.g., Company A) and sold to another (e.g., Company B). Company B builds a system consisting of the component supplied by Company A and other items may be proprietary to Company B.

Company A may have proprietary knowledge for the manufacture of the component but has in interest in licensing the right to build the entire system. A possible arrangement may be to have Company A pay Company B for the rights to build the system. Perhaps a lump sum is identified. During the course of the agreement, it may be decided that Company A will offer a discount on each component to offset the cost to license the system manufacture. If one looks at the transaction, it would appear to be a low-margin component sale, when in actuality there are a few transactions occurring simultaneously between the two companies. An example is in Figure 5-3.

This is only one example of how a transaction for the sale of products may occur. When analyzing a transaction, this must be kept in mind. When synthesizing a transaction, this may represent an alternative to create a better value on both sides of the sale.

3. Accounting and Finance Definitions

A. Finance definitions

The accounting and finance function have a specific terminology for evaluating performance of a company. Here we present some of these financial terms as they relate to new product development and operations.

1. Return on investment (ROI)

The return on investment is a measure of the business's performance based on the entire investment. In general, it is expressed as the ratio of the return divided by the investment. It places value on the amount of money invested and compares that amount with the annual funds returned as a gauge for each year.

2. Net present value (NPV)

The net present value is a measure of the discounted time value of money. It is based on the premise that money earned in the future is worth less than money today, due to inflationary factors. The NPV is used to evaluate the present value of future funds by reflecting them back to today's values at the discounted rate caused by inflation. The value is discounted each year by the estimated inflation rate of that year.

3. Cost of capital

The cost of capital is a measure of the corporation's cost to employ capital. This capital is used for equipment and can generally be the cost of a company's longer-term debt. If the company needs capital to operate and grow, this value is the measure of what it will cost the company to employ it.

Comparison of transactions			
	Actual transaction		**Expected transaction**
Sales price	$1,658	**Sales price**	$1,658
License fee	$300	**Cost**	$1,100
Cost	$1,100	**License fee**	
Net	$258	**Net**	$558
	Plus $300 toward license		

Figure 5-3. Comparison of transactions.

4. Internal rate of return (IRR)

The internal rate of return is also referred to as the time-adjusted rate of return. It is simply the actual interest yield of an investment program over the life of the project. It is computed by determining the discount rate that will equate the present value of the cash inflows generated by the program with the cash outflows required by the program. It is the discount rate that causes the net present value of an investment to be zero. The internal rate of return is used as a measure of the present rate of return on an investment, analogous to a time deposit account in a bank.

5. Net income

The net income is the amount of income that is left after all expenses in the business. It is the sum of material, labor, manufacturing overhead, operating expenses, and taxes deducted from the total income, and can be expressed as a percentage or as a dollar amount.

6. Owner's equity

The owner's equity is the amount of money the owner actually has after any liabilities have been subtracted. As the enterprise grows, the owner's equity should grow because this value represents the owner's value stake in the business.

7. Total assets

The total assets are the sum of cash, investments, accounts receivable (monies owed to the corporation), inventories, any prepaid expenses, and property plant and equipment.

8. Return on equity

Return on equity is a measure of the amount of revenue returned compared with the equity or ownership of the company. It relates the profits to the owner's stake in the business. The return on equity is a measure of how the annual return compares with the owner's equity in the business at that time. In general, it is expressed as the ratio of the annual dollar return to the shareholders' actual dollar equity in the business.

9. Return on total assets employed

The return on total assets employed is a measure of the return of the operations to the average total company assets within the specified period. It is calculated by dividing the net income and interest expense by the average total assets, factoring in the interest expense, which is adjusted for taxation. In this way, the measurement is an evaluation of the total return compared with the total assets, with no measure of how those assets are financed.

These are only a few of the many measurements that are used to evaluate an organization's performance. There are additional ones to measure specific items that circumstances might dictate.

B. Relating financial to actual events

From the development perspective, one of the most unacceptable accounting practices is to manage exclusively "by the numbers." There are instances in countless companies in which accounting renders judgments based solely on numerics without relating these to the operations of the business. This can be a mistake, just as evaluating a program solely for its contribution to technology with little regard to the numerics involved can be a mistake. The best way to manage the business is to evaluate both aspects.

To do this, the accounting function needs to become knowledgeable about products, the operation of the business, and new product development. In addition, development personnel need to become conversant with accounting principles, driving forces, and corporate concerns. Both groups need to accommodate one another's vocational and parochial interests.

For example, a lack of sales of a newly introduced product needs to be understood to effect corrective action. Both sides must understand whether the cause is a temporary delay in driving the introduction through the channel sales or is a permanent trend.

Figure 5-4 illustrates the financial analysis of sales progress and relates the numbers being reported to actual events. Understanding within the corporation is based on both of these aspects.

C. Product cost systems

The product cost system used by the organization should reflect accurate information about the product cost and provide management information to effect appropriate action. It is selected for use from an operations standpoint. There are three basic types of cost systems in use today: standard cost systems, job order cost systems, and activity-based cost systems.

1. Standard cost systems

The standard cost system differs philosophically from the job order cost system in that it is designed to set a specific rolled-up cost. The cost roll-up is the planned cost anticipated as a result of the product design process. Periodically during the production cycle, the actual incurred costs are gathered with the appropriate overhead allocations and absorptions included. These are then compared with the standard costs, and any difference is referred to as a variance. The variance is used as a prompt for management action to contain costs and align them with the standard. A standard cost system is illustrated in Figure 5-5(A).

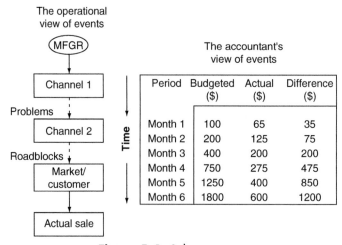

Period	Budgeted ($)	Actual ($)	Difference ($)
Month 1	100	65	35
Month 2	200	125	75
Month 3	400	200	200
Month 4	750	275	475
Month 5	1250	400	850
Month 6	1800	600	1200

Figure 5-4. Sales progress.

Standard cost summary					
	Standard cost				
	20XX	20XY	Variance	20XZ	Variance
Material (dollars)	50	45	5	60	−10
Labor (dollars)	35	37	−2	32	3
Burden (dollars)	110	120	−10	90	20
Total (dollars)	195	202	−7	182	13
			Unfavorable		Favorable

A

Job order cost summary						
	Description	Vendor		Cost each ($)	Quantity	Extented cost ($)
Material 1	Part 1	ABC		23	2	46
Material 2	Part 2	ABC		34	5	170
Material 3	Part 3	DEF		45	7	315
Material 4	Part 4	DEF		56	10	560
Material 5	Part 5	FGH		67	12	804
	Hours	Burden (150%)	Cost/ hour ($)			
Labor 1	10	15	50		Labor	1,250
Labor 2	12	18	37.5		Labor	1,125
Labor 3	2	3	24.5		Labor	122.50
Labor 4	3	4.5	12.6		Labor	94.50
Labor 5	6	9	78.45		Labor	1,176.75
Engineering time	24		55		Engineering	1,320
Miscellaneous time	2		30		Total miscellaneous	60
						Total cost ($)
						7,043.75

B

Figure 5-5. (A) Standard cost summary; (B) Job order cost summary.

2. Job order cost systems

The job order cost system is a cost-gathering mechanism that tracks the product through manufacture and posts accumulated costs in terms of material labor and burden. As costs increase, the job order cost system tracks the increase without reference to a maximum allowable cost or market pricing. As the product moves through the production processes, it simply gathers cost.

Job order cost systems are generally used with a cost-plus pricing strategy. The two approaches can get a company into trouble quickly from a profit and volume perspective. Such systems are, however, adequate in terms of posting the product's cost to record. A job order cost system example summary is shown in Figure 5-5(B).

3. Activity-based cost systems

The activity-based cost system generates a product's cost by focusing on the underlying activities that are necessary to produce the product. These underlying activities consume the resources. The activity-based system is appropriate when manufacturing direct costs shift away from labor dominance to material-dominated or other elements. With labor becoming less and less of a factor in the cost makeup, the manufacturing expenses normally burdened to labor become such a large percentage that it becomes difficult to generate a meaningful cost and understand the absorption of expenses.

The cost system employed will not materially make or break a program by itself. Its accuracy relative to its association to the operations is negligible, in most cases, compared with the other uncertainty aspects in a new development.

D. Costing a product as part of a project (quick test)

As part of the market investigation and the product presentation, a short-form test could be required to prequalify a program. This qualification is a quick test of the project's development cost, the product cost, and the revenue expected. It also incorporates means for evaluating prototype costs. This quick test can be used easily with a few data to determine project viability. Figure 5-6 shows a sample of the spreadsheet for such an evaluation.

In the example costing of a project shown in Figure 5-6, separate cost roll-ups are used for the prototype and the production units. Labor costs are determined by the number of minutes required to assemble the product multiplied by the cost per minute. Burden is calculated by the burden rate entered at the beginning of the chart multiplied by the direct labor content. The indirect cost is factored at base value. The material, labor, and burden costs are totaled for both the prototype and the production units.

The cost to develop the product is calculated by evaluating each of the engineering elements multiplied by its cost per hour. This value is entered at the beginning of the chart for engineering costs. Additional development expense is recognized in the next section and later totaled in the SG&A expense entries. The capital equipment is also factored into the payback calculation in the next section.

The next section takes the data on costs and allows a forecast for the prototype and the production units to be entered. The revenue per unit at the beginning of the chart is used to determine the revenue forecast. The expenses, product cost, and capital equipment costs are deducted to determine net profit. Although this sample is neither an elegant return on

Project development worksheet[1,2]					
Product costs					
Material					
Item	**Description**	**Prototype ($)**		**Production ($)**	
1	Drive 50 hp	1,870		1,870	
2	Enclosure	1,600		600	
3				0	
4	Amplifier	1,250		1,360	
5					
6					
7					
8					
9					
10					
Total materiai		4,720		3,830	
Labor			Dollars		Dollars
Direct minutes		480	360.00	480	360.00
Indirect minutes		120	90.00	120	90.00
Total labor		600	450.00	600	450
Burden			900		900
Total factory cost		6070.00		5180.00	

Figure 5-6. Project evaluation worksheet.

investment calculation nor an in-depth cost analysis appraisal, it is a quick calculator for determining the viability of programs under consideration.

E. The fallacy of the hockey stick

Whenever a program justification is prepared, there is always a tendency by the product planners to skew the forecast of product sales to the later years of the program. This is referred to as a back-end-loaded forecast or a hockey-stick forecast. The term *hockey stick* derives from the characteristic graphic appearance of the forecast.

For business development managers, it is critical to avoid justifying a program using the back-end-loaded forecast. If the characteristic sales volume is truly expected to follow this

Engineering					
		Hours	**Dollars**		
Application		24	1,320		
Hardware		16	880		
Algorithm		24	1,320		
Software		80	4,400		
Design spec.		8	440		
SPAF		1	55		
Manual		24	1,320		
BOM		24	1,320		
Cost roll-up		24	1,320		
Test data		40	2,200		
Routings		40	2,200		
Test proced		24	1,320		
Test SW		48	2,640		
Total hours		377	20,735		
Expenses				Capital equipment	
	Travel	4,000		Test fixtures	0
	Services	0		Tooling	5,000
	Miscellaneous equipment	500		Equipment	0
	Total expenses	4,500		Total capital equipment	5,000

Figure 5-6 (Cont'd)

path, then temper the forecast. If, however, a hockey-stick forecast is being used to justify expenditure and stall for time after development (as happens in some cases), be aware that it can have a devastating effect on the product development team, management, and careers. In many cases, a linear forecast can yield the same level of success as a back-end-loaded forecast from a net present value standpoint.

In the following example, the product price is valued at $500 each. The gross profit is 50% or half of the revenue. Forecast revenue values are generated for both the linear case of volume buildup and the back-end-loaded case of volume buildup. For both cases, the net present value is calculated based on the investment of $1.5 million and the gross profit stream generated throughout the five-year product run. Figure 5-7 illustrates this scenario.

Summary		
	Prototype	Production
Factory cost ($)	6,070	5,180
SG&A ($)	25,235	25,235
Tooling ($)	5,000	5,000
Net/unit ($)	8,500	8,500
Forecasted units	10	500
Revenue ($)	85,000	4,250,000
COS ($)	60,700	2,590,000
Gross profit ($)	24,300	1,660,000
Required sales for breakeven	12.44	9.11
Net profit ($)	($5,935)	$1,629,765

[1]Labor cost/minute $0.75; engineering cost/hour $55; burden 250%.

[2]Revenue each $8,500.

Figure 5-6 (Cont'd)

Fallacy of a hockey stick forecast							
	Investment ($)	Year 1	Year 2	Year 3	Year 4	Year 5	NPV
Price each ($)		500	500	500	500	500	
Volume of linear		1,000	2,000	3,000	4,000	5,000	
Volume of hockey stick		400	650	1,000	300	6,500	
Revenue of linear ($)		500,000	1,000,000	1,500,000	2,000,000	2,500,000	
Gross profit of linear ($)	−1,500,000	25,0000	50,0000	750,000	1,000,000	1,250,000	1,563,427
Revenue of hockey stick ($)		200,000	325,000	500,000	1,500,000	3,250,000	
Gross profit of hockey stick ($)	−1,500,000	100,000	162,500	250,000	750,000	1,625,000	808,425

A

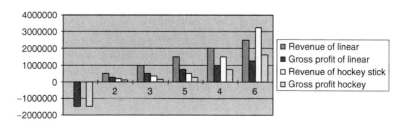

B

Figure 5-7. (A) Fallacy of the hockey-stick forecast; (B) Graphical representation of the comparison

In our example, the net present value generated from the investment and gross profit circumstances shows a higher value for the linear case than for the back-end-loaded case. There is, however, a more important point here. The forecast should be an accurate representation of the expected sales volume; it will be used in the program as a measure of success and as a justification for further funding. It is also important to show early progress and success to management. A forecast that is purposely back-end-loaded is a warning sign to management that the program must absorb much more uncertainty in the marketplace before the corporation achieves payoff.

Understandably, management finds it difficult to support back-end-loaded scenarios because time and uncertainty are the enemies of new product success. Management wants to know what magic will happen to generate such strong sales in years 3, 4, and 5 at a greater growth rate than in years 1 and 2. Management is justified in its position on this issue.

F. It's not development cost; it's factory cost and sales volume

In any development program that must result in volume sales to justify the investment, there are concerns about budgets, expenses, factory costs, volume of sales, and timing. The key to successful planning and product management is to focus on those items that will have a significant effect on the success of the program. As we will see in the next example, these parameters are not weighted equally in relation to their respective impacts. The example will show that volume of sales, factory cost, and timing are the key pressure points in a program. The development cost is less important if the market can support the projected sales volume. This is why it is critical to understand the market, the company's capabilities, and the company's effectiveness in securing orders. No amount of cost-cutting or other compensation can make up for lack of sales volume. Let's review the following examples (Figure 5-8). The examples are based on several assumptions:

- The factory cost is $250.
- The sell price is $500, yielding a gross margin of 50%.
- The net present value is the time-adjusted value of money reflected to the beginning of the project.
- Cases 1, 2, 3, 4, and 5 are decreasing in volume of sales.

As shown in the baseline example in Figure 5-8, the investment of $1.5 million is amortized with good return in Cases 1 and 2 and only marginally in Case 3. The decreasing sales volume affects the program significantly in Cases 4 and 5. This set of cases will serve as the baseline example showing the impact of sales volume reduction.

The Factory Cost Grows example illustrates the impact on factory cost increases as time goes on. This example shows immediate and deep impact on the net present value due to the increased manufacturing cost and fixed market price.

Case 3 shows the impact of increased development expense on the overall program. As shown in Figure 5-8, this has much less of an effect than the manufacturing cost increase example in Case 2. The third example, Development Expense High, supports the conclusion that if the market opportunity is real and the company can achieve market penetration within the anticipated time frame, the cost of development is a less important factor in the overall equation. Keep in mind that this conclusion applies only to development

Return on investment calculator								
Baseline example	Investment ($)	Year 1	Year 2	Year 3	Year 4	Year 5	NPV (based on 5% inflation)	IRR
Price each		500	500	500	500	500		
Cost each		250	250	250	250	250		
Case 1 volume	−1,500,000	1,000	2,000	3,000	4,000	5,000	$1,563,427	30%
Case 2 volume		800	1,600	2,400	3,200	4,000	$965,027	22%
Case 3 volume		600	1,200	1,800	2,400	3,000	$366,628	12%
Case 4 volume		400	800	1,200	1,600	2,000	($231,772)	0%
Case 5 volume		200	400	600	800	1,000	($830,172)	−17%
Gross profit case 1	−1,500,000	250,000	500,000	750,000	1,000,000	1,250,000	$1,563,427	30%
Gross profit case 2	−1,500,000	200,000	400,000	600,000	800,000	1,000,000	$965,027	22%
Gross profit case 3	−1,500,000	150,000	300,000	450,000	600,000	750,000	$366,628	12%
Gross profit case 4	−1,500,000	100,000	200,000	300,000	400,000	500,000	($231,772)	0%
Gross profit case 5	−1,500,000	50,000	100,000	150,000	200,000	250,000	($830,172)	−17%
Factory cost grows	Investment ($)	Year 1	Year 2	Year 3	Year 4	Year 5	NPV (based on 5% inflation)	IRR
Price each		500	500	500	500	500		
Cost each		250	290	330	370	410		
Case 1 volume	−1,500,000	1,000	2,000	3,000	4,000	5,000	$323,807	12%
Case 2 volume		800	1,600	2,400	3,200	4,000	($26,669)	4%

Figure 5-8. Return on investment calculator.

cost, not to the time required to develop the product, which is still critical to market acceptance and success.

For your specific program, you might want to try a few scenarios to experiment with the potential impact that each parameter might have.

G. Collateral costs: direct and indirect

The cases presented in this chapter are simplified for illustrative purposes. In actual situations, there are additional costs to take into account. Some of these costs are obvious and can be posted directly to the product line. Others are indirect and are considered elsewhere. Be sure to factor these costs into the calculation of sales return.

	Investment ($)	Year 1	Year 2	Year 3	Year 4	Year 5	NPV (based on 5% inflation)	IRR
Case 3 volume		600	1,200	1,800	2,400	3,000	($377,145)	–5%
Case 4 volume		400	800	1,200	1,600	2,000	($727,620)	–15%
Case 5 volume		200	400	600	800	1,000	($1,078,096)	–30%
Gross profit case 1	–1,500,000	250,000	420,000	510,000	520,000	450,000	$323,807	12%
Gross profit case 2	–1,500,000	200,000	336,000	408,000	416,000	360,000	($26,669)	4%
Gross profit case 3	–1,500,000	150,000	252,000	306,000	312,000	270,000	($377,145)	–5%
Gross profit case 4	–1,500,000	100,000	168,000	204,000	208,000	180,000	($727,620)	–15%
Gross profit case 5	–1,500,000	50,000	84,000	102,000	104,000	90,000	($1,078,096)	–30%
Development expense high	Investment ($)	Year 1	Year 2	Year 3	Year 4	Year 5	NPV (based on 5% inflation)	IRR
Price each		500	500	500	500	500		
Cost each		250	250	250	250	250		
Case 1 volume	–1,750,000	1,000	2,000	3,000	4,000	5,000	$1,325,332	24%
Case 2 volume		800	1,600	2,400	3,200	4,000	$726,932	16%
Case 3 volume		600	1,200	1,800	2,400	3,000	$128,532	7%
Case 4 volume		400	800	1,200	1,600	2,000	($469,867)	–4%
Case 5 volume		200	400	600	800	1,000	($1,068,267)	–20%
Gross profit case 1	–1,750,000	250,000	500,000	750,00	1,000,000	1,250,000	$1,325,332	24%
Gross profit case 2	–1,750,000	200,000	400,000	600,000	800,000	1,000,000	$726,932	16%
Gross profit case 3	–1,750,000	150,000	300,000	450,000	600,000	750,000	$128,532	7%
Gross profit case 4	–1,750,000	100,000	200,000	300,000	400,000	500,000	($469,867)	–4%
Gross profit case 5	–1,750,000	50,000	100,000	150,000	200,000	250,000	($1,068,267)	–20%

Figure 5-8 (Cont'd)

Cost assignment is determined by the culture of the organization. The manner in which costs are absorbed and posted to individual product lines can be misleading. The operative lesson is to be consistent. If comparable products do not include all of the costs, then do not post them to your disadvantage. You might want to keep track of them separately, however, for your own use.

DEVELOPMENT AND ACCOUNTING: TWO DIFFERENT DISCIPLINES

1. Comparison of Accounting and Development Operations

A. Motivation

Earlier, we discussed the difference between engineering and accounting and how they view operations. This is primarily due to the fundamental difference in their respective vocations. It is important to understand these differences as they contribute to the motivation and actions of both parties. Engineers are immersed in the work of absorption of uncertainty and problem solving. They look at uncertainty on a day-to-day basis and are quite comfortable dealing with it and overcoming it. They are also quite comfortable at overcoming adversity, failure, and reexamination. This applies to their programs, experiments, and systems.

Accounting and finance personnel are used to analyzing the result of uncertainty. They project and track the expenses, revenue, costs, quality costs, and progress in pure financial terms. They make dispassionate judgments about programs, time frames, results, revenue profit, and cash.

Neither vocation is enough to run a company in the absence of the other. The manager of the business development enterprise must mediate the two vocations and seek their cooperation to the betterment of the company.

B. Measurements of performance

Likewise, the measurements of performance for the two disciplines are quite different. The engineering vocation is measured on creativity, ability to contain factory costs, ability to procure and use tools to create new products, and accuracy of material and processes—all within the framework of meeting a specification of performance.

The accounting and finance vocation is measured by accuracy of funds flow, advanced information, trend analysis, and budgeting analysis as well as treasury functions. They keep the money supply flowing within the organization.

C. Organizational placement

The placement within the organization can also be different. Many times accounting and finance has a direct ear to the senior management controlling the purse strings of the program. The engineering vocation may not have the same influence. One word from accounting and it may take several hundred words from development to counteract.

The benefit derives when both vocations are placed on the business development team and measurements of performance include success of the business venture. In such cases, the cooperation of both can be a formidable force.

2. Accounting and Finance as Partners

A. Cooperating to get results

Even when they're on same team, the accountants' perception of new product development can be strictly financial. Consequently, they have empathy for neither the problems encountered nor for the means by which they are surmounted. The lack of concern for these issues can cause tension between the finance functional area and the engineering functional area.

It must be stressed that the cooperation of engineering and accounting is what sells management on a program. Therefore, make the financial team partners, not adversaries, in the development program.

B. Agree on expectations

The financial orientation evaluates things in terms of equations and checksums. This means that for every two pounds of invested effort, there is an equated result of some measure of progress. Unfortunately, actual development doesn't always occur in this manner. There are breakthroughs followed by little or no progress at all, then followed by further progress. The degree to which this process occurs is related to the uncertainty in the development program. Equations and checksums are not necessarily compatible with uncertainty analysis.

Because this equation orientation exists in the minds of the accounting team, there is limited understanding or tolerance for delays, failures, or restarts on product design. The perception is that the funds and effort were expended—where are the results?

To a large extent, this orientation is valid; there are times, however, when progress is neither linear nor forward. It is during these times that you most need the partnership of the finance people. Because they have one eye on the rest of the business and receive input from marketing on general market conditions, they make fiscal recommendations to senior management. They have the ability to influence funding or project termination. The operative lesson is to make finance and accounting your partners; they have the power to intervene to save a program during lean times.

C. Technological advantage

Because the development engineers are technologically advantaged relative to all others in the enterprise, things that are obvious to the development team might not be readily understandable to others. For example, at a certain point in a development when a roadblock can be removed with the procurement of outside services or capital equipment, a large expenditure required to make the breakthrough might be obvious to the technically oriented, but not to anyone else. This is where the trust relationship with the financial and other nontechnical people needs to transcend functional boundaries.

D. Narrative of similar programs

To enhance understanding, it might be helpful to cite similar programs that have experienced similar circumstances. In this way, a connection can be made between what you are requesting and a previous success story. Parallels can then be drawn, differences discussed, and resolution can be sought.

E. Business development's role: facilitate understanding

The business development manager must facilitate understanding between the financial community and the development community and to forge a partnership that will last through product introduction and beyond.

FINANCIAL AND BUSINESS MODELING

1. Financial Model for a Product Sale

Shifting our focus to external factors in new product financial matters, there are multiple models for sales transactions. These models are representative in nature and are not meant to cover all the possible scenarios, but they do give an indication of how funds flow with the sales transactions. The financial model for a product sale is quite different from a service sale or even a systems sale. Each has expenses and residuals placed in different time frames and financial exposure can be different in each. The following represents a financial model for a product sale going to the user via three ways: the direct sale, the representative sale, and the distributor or value-added sale.

A. Direct sales model

In a direct sales model, a sale is made to a customer. The factory purchases material and furnishes labor to build a product. The sales cost is incurred as the salesperson is securing the order. Other general and administrative costs are also incurred as a result of operations.

The product is shipped, and an invoice is generated and sent to the customer for payment. Payment, normally due in 30 days, can stretch out to 45 days. The company has carried the costs of this transaction for the 45 days, which helps explain why it requires cash for a product line to grow. As you increase sales, the cost to ramp up also increases. Figure 5-9 shows the cash flow for the direct sales model.

B. Representative sales model

The representative model is different from the direct sales model in that the bulk of the sales expense that is normally associated with the direct sales force is not incurred prior to sale. The representative advances these costs by virtue of being an independent businessman. The factory issues a check for the commissioned sale, generally after receipt of net payment,

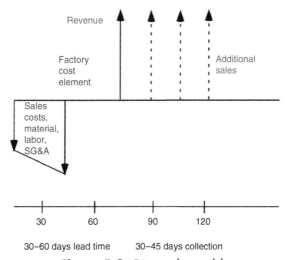

Figure 5-9. Direct sales model.

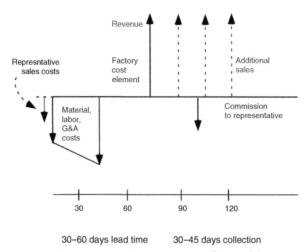

Figure 5-10. Representative sales model.

to ensure that representative-entered orders are real and collectable. Figure 5-10 shows the financial model for the representative sales model.

C. Distributor or value-added reseller sales model

The value-added reseller (VAR) model is different from the previous models in that an entire transaction occurs between the distributor and the customer. The manufacturer builds the products to a forecasted order and ships the product to the distributor. There might be other incentives between the distributor and the manufacturer, such as floor planning or rebates, but fundamentally, the distributor or VAR takes receipt of the product and resells it at a margin to the end-user.

The manufacturer receives payment on a predictable basis because the manufacturer and distributor work together throughout the term of their relationship so that uncertainty of credit with the customer is eliminated or reduced. The distributor must then rotate the product (sell, ship, and collect payment) through its organization at a rate of inventory turns that exceeds the bank's finance levy for business. The distributor then ends up with the margin for profit, as shown in Figure 5-11.

2. Financial Model for a Service Sale

The financial model for a service sale looks quite different, as shown in Figure 5-12. In the service sale scenario, there is no job or project specific investment. Looking at the graphic from left to right, an incremental investment is made in the labor hours doing the service and, hopefully, there is soon revenue coming in from that service. A larger service job may require more expenses, which are then offset by more revenue. The profit is the running difference between the expenses and revenue. There are also service jobs that are high-value to the customer, generating a lot of revenue from a small amount of service expense. Nevertheless, there are scenarios where the payment is immediate between the expenses and revenue. There are many specific types of cash

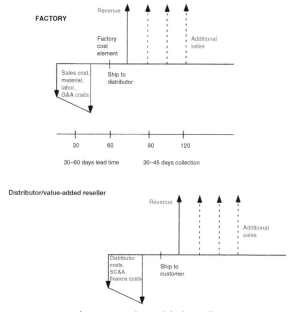

Figure 5-11. Distributor or value-added reseller (VAR) sales model.

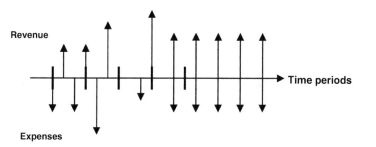

Figure 5-12. Service model.

flows in the service sector, but they are generally characterized by quick payment against labor expense.

3. Financial Model for a System or Project Sale

The financial model for a system sale is different from both the product and service model. This is because of two factors: the time in negotiating the system sale through to completion, and the level of absorbed uncertainty. Referring to the Figure 5-13 a sale like this consists of several items on the revenue side as well as the cost and expense side.

However, the system or project sale has certain elements of the product or service sale. There are initial sales expenses needed to get the order. With systems sales, there are generally factory sales personnel directly negotiating with the customer. After the order is placed, the factory usually requires a deposit for the system. Sometimes this amounts to approximately 10% of the total. The engineering begins and expenses in engineering labor

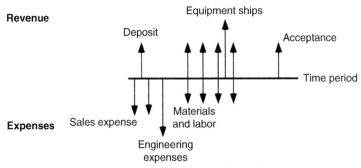

Figure 5-13. Systems sales model.

accumulate. When the engineering is completed, manufacturing starts to build the system. This is when inventory costs start to become significant. When the equipment ships, a larger portion of the contract is invoiced and collected. There can also be progress payments whereby, as expenses are incurred, invoicing is done to get incremental revenue throughout the duration of the project. After the system is commissioned and the customer accepts the system, the final installment may be billed. There are many different ways to characterize the systems sale. This is only one example.

RISK, REWARD, TIMING

The issue of risk/reward is widely known. One could assume a linear relationship in that the higher the risk, the higher the reward. In actuality, the curve may be more exponential in nature. This is because of the deployed assets both in personnel and opportunity costs in pursuing a risky venture. If the uncertainty is not resolved, huge amounts of money and time may be needed to implement the program.

1. Timing is More Critical Than Development Cost

A. Timing is indeed everything

In some strategic market situations, timing is the key element to success. It is more important to get into the game and sacrifice short-term profits than to miss being in the game. Long-term sacrifice of profits will have devastating effects on the organization and limit any future market response. To repeat an earlier statement, you must be able to deliver the product at specification at the market's cost and allow profits to amortize the development in *a market-oriented time frame.*

B. Success factors pertaining to business development

In terms of importance to product development, there exists a hierarchy of critical success factors. This hierarchy starts with the health and the size of the business, which determine the relative affordability for particular development programs.

Next, it is important to examine the profitability of the business to determine investment endurance. Can the organization carry off a long-term investment and fund it sufficiently throughout periods of uncertainty?

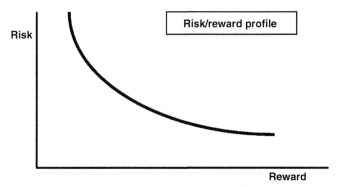

Figure 5-14. Risk/reward profile.

Equally important are management's spending priorities and habits. Is management consistent in what it spends funds on, and is there an established pattern of investment for the business? The answers to these questions determine the actual disbursable cash that management can support for a given program.

It is also important to determine the time frame for investment. Does the organization have a track record of investing and ensuring that investment through to completion? Is it averse to navigating uncertainty? Does management change direction in midstream? Knowing the historical pattern of corporate response sets the sense of urgency for the program. Finally, what is the organization's tolerance for error or failure, and what are the company's patterns of recovery?

Determine the enterprise's approach to cost management. How does the organization approach manufacturing cost roll-ups? How are costs contained?

These issues need to be explored and understood because at least some of them are likely to affect the program at some point during its execution.

C. Sliding windows of opportunity

We previously reviewed the sliding windows of opportunity for new products and saw that a marketplace opportunity does not remain in existence for a long period of time. Rather, it is available to capitalize on for only a short time and generally triggers activity by many competitors.

To be a long-term player, it is critical to be selective in choosing opportunities and to amass the horsepower needed to execute projects effectively. In years past, these opportunities evolved slowly and companies had more time to evaluate whether they wanted to go in a specific direction. Today, opportunities are intense but shorter-lived, with the spoils going to the competitor who can mobilize its forces most effectively.

D. The moving target

We also previously discussed leading the target in product planning. The development effort must be coordinated with business cycles and accounting's financial plan for the business. If we use the moving target analogy, then the bullet fired is the new product. The speed of the bullet's movement from the gun to the target is analogous to the speed of development.

The market moves a little faster each year, so product development must improve in speed to lead the target and secure a market share.

E. Comparative analysis of programs

One of the tools for deciding which program to fund is a comparative analysis of several programs. Comparative analysis allows dispassionate evaluation of the programs on a strictly financial basis of projection. It consists of estimations of investment and revenue and computes a return in a consistent, measurable way. Figure 5-15 shows three product programs with various investment scenarios and calculates returns for each.

In the first example, an investment of $2 million results in three possible scenarios for return. In this analysis, as in the other analyses we will review, the volume of product sold in the best, most likely, and worst cases is held consistent for each of the three products being evaluated, to give the reader a sense of the potential impact of volume and pricing on the return estimates. For Product A, as for the other two, the revenue is based on a price of $500 and a cost of $250. In these examples, the cost and price are held constant over the five-year product run.

The investment/return analysis for Product B, product cost decreases with volume and the price in the marketplace also tracks downward. In this case, the cost decreases at a greater rate than the price, with the result that margin dollars are enhanced. Also, the initial investment for Product B ($1.5 million) is less than for Product A.

For Product C, the initial investment is $1.0 million. The product price increases with each year and the product cost also increases; however, gross profit declines because the cost increases at a greater rate than the price, thus causing erosion in the margin. In this case, the combination of the product, profit, and investment generate a more favorable picture.

2. Preserving the Risk/Reward Ratio with Product Structure

With all the risk in business development, there must be a way to mitigate the risk. There is, and it is *platform management*. Not every new product will be a success. A slightly different version may be required to get a single market. Consequently, by making sure you are on the right product platform, you can recover from a missed opportunity quite quickly and generate a different version and still capture some market share.

When a company takes an incremental approach to the market, they can be successful. The generation of new and enhanced product features can keep a product fresh and always in front of the customers.

Another way to preserve the risk/reward ratio is to leverage development of versions off the platform with a partner. Sometimes referred to as an integrator, these partners can generate different versions for the lead company and allow them to manage the platform rather than generate product iterations.

3. Communicating Risk/Reward—Timing

It is critical to communicate the risk of a venture with the people charged with making the venture a success. These stakeholders are the keys to success and should have prior

Comparative investment analysis

Product opportunity A	Investment ($)	Year 1	Year 2	Year 3	Year 4	Year 5	NPV (based on 5% inflation)	IRR
Price each ($)		500	500	500	500	500		
Cost each ($)		250	250	250	250	250		
Best-case volume	−2,000,000	1,000	2,000	3,000	4,000	5,000		
Most likely case volume		600	1,200	1,800	2,400	3,000		
Worst-case volume		200	400	600	800	1,000		
Gross profit, best case ($)	−2,000,000	250,000	500,000	750,000	1,000,000	1,250,000	$1,087,237	20%
Gross profit, most likely case ($)		150,000	300,000	450,000	600,000	750,000	($109,563)	3%
Gross profit, worst case ($)	−2,000,000	50,000	100,000	150,000	200,000	250,000	($1,306,362)	−22%

Product opportunity B	Investment ($)	Year 1	Year 2	Year 3	Year 4	Year 5	NPV (based on 5% inflation)	IRR
Price each ($)		500	492.50	485	477.50	470		
Cost each ($)		250	240	230	220	210		
Best-case volume	−1,500,000	1,000	2,000	3,000	4,000	5,000		
Most likely case volume		600	1,200	1,800	2,400	3,000		
Worst-case volume		200	400	600	800	1,000		
Gross profit, best case ($)	−1,500,000	250,000	505,000	765,000	1,030,000	1,300,000	$1,640,903	31%
Gross profit, most likely case ($)	−1,500,000	150,000	303,000	459,000	618,000	780,000	$413,113	13%
Gross profit, worst case ($)	−1,500,000	50,000	101,000	153,000	206,000	260,000	($814,676)	−16%

Product opportunity C	Investment ($)	Year 1	Year 2	Year 3	Year 4	Year 5	NPV (based on 5% inflation)	IRR
Price each ($)		500	525	550	575	600		
Cost each ($)		250	285	320	355	390		
Best-case volume	−1,000,000	1,200	2,000	3,000	4,000	5,000		
Most likely case volume		600	1,200	1,800	2,400	3,000		
Worst-case volume		200	400	600	800	1,000		
Gross profit, best case ($)	−1,000,000	300,000	480,000	690,000	880,000	1,050,000	$1,775,064	46%
Gross profit, most likely case ($)	−1,000,000	150,000	288,000	414,000	528,000	630,000	$656,875	23%
Gross profit, worst case ($)	−1,000,000	50,000	96,000	138,000	176,000	210,000	($415,962)	−10%

Figure 5-15. Comparative investments.

knowledge of how the program is progressing and what the pitfalls are. Accordingly, it is important to communicate to the organization and all its stakeholders the status of the program and what is expected of each of them.

This is especially true when things may be going slower than expected, progress may be lacking, or initial revenue targets have not been hit. It may be tempting to shield the stakeholders from this news, but in most cases the worst thing is to keep it from them. Involve them in the good and bad, the ups and downs, and they will help you through to success.

CASH, TIMING, BUSINESS CYCLES

1. Cost, Volume, and Profit—Breakeven

A. Breakeven analysis

In any business, there are multiple costs associated with operations. Variable costs are a function of the volume of business, whereas fixed costs are a function of the business incurred costs without reference to volume. There is a point at which the profit of the enterprise is equal to the fixed cost. This is known as the breakeven point or zero-profit point. This is a crucial point to know even though no one ever wishes to operate in that mode.

As shown in Figure 5-16, the behavior of the fixed costs and the profit is similar to the equation for a line with a y intercept; for example, $y = m(x) + b$, where y is the profit, b is the fixed cost, m is the per-unit profit, and x represents the number of units. Assuming that the fixed costs are b, then the marginal profit per unit times the number of units equals b at breakeven. To determine the breakeven point, determine the value of x, the number of units, and where the total marginal profit equals the fixed costs.

In the example in the figure, the fixed costs are $150,000, the cost per unit is $275, and the sell price is $525. The breakeven occurs at 600 units. This is the point at which the incremental profit of $250 per unit multiplied by 600 units equals the fixed cost of $150,000.

B. Single-product company versus contribution to profit model

There is also a difference in the treatment and evaluation of profit between a large company with several product lines and a small company with one product or only a few products.

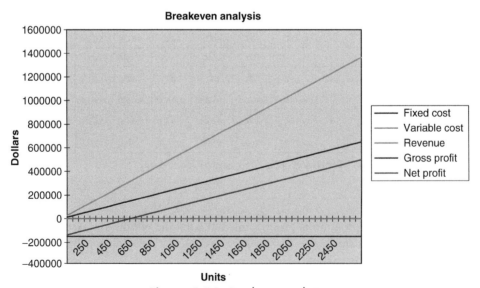

Figure 5-16. Breakeven analysis.

The profit motive is much more intense in the small company and is critical in the single-product company.

We evaluate the contribution to profit within the context of a large, multiproduct company. The absolute dollar profit value is less important than its overall contribution, pull-through sales, and strategic direction, even though it remains important to evaluate these products and the absorption of costs of their profits.

In the smaller company, there is necessarily little interest in percentages or global strategy; the issues are survival and profit dollars. The senior management and ownership of a small company generally do not respond to the discussion of what, to them, are abstract issues.

C. Every program is evaluated on its own

Regardless of the size of the organization, the key to long-term success for the product, the team, and the individuals involved is to have every product be profitable in its own right. In addition, ensure that the accounting system is set up to evaluate the various products uniformly and that the indirect and SG&A costs are absorbed according to their respective organizational activity bases.

There are many instances in which accounting has made historical changes in the treatment of products. Hidden costs or allocations levies shift organizational burden to higher-margin products in order to improve the financial appearance of others. If your organization is predisposed to this kind of treatment, costs could be posted to the new product to offset existing poor performers.

Each product should stand on its own merits in terms of profitability, and each should be managed by the correct financial data. Without these assumptions in place, incorrect conclusions can be drawn from the data and product decisions are not well-considered or well-executed.

D. Inventory: The hidden costs

We discussed the issue of accounting data being massaged to enhance the position of certain products. There is another issue in organizations—inventory—that affects the overall performance of a product line. Inventory is an indirect cost which, if managed improperly, can represent a significant burden on the corporation.

Most corporations are organized along functional lines, with product managers controlling specific products. Inventory is customarily considered to be the responsibility of manufacturing; however, this operational control generally defers to the product manager to facilitate the launch of the new product. If the product line is not well-considered or if the market shifts somewhat, the impact on inventory can be devastating.

The product manager needs to participate in inventory control and to operate within the framework of corporate goals given to manufacturing. In addition, product changes in response to problems can generate scrap, both in material and in labor.

2. Critical Unit Volume During Amortization

A. Amortization

"Yeah, all these plans are nice, but show me the orders!" If this has been said to you (as product manager) recently by a senior manager or by your banker, it generally instills an unsettling feeling. Whether we like it or not, this statement is rooted in fact and experience that reemphasize the critical need for a new product to become a viable business by generating unit volume. It is the volume that initiates and supports progress.

Volume means that a strong sales effort is occurring; it drives the unit cost down to where it has been targeted. It demonstrates the effectiveness of a marketing plan and establishes momentum. Fundamentally, volume creates the funds flow into the organization to begin to offset the investment in product development. A portion of each revenue element is used to pay back the organization for investment and therefore amortizes the development. There simply is no other means to offset these costs.

B. Don't misrepresent projected volume

The fact that unit volume is the initiator of payback is its own check and balance. If the development team overestimates the sales forecast, the organization will be unable to meet anticipated volume and consequences detrimental to the product will occur. If the forecast is too low, the organization will not approve the expenditure for the initial development.

A common danger in new product development and the team's zeal in qualifying the program is to overstate volume to sell the program to management. This practice can be a career-breaker for the team members. The best approach is to evaluate the opportunity realistically and forecast it based on the most likely scenario. If the program is viable, it will be approved and eventually will be successful. Remember that both management and the banking community look for promises kept, not promises made.

C. Examine return on investment: Unit volume pays for many errors

"There is nothing wrong with the organization that more orders can't cure." How often have you heard this statement? It, too, has a basis in business experience. A healthy amount of orders often can mask inefficiencies within an organization. In point of fact, a downturn in sales forces organizations to redesign and streamline their processes. There is, however, no amount of cost reduction, trimming of expenses, or austerity measures that can compensate for lack of orders, even though the common knee-jerk reaction to lack of orders is simply to cut costs. The key to a well-run organization is to keep healthy programs going, trim off poor performers, and keep a focused eye on order input to keep from having to operate in an austerity mode.

To reinforce this concept, look at the sample income statement in Figure 5-17. It shows the sales input side of the income statement. Net sales are at full value in this example. The material labor and burden are factored in on a normalized basis across the series of five examples of decreasing revenue and profits. The five scenarios represent a shrinking business.

The material cost is factored at 50% of sales, while labor is factored at 6.5% of sales. For simplicity's sake, the burden is 2.5 times the labor dollars. The SG&A expenses are shown as typical percentages of sales, leaving a net profit of 27.25% of sales. The expenses scale down as the business scales down, keeping the same percentage. This results in a fixed percentage of sales for the new profit with a decreasing dollar amount. To preserve the dollar amount of profit at $92,500 without decreasing the gross margin (as a result of volume reduction), the SG&A expenses would have to decline disproportionately. In the example given in the figure, the scaled-down expenses would be $117,000. To preserve the profit, however, these same expenses would have to total $84,625—a significant

Income statement					
	Case 1	Case 2	Case 3	Case 4	Case 5
Sales (dollars)	1,000,000	912,500	825,000	737,500	650,000
Return allowance (dollars)	0	0	0	0	0
Net sales (dollars)	1,000,000	912,500	825,000	737,500	650,000
Material @ 50% of sales (dollars)	500,000	456,250	412,500	368,750	325,000
Labor @ 6.5% of sales (dollars)	65,000	59,312.50	53,625	47,937.50	42,250
Burden @ 2.5 times labor (dollars)	162,500	148,281.25	134,062.50	119,843.75	105,625
Gross profit @ 27.2% of sales (dollars)	272,500	248,656.25	224,812.50	200,968.75	177,125
Sales expense @ 10% of sales (dollars)	100,000	91,250	82,500	73,750	65,000
Engineering expense @ 4.5% of sales (dollars)	45,000	41,062.50	37,125	33,187.50	29,250
Administrative expense @ 3.5% of (dollars) sales	35,000	31,937.50	28,875	25,812.50	22,750
Sales (dollars)	1,000,000	912,500	825,000	737,500	650,000
Total expenses needed to operate (dollars)	180,000	164,250	148,500	132,750	117,000
Total expenses required for profit of $92,500 (dollars)	180,000	156,156.25	132,312.50	108,468.75	84,625
Required reduction percent (%)	0.000	4.928	10.901	18.291	27.671
Reduction of sales		0.088	0.096	0.106	0.119
Reduction of expenses to preserve profit		0.132	0.153	0.180	0.220
Net profit (dollars)	92,500	92,500	92,500	92,500	92,500

Figure 5-17. Income versus volume and expenses.

difference. Figure 5-17 shows how this disproportionate relationship exists and how the impact of such a level of reduction can cripple a company.

The type of cost cutting just described affects staffing significantly. If the expenses are headcount-related only (an oversimplification for the sake of this example), the reduction rate will cause the enterprise to lose too many people, cutting staff below the levels needed for critical mass and for positioning the company in a strong development position in the future. In this example, we assume that every $100,000 of sales requires an additional staff member. Conversely, every reduction of $100,000 of sales revenue results in a reduction of staffing by one. At a $1 million sales company, there are 10 people. When sales decline to $650,000 according to this model, it would be necessary to fire 35% of the workforce!

In an actual company, eliminating headcount as drastically as is done on the spreadsheet example will irreparably damage the organization. Observe once more that little can compensate for unit volume.

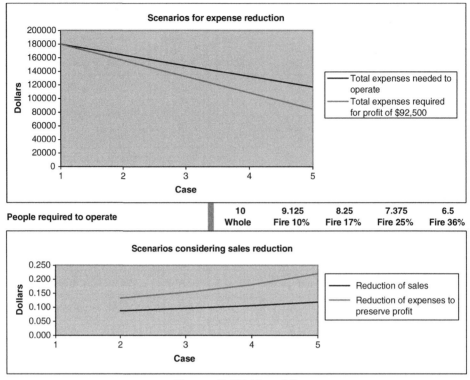

Figure 5-17 (Cont'd)

D. The importance of an accurate unit volume forecast

There is another, more self-motivated reason for the accurate assessment of volume forecast in new product development. The organization will be less than kind to you if the forecast is skewed to the high side, especially if the initial numbers are not as aggressive as originally anticipated. In addition, whatever mentoring might be in place for you with senior management will soon fade in the wake of an overrated forecast.

The forecast of volume and the profits realized from them are factored into the budget. These funds will be used for other programs, so they must be achievable. If the funds are not realized, the budget process falls apart and follow-up investment will suffer. Worse yet, management's confidence will be shaken to the point that follow-up funding will be more difficult.

E. Feeding on scraps rather than owning the market segment

The need for timeliness in new product introductions is a given. The market dynamics dictate that the first company to market usually is in the best position to reap the rewards of that market. If you miss the timing, you could end up serving a segment of the market you have no interest in serving, as shown in Figure 5-18.

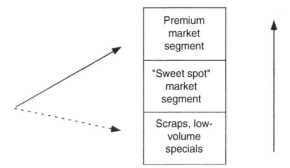

Figure 5-18. Market segment.

In this illustration, the initial target is for the premium market segment, represented by the solid arrow pointing to this premium segment. If development is lethargic, the dynamics of the marketplace keep the market opportunities moving. The market in effect moves vertically past the manufacturer's target if the manufacturer's development is late. In such a case, the dashed line represents the manufacturer ending up with low-volume specials, completely missing the more desirable segments.

F. Focus on unit volume and costs

Previously in this chapter, we discussed the concept of amortization as the streams of payments required to offset an investment. Ideally, the stream of payments (factoring in the temporal value of money) should exceed the original investment and incrementally add to the retained earnings. A company must have a stream of profit-generating revenue to amortize the development investment. Ensure that all of the expenses, costs, and investment are coordinated with the revenue and profit to guarantee that the program is on sound financial footing.

3. Generating Cash and Profit

A. Impact of route to market

The route to market can have a significant impact on the overall program. As a new product manager, your job is to maximize the profit for the company. Consequently, you must evaluate the tangible exchange that will take place between the company, the market, and the route to the final customer. The worth of the roles they play and the value given to those roles must be evaluated.

Pinpointing the risk and the party who endures the burden of warranty both enter into the analysis. A route that can actually deliver customers and volume sales that could not be delivered otherwise is worth more of the profit margin than a route that is merely in the opportunistic place at the appropriate time. Figure 5-19 shows the impact of the route on profitability.

In the example shown in the figure, the product has a normalized cost structure as follows for all three cases. In Case 1, the channel cost is 15% of the sales price of $120, or $18. This

Impact of channel on profitability							
Sales price Dollars	Material Dollars	Labor Dollars	Burden Dollars	Factory cost Dollars	Channel Cost	Channel Cost Dollars	Gross profit
120	50	6.5	14.75	71.25	15%	18	30.75
120	50	6.5	14.75	71.25	20%	24	24.75
120	50	6.5	14.75	71.25	25%	30	18.75

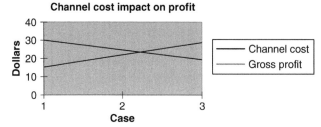

Figure 5-19. Impact of channel on cost.

leaves the manufacturer with a gross profit of $30.75. As the channel cost increases, the gross profit for the manufacturer declines to a crossover point, as shown in Case 2. A further reduction causes an inverted situation in which the manufacturer is making less on the product than is the channel taking the product to market. This is undesirable because the manufacturer bears most of the risks, considering the warranty and the initial investment.

B. Impact of economic cycles

The impact of economic cycles is significant as regards the development of new products and new business. The ebbs and flows of the larger economy affect the profitability and cash movement of a company. A well-run program can be affected if corporate finances are poorly managed. In addition, there are several points in a cycle in which the company is cash-rich and others in which it is starved for cash.

Figure 5-20 illustrates this point. In an economic downturn, sales of products (normalized as shown in the sales index values) are lower than previous months (T1–T8). This is a situation in which the company operations are throwing off cash, as shown by the values of the row labeled Cash (Activity). The values in this row are computed by subtracting the T values in the Index (Normalized) row from the previous T value, yielding a measurement of the cash flow. Accounts receivable is collecting for previous shipments at the higher rate. Meanwhile, accounts payable is paying out at a reduced rate.

Cash is plentiful in this section of the cycle. At the lower level of business, management has throttled back on expenses and programs to maintain some level of profitability at the reduced levels.

As the business increases again, coming out of the downturn in the cycle (T9–T15), inventory must be purchased at a higher cost than the lower steady state, which increases

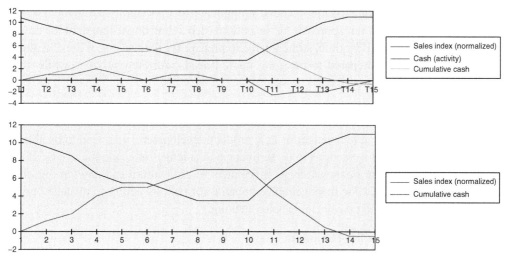

Figure 5-20. Cash movement in an economic cycle.

accounts payable. Products are built at a higher cost and accounts receivable are growing because of the increased shipments. This dynamic drains cash, as shown by the negative numbers in the Cash (Activity) row. The normalized cumulative cash flow is shown in the third row of the table.

The graphic representation of what is occurring also appears in the lower part of Figure 5-20. This is not a pure representation of a cash flow example but is merely intended to stress a point about economic cycles and the company's ability to fund programs during specific economic phases.

C. Insolvency after a recession

As you view the charts in Figure 5-21, it is easy to see how recovering out of a recession or even a growth spurt can strap a company for cash. The finances of the company are most at risk during these times, and a watchful eye will focus on expenses. Many firms become insolvent during these times because of cash requirements.

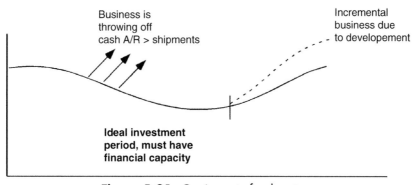

Figure 5-21. Coming out of a downturn.

If, however, the company has the financial strength and an opportunity worth the risk, the bottom of the cycle or a point approaching the bottom (depending on the length of the development cycle) is an excellent time, strategically, to prepare for the rise out of the downturn. Your company will be prepared to meet the market needs with new product, while your competitors struggled to survive the downturn.

D. Read management well to see how fast it reacts

In order to be an effective manager of new product development, you need to be able to read senior management's reactions to these financial challenges in economic cycles. If the company can tolerate the financial stress, investment in a downturn can pay out big dividends. You need to assess the management's appetite for risk, its ability to manage during a downturn, and its staying power in decision making.

E. Watch the health of the business

Like it or not, the new product manager needs to keep a watchful eye on the general health of the business during the project. The financial pressure points in a project (e.g., appropriation requests and equipment procurement) need to be handled within the framework of the finances of the business as a whole. For example, do not approach senior management for funds in the same time frame in which the company is reporting a loss. You can improve your chances by letting the bad news of a loss wear off for a few days and present your request later, when it might be more palatable.

F. Consider business development as a discretionary expense

We have discussed the overall strategic plan of the business and the need for new product development to be an integral part of the operation. In most companies, however, new product development is considered a discretionary expense. It is viewed this way because of the tendency toward shorter-range focus on profitability, which makes development subject to the nuances of quarterly reports. If there are going to be reductions in program expenditures, it is important to point out to senior management the impact of these reductions. Often these decisions are made in a vacuum and neglect previous business assumptions that have been made. The impact of previous promises is not readily apparent. Furthermore, the budget might not be updated to account for the reduction in future income due to the current reduction or delay.

It is also important to understand how management reacts to financial pressure within the framework of a product development. Is management prone to terminate expenses unilaterally at the first sign of bad news, or will it tend to follow through on commitments previously made to a new product?

G. Importance of economic analysis and company reaction to external forces

There are two factors that management needs to correlate in a downturn. The first is the general health of the industry and the economy in general. The second (and more important) factor is the company's performance, which encompasses its plans for the future. The two factors do not necessarily follow each other, as shown in Figure 5-22.

Does management react with mild concern at the first sign of a downturn or does it move directly into the panic mode? If panic is the first reaction, the impact of lost revenue will be

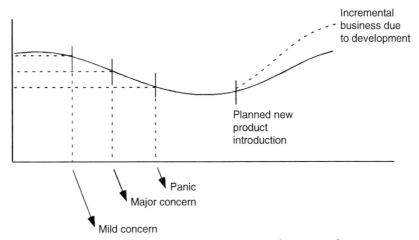

Figure 5-22. Company reaction to external economic forces.

more dramatic. It is management's job to react with prudence as the economy fluctuates; more important, management should react to both the health of the economy and that of the company itself. It is the program manager's job to be an advocate for project continuity.

H. There is no such thing as one bad quarter

Every company has a certain amount of inertia associated with it. The link between the financial reports and the field sales personnel is not direct, and it is unfiltered by third-party interpretation. This is why initiatives take some time to pass through an organization.

This phenomenon is the reason why there is no such thing as a bad quarter. By the time the first report of bad financial news reaches the management of the company, the company in aggregate cannot react instantly to effect a change in the trend. Management can initiate action but the action takes time to execute. Field sales personnel appointments are set up, focus is directed, but the final effect of the focus—the purchase order—does not occur immediately. There is generally a lapse of time between sales cultivation and securing orders. Therefore, while this redirection is occurring, the next quarter is passing the company by.

I. Launching in the eye of a buyout

A buyout is one of the most exciting aspects of a business development manager's job, and it occurs in rare cases. There was a film some years back about the Grand Prix racing circuit wherein the lead character was interviewed about his success in racing. He described how he wins races: "When I see an accident off in the distance, I accelerate as a first instinct, because others decelerate as their first instinct." It is not a very nice way to win, but it sometimes gets results.

There is a parallel between this racer's strategy and launching a new product initiative. While news of a buyout or change in ownership of a company circulates, a paralyzing effect occurs among the people involved. In the interim, very little progress is made; time is spent commiserating and guessing. This is the perfect opportunity to differentiate your team to the new ownership. The new management will be looking for this type of leadership. In a larger company, maintain steady progress and continue with the plan. If your management is paralyzed, take time to explain the impact of no action. In a small company, if you are in control, accelerate your team's efforts—it will pay off in the long run.

J. Dealing with the new owner

After the new ownership is in place and the dust has settled, there could be some cost cutting. You might be placed in a position to requalify and rejustify your program. Assumptions made before the change in ownership might no longer be valid to the new management or even to the same management under new direction. Be prepared to fight even harder than before to preserve the program.

Get in tune with the new ownership's direction, and focus and tailor the program's presentation accordingly. Also, prepare alternatives to the program; the purpose of such alternatives should be only to enhance the original plan or to align it more closely with new company objectives. The important point is to move ahead according to the plan in concert with the new ownership rather than rethink the tactics or strategy midstream to appease new management. If you fall prey to this latter temptation, you will get caught short in delivering previously stated goals with new assumptions overlaid on them.

4. Profit in Backlog

A. Monitor the 90-day and 180-day backlogs

Keep tabs on the business level by watching out for the 90- and 180-day backlogs. These two numbers, along with the calculation of profit in backlog, can give you a sense of the general health of the business.

The profit in backlog is a detailed calculation of the various products in the backlog, their cost structures, and the amounts of gross profit they generate. If the accounting system is set up with allocations for SG&A expenses, then an SG&A expense percentage can be entered in the appropriate space for each product category. The funds for development that you require in the six to nine months that follow will come from this backlog calculation.

Figure 5-23 shows an example of a profit in backlog calculation. The 90-day or 180-day totals are placed in the chart for analysis in addition to specific product information. The table shows the total backlog numbers for each of five different products. Also shown are the prices for each product along with their respective cost structures. SG&A costs are entered as a percentage of sales, and net profit is calculated and totaled. This type of analysis can also be used for the comparative analysis.

B. Forecast of the general immediate health of the company

The 90- and 180-day comparisons with previous backlog figures on a rolling basis can provide insight into trends or into critical changes to the business. Depending on the lead times, the 180-day backlog can be significantly smaller than the 90-day backlog; however, a comparison of the present 90- and 180-day backlogs with the previous measurements (for example, 30 days previously) can foretell certain trends.

EXTERNAL ECONOMIC FORCES

1. Economic Factors When the Product Rides the Economy

When times are good, all is well. When times are bad, we look to the new business to develop the revenue. There are distinct differences in assumptions when the economy is not

Profit in backlog						
Product category	Product 1	Product 2	Product 3	Product 4	Product 5	Total product
Backlog units	30	23	56	76	87	
Price net each ($)	6,000	7,500	4,500	9,000	1,275	
Backlog ($)	180,000	172,500	252,000	684,000	110,925	1,399,425
Material	0.275	0.375	0.3	0.325	0.275	
Labor	0.085	0.09	0.075	0.065	0.05	
Burden	0.2125	0.225	0.1875	0.1625	0.125	
Factory cost	0.5725	0.69	0.5625	0.5525	0.45	
SG&A (%)	28.5	27	32	34	30	
Profit ($)	25,650	6,900	29,610	73,530	27,731.25	163,421.25

Figure 5-23. Profit in backlog.

growing. Our earlier discussion of the seasons of change also impacts our ability to be successful. When the economy is not expanding, consumers and users are focused on necessity and cost reduction. There is little interest in developing markets and the money supply is tight, forcing competition in business development. These are times when new business models and cooperation amidst consolidations may occur. There are four characterizations that a market can have based on the economy alone. They are growing and expanding, developing a new market, experiencing a shrinking market, and stable. Figure 5-24 outlines some of the dynamics.

2. Economic Factors When the Product is Not Directly Related to the Economy

There are cases where the business model and product sales are relatively unaffected by external economic forces. In these cases the demand is such that the consumer or end-user

Market characterization				
	Price Pressure	Acceptance of new technology	Competition	Company's risk aversion
Growing and expanding	Tolerant	High	Less	Low
Developing a new market	Tolerant	High	Less	Measured
Shrinking market	High	Low	Acute	High
Stable market	Downward	Average	More	Medium

Figure 5-24. Market characterizations.

will buy the product within a very wide price band. When this happens, an increase in price will not result in a drop in volume and, more important, a reduction in price will not generate additional volume. In these cases, command a higher price and make the appropriate profit. This inelastic demand is usually short-lived because other players will enter the market. The other scenario is when the product is a necessity and there is little competition.

The important thing to remember is that nothing is forever and that, just as we had the seasons of change in market behavior, the market landscape populated with the competition can also change.

PROGRAM CONTINUITY

1. Cost Impact of Lack of Continuity

A. The damage caused by lack of continuity

There is a cost impact due to lack of continuity within a product development program. A certain degree of efficiency is developed in a team's joint effort. Constant stops and restarts cause delays and also increase costs, and inefficiency settles into the team. This inefficiency manifests in two ways. The first is a tendency for the team to slow its progress by working at a measured pace. An attitude of "Don't exert too much effort, they will change their minds again anyway" becomes commonplace. The second is due to refocusing the team on the latest project at hand, a process that is similar to the setup time for fixtures that must occur before progress can be made.

In addition, development personnel are not light switches. They do not start and stop programs as swiftly as the senior management might desire. They will generally complete a certain phase of a program and document it, even though it might have been placed on hold. This is to force a natural break in the program so that the program can be picked up at a later date.

As you work with more people and more projects that involve technical personnel, it will become evident that continuity is the best thing a manager can provide to the team. It is the manager's job to clear pathways and provide consistent direction for the development team.

B. Momentum, once lost, is difficult to rebuild

The issue of project management will be discussed in Chapter 7, where we will see that time lost in a development effort, given the constraints of the development program, cannot be made up, worked around, or reorganized around. It simply is lost. We can delude ourselves or pacify senior management by restructuring the program, shortening task times, or adding more personnel to execute tasks; however, even if the program was laid out well initially, lost time cannot be recovered. Don't lose it in the first place! An effective new product development manager will foresee the potential for time loss and act to prevent it.

C. Loss of trust between development and management

An often-underestimated effect of redirecting priorities for development people is the loss of trust or confidence that can occur between the development team and senior management. Too many swift moves can cause the development team to question management's sincerity, decisiveness, and commitment to the program. If you are a senior manager, keep this in mind because you might be the one causing problems regarding lack of focus.

D. Loss of continuity: slower results and poorer quality

A final thought on the impact of lack of continuity on cost is that, in corporations, decisions are often made in a linear fashion with justification rooted in equations, relationships, and schedules. Unfortunately, we as managers forget that it is people who execute programs, it is people who make progress, and it is people who implement our visions. It is human nature that lack of continuity causes poor quality of work from those people who are affected. A program that is interrupted several times will have poorer quality than one that is pursued from start to finish. There are details missed during the transitions, and frustration can lead to inaccuracies. Over the long run, it pays to establish a pace and preserve consistency of purpose throughout the program.

2. Halfway Versus all the Way

A. Don't fall for the halfway commitment

Management often deals with the uncertainty of new product development by only partially committing to a program. Half-hearted commitment gives management a false sense of security about the program because they are deluded into thinking they are mitigating risk somehow. Fundamentally, there is no way to mitigate risk after the scope of a program has been defined and the market opportunity has been identified. A halfway commitment merely places the risk of failure with the development manager. If you are this manager, take care in accepting this assignment because senior management has severely hampered your ability to succeed. Furthermore, the halfway commitment is not a fair treatment of the diligence your team has performed in identifying the opportunity and scope of the product. It is a prescription for failure. This prescription is not only limited to the development effort; it includes you, as well.

B. Dealing with management's fear of the unknown

The real issue in the application of the halfway effort or resource allocation is management's fear of the uncertainty of new product development. There are several reasons for this uncertainty. One is that management might once have funded a poorly researched program, which then failed. Now management has become very conservative. Another reason is that your program is off strategy—too expensive for the financial condition of the company. If you have followed the precepts presented in the previous chapters, however, you have accounted for these issues.

The final reason is that senior management might be risk-averse. In such a case, there is little that can be done other than to re-present the program with more reassurances. The risk aversion might be so strong that management is paralyzed, in which case there are more severe problems in the enterprise than funding of the new product development.

How often have you heard, "Here is some funding to get started, and then we will see how it goes?" This might sound comforting if you are in a financial firefight to get funding for a program; however, do not settle for it. When the going gets rough in the middle of the program, your senior management might not be committed to going the distance to complete the program. There can be no solace in this type of funding because your acceptance underscores and tacitly approves of management's reluctance to make a command decision and see it through.

C. A well-documented, well-researched program deserves full funding

If there is a single thought you need to take away from this chapter it is that a well-researched program that is on strategy within the constraints of the business deserves to be funded properly. It is unfair both to the involved management personnel and to the development team to string them along or signal a false sense of commitment. If you cannot garner full support at this critical juncture, what does that say about the prospects if the program runs into trouble?

3. Strategic Impact of Halfway Efforts

A. Loss of the competitive edge

The arena of new product development is a race against both time and competitors. It is a race against time from the perspective of cost and return of funds to the corporation, and a race against the competition to reach the market first so as to retain market share. When qualifying programs for the development of new products, loss of competitive edge can occur if a firm financial commitment is not made. These halfway efforts serve to water down the programs and lead to their ultimate failure. While your company is vacillating financially over a program, others are pursuing programs to completion with increasing efficiency. Your company will get left behind. This situation arises frequently as a result of management wanting to accomplish too much in the way of development. Too many programs are started, and none end up with the support to execute them.

B. Morale

Another circumstance to beware of is the loss of morale of a development team that has been supported only halfway. A team needs to complete a program as a group. If the priorities are shifting and the finances are a constant source of roadblocks, the team will lose heart in its efforts. It is very difficult to rebuild this enthusiasm with the same management in place. The team's morale must be nurtured during transition times to preserve the energy required to navigate the technical obstacles of the program. It is the role of the manager to insulate the team to some degree from the dynamics of the day-to-day operations and priority shifts that take place.

C. Achieving critical mass of effort

There is also the issue of critical mass of effort. The easiest way to allow failure is to fracture the team's effort. To achieve steady progress and momentum, a certain critical mass of effort must be maintained. It is what causes the team to progress because each member of the team is interdependent for results. If there is no critical mass, individual team members take on more and more peripheral activities to the point where they think they are handling the bulk of the program. It is at this point that the process of momentum-building breaks down. The members begin to work at their own paces because they perceive that they are responsible only to themselves, not to the momentum of the team.

D. Time lost

The loss of momentum causes time loss in the project and all of the dangers associated with a late product introduction. The accounting function can actually create failure within the company by trying to control expenditures too tightly. The key is to create balance, select the programs to execute, and fund them properly. Don't let management start too many programs and then allow them all to fail.

E. The fallacy of "if it's a good program, it will survive"

The notion that "If it's a good program, it will survive" projects a certain romanticism about product development. The most appropriate goal, however, is to create and foster camaraderie within the team to meet deadlines and market needs, not to place them in competition with other teams. An approach that places each team in competition for survival with the others misdirects the focus inwardly into the organization and away from the marketplace and customers.

STRUCTURE OF FINANCES

1. Internal Financial Contribution

Financing a program can take may different forms. The most straightforward method and one rooted in recent history is the direct funding approach. The company funds the development of a new product with its own resources that come from retained earnings.

Shared development resources is another means. The matrix management approach can be a good thing for a project because it allows the team to draw from available resources. However, it has serious drawbacks when it comes to prioritization and commitment. The structure of the team will be covered in Chapter 6. The matrix setup is appealing to may companies because, in this approach, a pool of talent with specialty capability can be used to address various parts of the program.

2. External Financial Contribution

Depending on circumstances, one may choose to borrow funds to execute the program and place the liability in long-term debt. This is a convenient way to shroud the program from potential onlookers in a publicly held company. Another way is to take the funds and expense them. Still another way is to use available funding from the government in the form of grants to get some of the work done.

3. Contribution in Kind

Contribution in kind is especially useful in a partnership whereby one of the partners contributes assets in the form of intellectual/property, equipment, inventory, or some other valuable asset. This is popular in international cooperation where one partner is in a country that demands certain investment with the other partner. As an example, Partner A is setting up a partnership with partner B in Asia. The target country requires certain investment to create the joint venture. Partner A could contribute know-how and goods at full market value and get "credit" toward the contribution. In these cases there is generally latitude in defining the goods and know-how.

APPROPRIATION

1. Getting the Appropriation Request Approved

A. The critical juncture for a business development activity

The appropriation request marks the point at which conversation ceases and commitment begins. At the beginning of this chapter, we outlined a philosophy of viewing accounting and financial departments as partners rather than adversaries in the new product development process. The appropriation request is the milestone for making that partnership work.

The issue of funding cannot be ignored, given a lower priority, or diminished in importance in any way. If the tone of the rhetoric sounds like brinkmanship, to some degree it is. All of the previous effort, research, investigation, and planning lead to this point. Your team has the right to drive the decision to load the accounts with funding.

Any action short of funding at this point is merely conversation. It is your role as business development manager to navigate the distinction with senior management. Because funding marks a pivotal point in the approval process, management's true intentions become apparent at this juncture. This statement is made not to encourage confrontation or animosity between development managers and senior management but, rather, to drive decision making and back up that commitment with corporate resources.

B. Homework done well will pay off at this point

The degree to which you prepared the research and planning is the degree to which you can be vocal and aggressive in securing funds. If your team did a poor job, you shouldn't expect too much if the first attempt at financing fails. If, however, you have prepared your case well and if the opportunity is real and fits with the direction of the organization, be vocal and passionate in securing funds. This is not to say that you should be abrasive, but you should demonstrate to senior management that you are the program's champion and have the commitment to see it through to completion.

C. What to do if a program is not approved

If you do not get the financing you are seeking, what do you do about it? There are some alternatives. You can scrap the program, broker the idea to someone else with the corporation's support, or finance it by alternate means. In each case, the onus will be on you to resell the opportunity to the appropriate party.

The idea can be brokered to another company; however, be aware that the other company might need to secure some of the people on your team. The other company should outline a quid pro quo arrangement because you are turning over an opportunity that your firm spent considerable time in researching and planning.

The alternate financing approach might be best served by creating a joint venture or brand-label arrangement with another company. This would allow your team to drive the development or, at a minimum, to participate in it. It offers a way to get started in the marketplace and gain experience with the business.

D. Leading the next finance stage

There is also the issue of securing financing as the program progresses. At each stage, the work of securing funds for the next stage of the project begins. All too often, the work of securing financing is left until the last moment when the decision actually takes place. The work of securing the funds needs to be done long before the decision-making event. It is the role of the program manager to retain management's interest and support throughout the entire program, not just at the funding stage. To that end, if you are in a program presently, funding for the next stage must be planned for now.

E. Create the financial momentum

The goal is to create the financial momentum to keep funds flowing at a rate equivalent to the funds currently being expended. This process needs to be financially correlated to the progress of the development to obtain management's support.

CONTINUOUS APPROPRIATIONS

1. Reappropriation

When a new project is started, it is best to determine the entire funding requirements. Many times it is tempting to state only the near-term requirements and leave the future requirements on their own. In reality, for a program to be successful it is best to know up-front (as best as you can) what the actual requirements for the entire round of funding will be. By doing this and communicating this to the organization, one can assist finance personnel in planning cash management and assessing the program's impact on the financial health of the business.

When the time comes for reappropriating funds to overcome adversity or to fund the next stage, it will be easier. This is because finance has been made aware of the issue all along and can best judge the impact of the event. This will support the continuity of the program.

2. Pulling the Plug

If there is justification, it may be necessary to discontinue the program altogether. If this is the case, it is best to confront it with the finance personnel and present your findings together to management. This will enhance your standing in the firm as well as demonstrate fiscal responsibility to the rest of the group. In some cases, management may decide to continue the program in spite of the problems, and you will have the appropriate funding. If not, chances are the program would not be successful and would be starved for funding if you insisted on continuing. It is best to face up to the issue, make the recommendation, and move on to another program that will benefit the company.

SUMMARY

This chapter was designed to provide you with a basic background in accounting and finance to create understanding about the driving forces. In other words, we looked at a project from the accountant's perspective.

A variety of information and specific tools for analysis and planning were developed and presented. Definitions pertaining to finance and accounting were reviewed, and the fundamental differences between accounting and engineering operations were discussed.

Financial models for product sales were presented, along with a review of cost systems and how to estimate an entire program. A discussion on project and cost estimating, factors associated with the timing of future sales, and a sample of a comparative analysis for projects were also presented. We also looked at risk management, cash timing, and business cycles.

Following these items were discussions pertaining to the generation of cash and profit and the financial cycles of a business. We also reviewed the external forces and what some of the economic driving forces are. Finally, we discussed the requirement for project continuity, the structure of finances, and appropriations.

With a new product concept, a development plan, a business plan, and the funds to execute it, you are now ready to start out on the venture. This is the focus of Chapter 6: Starting Out.

STARTING OUT

LOGISTICS AND PROVISIONS

Now you have the funding and are ready to start the program. The process of doing the program will come down to four major variables that must be managed. They are funding, time, people, and technology. As we will see, each has its own level of uncertainty that must be managed to get the program done on time, correctly, on budget, and with a team intact.

1. Absorption of Expenses

One of the first things on the business development manager's agenda is to organize the expenditures and expenses. Accounts should be set up and mechanisms for withdrawal from those accounts should be in place. To keep finances tight, the approvals' protocol should be set in place and communicated to all parts of the team.

2. Absorption of Equipment and Training

An often-underestimated element in program development is the time required to assimilate new equipment and software tools. Many times managers count on the productivity increases available with these tools but fail to understand the impact of cost and time to become proficient. Almost any new software will require training, which costs funds and takes away precious development time. The manager must factor in the time needed to become proficient enough to leverage the tool on the particular project.

3. Absorption of Technology

This is the most significant issue in a project. The problem starts with the level of expertise within the group for the particular technology. Just because the technology exists doesn't mean that the group is conversant with it. In addition, technology in its raw state may not be readily usable. This is especially true for acquired technology or technology that has been purchased from another party. The term *diffusion index* can best describe this dynamic. The tribal knowledge cultivated over years in a company cannot be diffused into a fresh team starting from a zero base. This takes time.

The transfer of technology is a contact sport and, as such, must be done from person to person. If the package of technology is sold to the target company and you have to use it to make a product line, your team may not know everything there is to know about the technology to make it deterministic and readily integrated on a platform. In these cases the manager must make the team as well as senior management aware of the issue. In order for the business development model to work, the tools, parts, and pieces must have a clear pathway to the model.

Many times the acquisition of the particular intellectual property cannot be readily duplicated in the acquiring company. This necessitates training, collaboration, and extra time and funds to get to the level that was assumed in the beginning. All this is happening while the marketing clock is ticking.

4. Identifying the Requirements

A. Identify the resources

What are the resources required to initiate and complete the program? What is the mix of internal and external expertise that will be needed? What are the capital resources needed and when will they be required? Is there a learning curve for the capital equipment? All of these essentials must be sequenced and coordinated.

As we will discuss in this chapter, the program must be laid out and the resources identified and sequenced based on program activities. It is best to start with the project activities, sequence them, and generate the list of requirements to accomplish the task.

B. Planning the requirements

Alterations in corporate finance philosophy have often forced changes in the procurement of equipment and resources. Without external controls, most managers probably would spend all their capital expenditures in the first quarter after the start of a new fiscal year. This tendency could be due to shortsighted senior management and its changing preferences for expenditures and programs. There is a natural tendency to gather the resources even before they are needed in order to ensure availability.

In the ideal situation, not all the resources are required on day 1 of a program. Instead, they need to be scheduled to minimize financial impact and to allow the corporation and the team to absorb them and use them effectively. To the extent that the company can make a commitment and see it through, plan and execute the resource procurement accordingly.

C. Organize the program tracking chart

There is only one systematic means to identify the resources and equipment needs for a program: break the project down into its component stages and identify the subtasks for each stage. Each task will require personnel resources and, possibly, equipment resources. Note the requirements for each type of resource under the task. If you have laid out the program properly and the sequence is correct, the resource and timing will also be correct, as illustrated in Figure 6-1.

D. Graph the demand

The next step is to graph the manpower and resource requirements during the run of the development. For example, Task 1 might require three people during its duration, whereas Task 2 might require five people. The team needs to be staffed accordingly to meet the demands of the tasks, given the time frame established. Supplying fewer personnel than are required will result in extending the execution time of a given task.

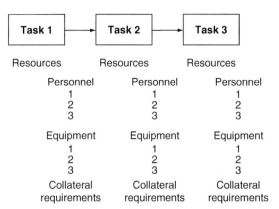

Figure 6-1. Resources.

In a similar manner, the equipment needs to be sequenced and procurement and commissioning schedules can be put in place. Integrating the equipment might also require its own personnel or they simply may be part of the development team. All of these elements need to be considered for assessing the requirements of the program.

E. Determine tools required

One way to keep the program on track is to determine the tools that the development personnel need early in the program. Often these tools are decided upon only as the development progresses, which results in building inherent delays into the program. Procurement delays, funding delays, and integration delays can occur.

A frequent excuse for the delay in the selection of these development tools is their continuous evolution and the desire to be on the latest platform or have the latest version of equipment or software. In reality, the delay incurred by not planning ahead often far outpaces the loss in efficiency due to using less than state-of-the-art equipment. Consequently, have the team identify the equipment and resources, and procure and integrate them into the organization so that they are truly available at the time they are needed.

F. Define the levels of expertise required

Similar to determining the tools required, you need to determine the level of expertise required of the team members. Be sure to get the requisite talent in place, as a minimum, before starting out; otherwise, you could end up *searching* for talent when you should be *executing* tasks with that talent, hence delaying the program.

G. Establish the pathway for tool development

One part of the identification and procurement process is a plan to integrate and diffuse the new technologies and methods into the development team. Regardless of their specialty, each member should be on a pathway toward certification in the technologies brought into the company. Following this strategy will go a long way toward giving management the flexibility to assign personnel to various tasks. Eventually, every member should have a broad repertoire of skills. Figure 6-2 illustrates such a pathway.

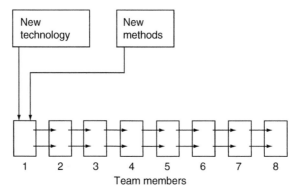

Figure 6-2. Pathway for tool development.

The new technology and methods are taught to Team Member 1. After the work content for the task is complete, that team member transfers the technology to Team Member 2, and so on, until all of the team members are conversant with the technology and methods. This strategy provides the manager with the utmost flexibility.

H. Review process questions for planning

There are several planning questions that must be considered as part of the process of identifying resources. At this point in the development of the program, most elements of the business plan should have addressed these issues; however, the following points can be used as a reminder for the program manager:

1. Write the planning statement

What is the desired result or outcome? Be very specific about this. What are the measurements of success? What do we want to accomplish? Who, what, when, where, how, and why?

2. List the action steps in the plan, identify timing, and assign responsibility

Develop the chronological list of action steps to be accomplished. Who is accountable for each task and what is the realistic starting and ending time for each task? Also, list and correlate the interdependent tasks at this time.

3. Review each step in the plan and identify the critical areas

Which action steps are the critical impact tasks? Where is uncertainty associated with a task or group of interdependent tasks? Where are there potential gaps or overlaps in responsibility?

4. Identify and list likely causes of key potential problems

What are the specific influences or events that can cause these potential problems? Test each problem listed to determine whether it represents a cause-and-effect relationship with the corresponding potential problem.

5. Develop preventative and facilitative actions

What can be done to prevent each likely problem from occurring? What action should be taken now to facilitate the opportunity or to prevent the obstacle that could stand in the way of the opportunity?

6. Design contingent actions

What contingent actions should be planned for each potential problem so as to minimize the impact? If the potential problem occurs in spite of preventive action, what must be done to mitigate the circumstances?

7. Identify and list contingency alarm points

This is probably one of the most important elements of planning—to understand clearly the points at which you are in trouble and to set alarms for them so corrective action can be taken. Often a program gets into trouble and the team is deeply immersed in the issues. If proper alarm points were used, precious time and energy could have been preserved to focus on corrective action at an earlier stage.

ORGANIZATIONAL DYNAMICS

As we delve into the business of building a functional team and group to execute the program, the role of organizational dynamics comes into play. There are two embodiments of this interplay that we need to concern ourselves with, namely, the dynamics between the departments within a firm and the dynamics of departments across two separate firms.

1. Dynamics of Departments within the Firm

The biggest stumbling block to success in an organization is interplay between departments and personnel. If we do our management jobs correctly, the business development team involved with the project and the participants within the department are sufficiently motivated and compensated to accomplish the task. The issue of linking department objectives is also critical. The program will be able to gather momentum if there are linking objectives where the departments having members installed on the business development team are evaluated with success of the program as a critical measurement parameter.

This issue has such an impact on a program that it cannot be overstated. The flowing is a summary of various parts of a business development process that outlines the natural dynamic that occurs and what needs to be done to chart the pathway toward success.

1. Product definition

- Natural organizational dynamic
  - Loose definition—wide, broad specifications contributing to higher cost structures
- Charting the pathway to profit and customer satisfaction
  - Tight, targeted definitions with specific volume customers identified and engaged

2. Business case development

- Natural organizational dynamic
  - Broad market assumptions
  - Little regard for timing
  - Little discipline in costs structures
- Charting the pathway to profit and customer satisfaction
  - Tight, targeted market assessment
  - Market dynamics versus steady-state analysis
  - Contingency plans built in

3. Integration into the corporate goals and objectives

- Natural organizational dynamic
  - Topline focus in opportunity evaluation overshadowing of impact on secondary operations
  - Charting the pathway to profit and customer satisfaction
  - Focus on sustainable advantage and integration into entire income statement

4. Assembling the supporting cast and team members

- Natural organizational dynamic
  - Personnel selection based on availability
- Charting the pathway to profit and customer satisfaction
  - Selection based on talent, skill set, drive, and dynamics

5. Executing the actual development

- Natural organizational dynamic
  - Slow start, nonoptimized approach, early delays
  - Playing catch-up game, project logic affected
- Charting the pathway to profit and customer satisfaction
  - Hard start, paced execution, logic of program unaffected by delays
  - Sense of urgency imbued into the organization

6. Integrating of the manufacturing development function

- Natural organizational dynamic
  - Manufacturing integration is planned after product configuration or, worse yet, after development
- Charting the pathway to profit and customer satisfaction
  - Manufacturing integration is planned at conception and along with each phase of the program

7. Launch

- Natural organizational dynamic
  - Fit the launch to a standard launch package
  - Measure too early and deny nonsuccess
- Charting the pathway to profit and customer satisfaction
  - Fit to the opportunity

- Initiate, measure success factors early
- Manage to success

8. *Product management*

- Natural organizational dynamic
  - Passive "certificate of deposit" mentality and approach
  - Disconnected from market and "professionally managed"
- Charting the pathway to profit and customer satisfaction
  - Active, dynamic, connected to marketplace, customers, and competitive moves

9. *Cost optimization*

- Natural organizational dynamic
  - Value management internal, insufficient
  - Customer input obtained in a vacuum of the marketplace dynamics
- Charting the pathway to profit and customer satisfaction
  - Strict, regimented evaluation of value
  - Interdisciplinary approach
  - Market reassessment of original package of values

10. *Aftermarket operations*

- Natural organizational dynamic
  - Margin cash cow orientation, clip coupons
  - Steady price increases
  - Disconnection from the value chain comprising the value package
- Charting the pathway to profit and customer satisfaction
  - Proactive selling, focus on life cycle costs for customer

11. *Continuous competitive assessment*

- Natural organizational dynamic
  - Internally generated, biased, corporate bravado
  - Additional features = additional price
- Charting the pathway to profit and customer satisfaction
  - Internally and externally generated, objective and painful assessment
  - Introspection, listening rather than talking

These assessments are critical and harsh, but in many cases accurate.

2. Dynamics of Departments Across Firms

This situation is even more difficult because the program success is not directly fed back to the two companies. The companies are potentially at different stages of development, have different pressures on cash requirements, and, most important, have different philosophies on how to staff a program. For example, if you have a small firm cooperating with a large firm, the large firm is somewhat at risk. The smaller firm can react to a market opportunity and may choose to do so. In this case, critical talent may be diverted from the primary program to chase a different opportunity. This causes discontinuity for the other partner and the project. The dynamics of a sequential project are such that any interruption can cause irreparable damage to the schedule. Therefore it is imperative that both contributing teams from each company do in fact meet their goals on time and stay focused.

3. Assembling the Team Members

A. Canvassing and interviewing team members

The technique of putting a team together to suit the needs of a program is an acquired skill. It takes several years of operating in a team environment to understand the potential pitfalls and the ingredients for success. The elements discussed in this section are not totally inclusive of all issues in canvassing and interviewing team members; however, they do represent key points to keep in mind.

Personnel assigned to a development program are measured by the combination of two attributes: intellect and energy. Intellect is the knowledge to create, design, analyze, and synthesize solutions. Energy is the human drive to apply that knowledge to effect progress. Either attribute by itself is worthless in a development person, but the combination of the two is what makes things happen. A person with intellect but no energy to apply that intellect will not move a program along. A person with only energy and not enough intellect doesn't have the tools to do the job.

In our experience, however, energy, interest level, and drive can offset many shortcomings in experience and intellect. Never underestimate the capabilities of a person who wants to do something, and never count on a high intellect as a sole means to an end.

To accurately assess a potential team member's ability to execute a program, you need to conduct performance-based interviewing. Performance-based interviewing is a technique whereby the candidate is given the opportunity to draw parallels between his or her past accomplishments and the proposed program for which he or she is interviewing. It focuses on specific skill set requirements of the position and looks for those skills demonstrated in the candidate's previous work.

Performance-based interviewing is an objective, dispassionate method for assessing skill level and measuring it against the job requirements. It allows an objective selection between two individuals based on their anticipated contributions to the program.

B. Makeup of the team

The makeup of the team is critical to the team's success. Certain personalities can enhance success or contribute to failure. Members with exceptionally strong personalities can undermine the team leader. Infighting can erode the sense of purpose. The selection and complexion of a team must be geared toward the program at hand.

Each team needs a blend of creative people and pure implementers. There should be individuals with enthusiasm and also naysayers. Each type complements the other to create balance in the decision-making process. If the program requires navigation of uncertainty, focus the team orientation toward creative people. If critical time and cost issues are the focus, make use of implementers to deliver results.

C. Aligning the team

The alignment of the team is crucial to get off to a productive start, but teams do not arrive in-place and aligned. In fact, all teams when first assembled are disjointed, fragmented, and somewhat ineffective. They are a loose collection of people with differing goals, agendas, biases, and allegiances.

It's the manager's responsibility to focus the group on the goal and accomplish the result. Along the pathway toward the result, the manager needs to reinforce the interdependency between the team players and within the team as a whole. As this goal is attained, remind the team of the basic tenets that made them successful. The manager needs to drive the team culture from these tenets, which is why it takes time to construct an effective team. In some cases, several projects might be required to cement the team together. Any senior manager who proselytizes about teamwork and who expects a fully functioning team in place on day 1 probably has never put a team together.

D. How a team is made

There is value in cementing team consciousness by team success; however, nothing cements team relationships as much as the process of the team overcoming a problem or an adversity together. For the manager, keeping a team together during these times can be especially challenging, but the manager always must overcome the desires of the individuals and place the needs of the project above personal preferences.

Although members might want to dissociate themselves from the program, the manager must force cooperation that produces results. Later, the manager needs to remind the team of what works and what fragments a group, and how fragmentation can destroy a program. At the successful completion of the project, celebrate with the team and reinforce that commitment pays off.

E. Development engineers apply the technology

The role of the development engineer involves a dichotomy in required personality traits. In one sense, the development engineer needs to be a creative, freethinking individual to overcome the uncertainty inherent in a program. In another sense, the engineer needs to be structured in implementing the solution to the problem. Most engineers are quite comfortable in the creative role because they view resolution of the problem as a challenge, but many are averse to mechanizing it so that it can be absorbed by the organization. For many engineers, the latter task simply pales in excitement and satisfaction compared with the problem-solving aspect of the project.

These two apparently opposing personality traits are very difficult to find or cultivate in one individual. Many times it is easier and more expedient to implement these dual functions with two people, each of whom possesses one of the required traits.

The needs of the program at any point in time define which traits are required in the development people. Often the need for freethinking and creative solutions to problems is greatest at the beginning of the program, whereas the more rote activities of implementation take precedence toward the end. The development people need to seamlessly flex from one process to the other; when this is not possible, the program must be structured so that different personnel take over as the program changes.

F. Get the requisite talent

As a basic tenet of assembling a team, be sure to secure the requisite talent at the beginning of the project. A common theme throughout this book is that time is a competitive weapon. This being the case, you want to minimize lost time by securing talented people at the outset. Otherwise you will be spending precious time getting the personnel up to speed instead of making progress on the project.

Many times, senior management attempts to influence the selection of team members as a means of containing expenses for salaries; however, this practice sets the project up for delay. The team becomes comprised of trainable personnel rather than experienced personnel. You have learning curve issues and experience curve issues to contend with. To understand the impact of this type of team selection process, consider a mistake that could be avoided by the selection of experienced people. In the long run, you will save expenses by not committing the mistake of selecting trainable personnel, and you will save even more valued time.

G. Obtain team members with drive

Secure personnel with drive and it will pay off well in the long run. This strategy also allows the manager the freedom to manage the program rather than expend energy on incentives for the group to perform. We have found that it is quite helpful to select personnel who have something to prove in life. The challenge for the manager is to channel that desire and make it consistent with company and program objectives.

Team members with something to prove are driven to perform. They break down barriers and navigate uncertainty better than most people because they are focused on the result. All too often, teams are composed of personnel who are not driven and who have few personal goals that are linked to corporate goals. The problem is that these people do not contribute energy to the group and actually cost energy to extract minimal performance.

H. Temperament of the team players

In new product development, remember that results are the only things that count. Senior management is appreciative of all the storms the team might have gone through, and your navigation skills might earn you recognition or a better assignment someday. Ultimately, however, management is only interested in whether you brought the ship in. Results are the only tangible measurable; consequently, the team must perform.

Developing a new product places the team under pressure, and tempers can flare occasionally. This is why it is a good idea for the team's manager to understand the temperament of the team. Individuals respond in a variety of ways when placed under pressure. It is important to recognize this fact and to mitigate the effect of conflicting personality profiles within the team. Members of the group need to be reminded that time, not their colleagues, is the adversary and that the needs of the program outweigh individual needs. There are many ways of profiling personalities and many recommended methods for managing them. Our focus here is not on the mechanics of people management; rather, reemphasize the objective of the team and manage toward that goal—to execute a program in the best way possible, to meet the needs of the corporation in the time frame allotted while cultivating an atmosphere of mutual respect and cooperation, and to provide a vehicle for individual personal and professional growth.

There can be occasional conflict; that is acceptable as long as the conflict is constructive toward the program. The team leader must manage conflict. Each member can have a dissenting view. The views may be discussed, voiced, and argued. In the final analysis, however, it is the manager who needs to make a command decision. It might not be popular but it is the group decision. In addition, the manager's job is to monitor behavior and immediately dismiss a member who engages in destructive conflict for the purpose of delay or for political motives. This doesn't mean dismissing a dissident, the bearer of bad news, or someone who is passionate about his or her work. It does mean removing personnel who knowingly do not operate in the best interests of the program. If this activity is taking place, action must be taken immediately. Failure to do so will undermine the manager's leadership and delay progress for the whole group.

If a team member is not doing what they are supposed to be doing, it is important to correct the situation. Managers often see this activity and try to legislate the member's behavior or, worse, legislate the entire group's behavior. In these cases the business development manager should focus the team on the tasks at hand and manage to the program requirements rather than trying to effect behavior modification in the vacuum of the program requirements.

I. Dynamic interplay

The management of a development team is a dynamic endeavor. Assumptions change, situations arise, obstacles present themselves, and attitudes change. The manager needs to watch diligently, listen, observe, and redirect and reinforce behaviors to facilitate the completion of the project.

4. Apprenticeship and Mentoring

A. Focus on personnel development

Traditionally, the two practices of apprenticeship and mentoring have represented the means to transfer technology from the experienced, sage, and wise members of the workforce to the younger, inexperienced, and naïve members. This strategy has worked historically in industry and could be effective today if not for the rapid change in technologies. Even mature members of the workforce need to expend a portion of their day learning the new technologies and methods. The focus has shifted to training new members in basic requirements for a lifetime of continuing education. If apprenticeships historically involved a fixed time to practice and learn a fixed body of knowledge, then today the body of knowledge is ever-increasing. Consequently, we need to give the new members the tools to learn and then get them into the productivity race.

On the professional side, mentoring is a traditional method to augment a new member's education with the wisdom of the elder statesman of the corporation. Mentors generally provide advice regarding career development, pathways to success, and elimination of roadblocks. Mentoring is effective today, even in the wake of corporate hacking away at higher-level managers during cutbacks. Unfortunately, in some organizations the use of mentoring has been replaced with self-preservation; in a healthy situation, however, it remains an effective means for educating the younger members of the workforce.

B. Growing the young workforce

The growth and assimilation of the young workforce is almost a sacred corporate trust. Through mentoring and the transfer of wisdom, the culture and values of the organization get passed from management to the new hires. If a company is going to grow, the integration of the value system and the culture must be spread throughout the organization. If this is not done, the enterprise will eventually be destroyed because the basic value system and frame of reference system are corrupted. As in life, take care in growing the young; the results will have tremendous positive impact. The business development manager needs to strike a balance for the team member between their own development and the needs of the company.

C. Cross-fertilizing the experience base

Previously in this chapter, we discussed the assimilation of new technology by different members of the team. It is important to have a coordinated plan for cross-fertilizing the technology and skill sets of the members. The ultimate test of cross-fertilization of knowledge and skill sets is to be able to reassign members in mid-task and to have the members still be able to complete those tasks seamlessly. Pursuing this strategy yields the maximum flexibility of the team and, equally important, creates loyalty and enthusiasm within the group because no one remains vocationally stagnant.

D. Concept applies to all disciplines

The practices of apprenticeship and mentoring are valuable for all disciplines of a development group. They apply to professional, vocational, and technical disciplines. Consider how apprenticeship evolved historically, as the transference of knowledge to young employees who had to be taught the procedures and methods for a task or series of tasks. It was instruction on the mechanics of the job. After the apprenticeship period, the individual was expected to demonstrate a certain level of skill, which was further enhanced throughout the individual's career. This concept is still valid today for all disciplines. Individuals new to the organization need to be taught the values and mechanics of the job to which they are assigned.

In contemporary business practice, enterprises tend to give new hires a small dose of instruction and then send them on their way. More popular in the United States than in Europe, this practice can generate waste and a frustration on the part of both individuals and management. Managers simply need to dedicate the resources to cultivate new individuals in their professional and vocational roles.

The practice of mentoring is less about mechanics and more about work style. The workplace is increasingly diverse and the style of management can affect program outcome more significantly than in the past. In current practice, turnover in the managerial ranks increasingly points to the need for mentoring. Mentoring imbues the individual with the value system of the corporation. It teaches the individual how to conduct himself or herself within the constraints of the values of the organization. Properly implemented, these values are tremendously more powerful than any energy a single manager can expend.

E. The case for both apprenticeship and mentoring

In the contemporary development environment, there is a practical need for both apprenticeships and mentoring to cultivate new personnel in the professional and vocational areas. The objective is to create a group of team members out of individuals who are well-trained, who operate within the boundaries of the company mores, and who are motivated to improve.

PROJECT/TEAM FORMAT

1. A Statement on Teamwork

A. Teamwork

The term *teamwork* is often misused as a rallying cry for a disjointed organization. It can be a euphemism that managers tend to overuse when the goal is to coalesce and galvanize a diverse group of people to achieve a common objective. In reality, a team is a work in progress, in which the objective focuses the actions of the individual. In a team, the needs of the corporation outweigh the needs of the individual. A team is a forum for individual contribution to an aggregate goal.

B. Understanding the concept of a team

Do not fail to understand the concept of a team as a work in progress. Do not rely on the "teamwork speech" and expect to have a disjointed group of people with differing agendas suddenly perform as a cohesive unit. The concept of teamwork needs to be nurtured, reinforced, and supported on a day-to-day basis. It is integral to the entire program, not just to its beginning stages.

C. Teamwork doesn't happen naturally

Teamwork doesn't happen naturally. Groups are composed of people who often have differing agendas and different perceptions of urgency. The manager must convert the energy of the individuals into directed energy focused on group progress—a very difficult task. The key to achieving this goal is to foster group progress with individual recognition and group rewards.

D. Leaders galvanize the group to perform

The development manager must shepherd the group to perform. This means consistent refocusing of individual efforts to the stated objective. The leader of the group must be respected by the group, not solely for technical capability but also by virtue of being the guiding force in executing the development. The leader must provide the tools, the resources, and the uninterrupted time to allow progress to occur. The leader must verify that the progress made is appropriate to the initial objective and also ensure that the original objective is still valid. The group leader must foster enthusiasm, provide continuity during trying times, and protect the momentum of progress toward development.

If you are the leader of a development group, you hold the primary responsibility for success or failure of the group. You (along with the team) will take the credit for success, but any failure will be yours alone. It therefore is in your best interests to understand the dynamics of a group of people and to become proficient at managing them toward a goal.

2. Selecting the Format

A. Types of organizational formats: Functional, project, matrix

The issue of which organizational format to employ in a program is a critical one. Given ample resources and reasonable commitment, almost any of the three commonly accepted formats can work; however, in actuality, there are never enough resources or time to execute a program ideally. Therein lies the importance of which format is used and how to maximize the use of the development personnel. The three basic types of project organization are as follows:

- Line or functional form
- Pure project form
- Matrix form.

In the *line or functional form*, a group of projects is assigned to a functional manager. The members of the team report to the manager in the standard hierarchical structure. This form of project organization is illustrated in Figure 6-3(A).

The *pure project form* of organization is one in which each project is self-contained and resources are assigned to a single project for execution. Members do not work on anything other than the project. The traditional skunkworks is the ultimate in pure project format. It is illustrated in Figure 6-3(B).

In the *matrix form* of organization, development personnel are located in a resource pool. Assignments are given to the pool and talent is assigned based on availability. Section managers in charge of specific technologies or skill sets provide line management. This format can cause conflicting priorities between section (line) managers and project leadership. This form of organization is illustrated in Figure 6-3(C).

B. Perspectives on organizations types

1. Line organization format

The line or functional form of organization is an effective format in that one manager for all the projects directly applies the resources to the programs. Compromises will have to be made by strategizing on the manager's part; however, if the members of the department are flexible enough, this can also be effective.

The line organizational format is another relatively good method for applying resources to a program. There is, however, a drawback in this form of organization in that engineering talent can be shielded from the customers and corporate strategy by virtue of the bureaucracy of a line-organized department. Often referred to as monolithic, this organizational

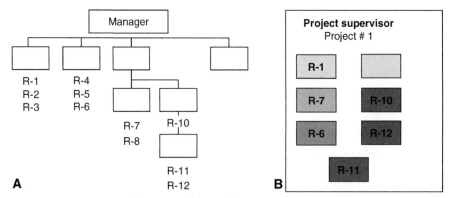

Figure 6-3. (A) Functionally managed form of project organization; (B) Pure project form of project organization.

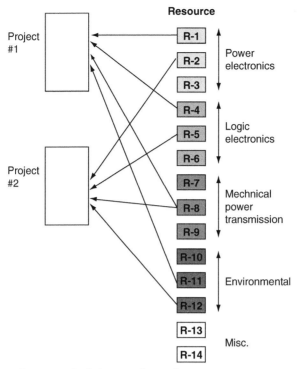

Resource

Power electronics

Logic electronics

Mechnical power transmission

Environmental

Misc.

Figure 6-3. (C) Matrix form of project organization.

form can execute programs with proper management. From a personnel perspective, it is easier to lose someone within the organization, hence the use of personnel resources is not as good as in the matrix form.

2. Pure project format

The pure project form is probably the most efficient organizational form for executing a program. It focuses the development horsepower directly on the task at hand with no compromises. There is single-minded focus, which helps a team to be highly effective.

The pure project format is one of the best formats to employ when program execution is critical. This format places resources exactly where they are required to ensure the application and management of resources with respect to project timing and results. In terms of allegiance and ownership of the program, this format is one of the best choices. It has disadvantages in terms of flexibility of resources, which is simultaneously part of its appeal. Flexibility in terms of the application is in conflict with this type of format.

3. Matrix format

The matrix form of project organization is the most comfortable for senior management to support throughout a program because the ratio of projects to people seems better. With this form, senior-level managers have a tendency to stuff projects into the pool at will, with little regard for the effect. Resources seemingly are endless. Unfortunately, this format is used when resources are actually thin and there are many projects that have to be executed simultaneously. Consequently, programs become resource-starved through reprioritization and infighting between project managers who are competing for resources and expertise. The matrix form is nonetheless a very commonly used format, even if the company's formal structure shows something else.

The matrix form of organization is sparing in terms of the use of resources. It makes use of the proper personnel for each segment of the project. Typically, personnel are used to their full potentials. The weakness with the matrix form manifests in two ways, both stemming from the same root phenomenon. Program allegiance is generally weak in these systems because personnel become "experts" in a specific discipline and choose their favorite program to concentrate on. It is thus difficult to measure the application of talent in uncertain situations and to assess their diligence. Accounting for their time is difficult, as is focusing horsepower on the tasks at hand. Project contention issues also arise, making multiple projects management far from easy.

No single format is optimal for all types of development. One must evaluate the enterprise philosophy, the temperament of the resident talent, the program needs, and the ability to apply resources to the various programs.

C. Selecting the right form

If you have the flexibility to design an organizational format, select the one that best fits your needs and the organization's level of flexibility. If you cannot change the format, understand the weaknesses in the one you must live with and bolster those weaknesses to serve the interests of the project. You might want to take the time to project potential conflicts in a program according to each organizational format and to generate different scenarios. By so doing, you can gauge the impact of organizational structure on a program.

Sometimes it is very difficult to change entrenched organizations from one pure form to another. Enter the dotted-line relationship. A dotted-line relationship is generally a sign of two things: A weak organization and an organization in transition. If your company needs to transition from one format to another, you might want to make use of the dotted-line relationship, which also assesses behavior of the individual reporting to two people. Keep in mind, though, that it is not fair to the individual to let the format go uncorrected indefinitely.

D. Making do with existing organizational formats

Sometimes there is no simple or practical way to modify the existing format of a team. Remember an earlier observation: Almost any system or format will work as long as it is managed properly. If the format cannot be changed, work within the parameters at hand. Deliver demonstrable results to senior management first, then negotiate organizational changes. This strategy is especially important to follow if the manager is new to the organization.

E. Meet the program needs, not organizational preferences

One final thought on the subject for organizational formats: Let the needs of the project drive any organizational format changes, not the other way around. If the project requires a certain format, modify the organization to that format. Do not try to meet the political needs of the organization with the program; the program simply cannot be compromised.

3. Key Players and Backup

A. The parameters involved

There are several parameters involved in dealing with team members. All of these factors need to be balanced to meet the needs of the individual and also the program and company needs. The following is a listing of the items a manager has to influence for an individual team member:

- Salary
- Expenses
- Motivation
- Training
- Pathway
- Career development
- Maturity of the individual, both vocationally and interpersonally.

Dealing effectively with all of these issues is the key to unlocking maximized effort.

B. Every team has key players

Individual contributions can vary from one member to another. No one contributes exactly the same amount of value to the project each day, week, or month. In a coordinated sequence of scheduled events, however, the contributions of everyone are required to move the program forward. Furthermore, failure of one member (regardless of how small the failure is) can affect the program. Using the example of a race car that loses the race because an 89-cent part fails, the lowest-level contributor can take the program off schedule or affect the time line.

The key contributors in a program can be recognized for their contributions but must be placed in perspective. No one player can be larger than the effort of the team or the program as a whole.

C. Not getting held hostage

It is vital for a manager to prevent being held hostage by the key players in a program. Given their talents and critical importance to the program, it behooves management to develop alternatives and strategies for dealing with these resources.

D. Prima donnas not allowed here

Dealing with team members who have prima donna attitudes is a difficult and tedious management task. Managers certainly can afford to absorb the effects of these attitudes when they are directed at them personally; however, the real danger is in the demoralizing effect these people have on the rest of the group. They have a tendency to bully their way around, discount the ideas of others, and align support along polarized positions that are not necessarily in the best interests of the program.

Fundamentally, these people have to be dealt with in a firm and fair manner. If necessary, they might need to be removed from a program. Keep in mind that a single project does not generally make or break a company. Because the main objective is to execute a program and to cultivate a functioning team, prima donnas have no long-term place in that team.

E. Developing strategies to cover loss of key people

There are several tactics and strategies that can minimize the damage caused by a disruption in applied talent to a program. These include cross-training personnel to diffuse key knowledge among several players; using the key players as trainers of the balance of the group; and continuously evaluating personnel for upward mobility for training. Select understudies for each of the key members and ensure that they have at least half of the knowledge base of the person they are backing up, as illustrated in Figure 6-4.

The use of understudies will ensure some coverage in the event that a key member is lost. Such a strategy is also a good means for development planning of the team members. Each member so trained is effective in reducing the dependence on the prima donnas and also sends a clear message to them that their behavior will not be tolerated for any period of time.

F. Overcoming the crisis: Trading ground for time

If for some reason you find yourself running a project with a high-risk key player, you might need to endure using that person until a backup plan is exercised. This means keeping this person on the team even though he or she might be disruptive and bad for the group. The manager needs to hold the group together until a replacement is found. Then the personnel action can be taken and the values of the group reset to the priorities of the project.

MANAGEMENT REPORTING

1. Management of the Team

A. Systems for program management

Many systems for program management are offered in various forms. Fundamentally, they are based on two basic types of project tracking methodologies:

- Gantt chart
- PERT (Program Evaluation and Review Technique) chart

The Gantt chart is represented by a list of activities along the y-axis and a time line along the x-axis. The x-axis indicates the start and stop time for each activity. This is shown in Figure 6-5. This chart is a quick and easy way of starting out a program that is relatively simple and does not require interconnection of events. For programs that require this interconnectivity, the PERT chart may be a better choice. In this chart, there is some logical organization of the tasks, but not to the degree that is found in the PERT chart.

The PERT chart shows a sequential connection between the separate tasks. Each task has a point of connection for starting and a point of connection for completion. The interconnection

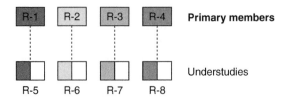

Figure 6-4. Program understudies.

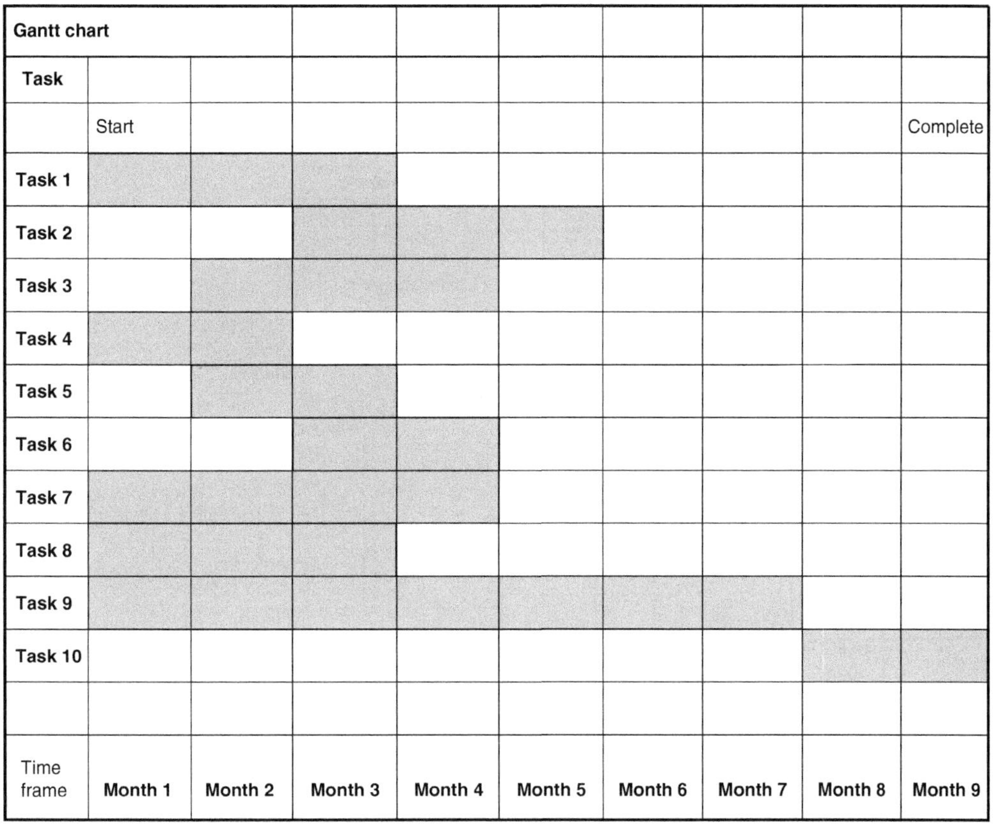

Gantt chart									
Task									
	Start								Complete
Task 1									
Task 2									
Task 3									
Task 4									
Task 5									
Task 6									
Task 7									
Task 8									
Task 9									
Task 10									
Time frame	**Month 1**	**Month 2**	**Month 3**	**Month 4**	**Month 5**	**Month 6**	**Month 7**	**Month 8**	**Month 9**

Figure 6-5. Gantt chart.

of the points represents the logical sequence of activities. The PERT chart is the more accurate method of program management, as it accounts for task listing, timing, and interconnection of events.

The PERT chart outlines both the tasks in a project and the interconnections between task start and end dates. This constitutes the logical sequence of tasks for the program. In addition, the critical path is defined—the shortest pathway from start to completion without compromising the integrity of the logic. An example of a PERT chart appears in Figure 6-6.

As the figure shows, each task is interconnected to other tasks to formulate the logical sequence of task completion. In this example, the second and third tasks can be executed in parallel time frames with the tasks immediately below them. In addition, the last task cannot be started until the previous tasks are complete.

B. Model for program management

The management of a program is an interdisciplinary, widely diverse activity. There are issues to manage that range from creative specifications to communication and security (both internal and external). Figure 6-7 is a model showing the diversity of management issues involved in a project, in a format that can be analyzed and optimized.

The overall project consists of six basic stages. They are as follows:

Stage 0 Market assessment and internal assessment
Stage 1 Prototyping

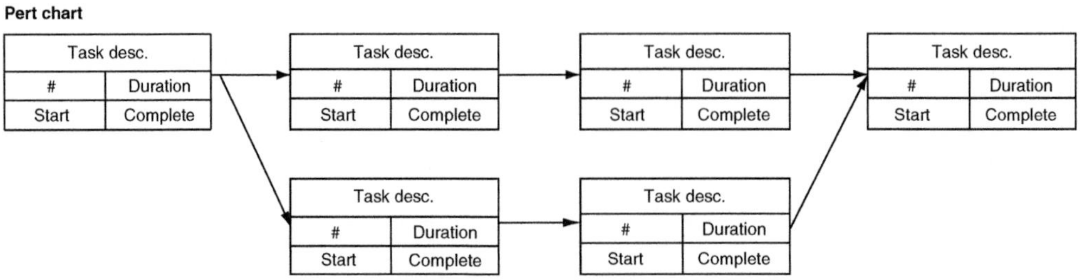

Pert chart

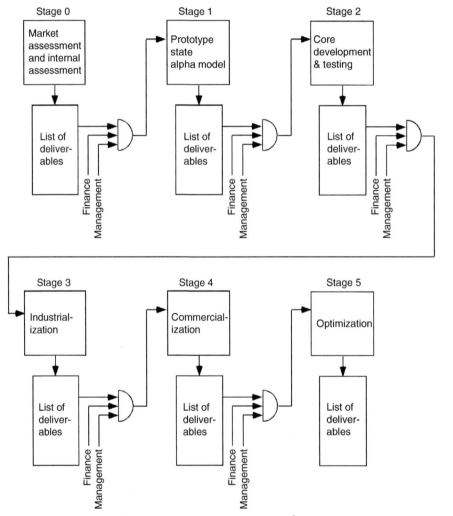

Figure 6-6. PERT chart.

Figure 6-7. Project management format.

Stage 2 Core development and testing
Stage 3 Industrialization
Stage 4 Commercialization
Stage 5 Optimization

The first stage is the assessment phase, which involves internal introspection of the company and direction and an external look at product opportunities. The next stage is the prototyping stage. It leads to the physical embodiment of a product concept as a starting point. The third stage is the core development of the product. This is the stage in which the product is developed and testing takes place.

Next is the industrialization stage where the product is made manufacturable and the manufacturing processes are developed. The fifth stage is the commercialization stage where the product is rolled out to the marketplace and units are placed in the hands of consumers. The final stage is the optimization stage, when feedback and improvements are incorporated.

Throughout the entire program, each stage has all the required deliverables, finance, and managerial approvals for proceeding to the next stage. Program management is the coordination and the execution of all of these tasks in an effective and timely manner.

C. Internal, external, and combination resource management

Not all of the resources to be managed are internal to the organization. Besides advice gathered from vendors and others, there could be an arrangement by which outside resources are required because they are not available within the constraints of the company. In such cases, there must be a contractual means for obtaining the resources.

The program manager must have the power to effect results for the program. Simply stated, this influence needs to extend to all areas having an effect on the progress of the program.

D. Managing uncertainty in a program

Oxymoron is the term that first comes to mind when referring to the management of uncertainty. By its very nature, uncertainty is "uncertain" in terms of degree, duration, severity, and extent. To suggest that it can be managed in some deterministic, scientific manner is somewhat naïve. Rather, the idea is to mitigate the impact of uncertainty and all that it brings with it. Take care at the outset of a program to identify the items that are known and understood and those items that are fraught with uncertainty. Channel energy to understand and clarify the uncertain items as soon as possible.

E. The dynamics of schedules, timing progress, and slippage

The impacts of schedules and slippage in the schedules can devastate a program. What seems like a small slippage in a schedule can accumulate to huge delays in product introduction. Unfortunately, many teams feel that slippage in a schedule can be made up as the program progresses. This is rarely the case. In addition, there could be a tendency to try to make up for lost time by changing the logic of the program to work around delays. Although this goal is admirable in terms of trying to preserve program timing, this practice can introduce oversights and poor compromises.

To help you gain a sense of the degree to which slippage can affect a program, Figure 6-8 shows a sample project consisting of six stages. Each stage has a scheduled duration. As the project exits Stage 0, the team incurs a one-day delay. In a 20-day activity, this represents a 5% delay. Because this is the first stage, the cumulative delay is also 5%. The use of the cumulative measure is the percentage of that stage's total duration that represents the accumulated delay.

Impact of delay							
Item	**State 0**	**State 1**	**State 2**	**State 3**	**Stage 4**	**Stage 5**	**Total**
Scheduled Duration	20	20	26	30	24	26	146
Schedule Slip	1	2	2	3	2	2	12
Cumulative Days	1	3	5	8	10	12	12
New Duration	21	22	28	33	26	28	158
% of Duration	5.00	10.00	7.69	10.00	8.33	7.69	8.22
Cumulative %	5.00	15.00	19.23	26.67	41.67	46.15	8.22

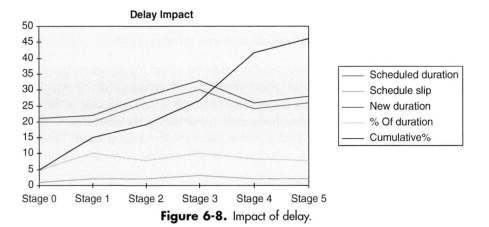

Figure 6-8. Impact of delay.

As the project exits Stage 2 (another 20-day duration), two days are lost in delay. This represents a 10% delay for that isolated stage and a cumulative delay of 15% dragged into Stage 2. Progressing through the example, a few days here and a few days there add up to a staggering 46% delay dragged into Stage 5. This means that 46% of the time originally budgeted for Stage 5 of the project is delay. It is impractical to catch up within the stage you are operating in and get back on track for each stage. To bring the program back into alignment with the original schedule, this amount of time would have to be added to budgeted time planned for Stage 5. Looking at the project as a whole, the overall delay represents 8.22%, or 12 days. The point to emphasize is that a project simply cannot catch up on a delay. Once the time is gone, it's gone!

F. Managing for results

What are the practices that the group and the company need to follow to become an effective development team? There are distinct practices that will result in distinct goals for the company. The summary chart is presented in Figure 6-9.

As shown, one key practice involves the training in and use of a common vocabulary throughout the enterprise. This assists in yielding a corporate goal and in cultivating effective

Managing for results				
	Effective and uniform communication	Interdepartmental linking structural objectives	Development program timeliness	Optimization of corporate resources
Common vocabulary	X	X		
Common methodologies for evaluations and decision making	X		X	X
Process for establishing priorities		X	X	X
Decisiveness and consistency of purpose		X	X	
Realistic planning and expectations			X	
Team structure and authority	X		X	

Figure 6-9. Managing for results.

and uniform understanding of goals and issues with the development project and the business. It also assists in linking the goals of each department.

The next practice is the development and use of common methodologies for evaluations and decision making. Working to reduce misunderstanding and false starts and progression down the wrong pathways, this practice establishes a uniform basis for decision making. This has a net downstream effect of promoting program timeliness. The use of corporate resources is also a collateral benefit, as common values and goals tend to encourage placement of resources to the proper tasks.

Establishing and supporting a process for determining priorities is the next desirable practice. It assists in the structural linkage of objectives by ensuring that each department has the correct priorities. This practice encourages development timeliness and the most effective use of corporate resources.

The next desirable practice is the corporate operational trait of decisiveness and consistency of purpose. This has tremendous benefit in cementing interdepartmental linking of objectives and the timely execution of a program.

Next, realistic planning and expectations bring about benefits in terms of helping the organization to be timely in its pursuits. Unrealistic expectations and planning done through rose-colored glasses create intrinsic delays in the execution of the program.

Finally, team structure and authority have a direct impact on both corporate-wide communications and program timeliness. If the team is a functioning unit with effective management and clear goals, the organization will be plugged into the program and the product will have a much better chance at a timely introduction to the marketplace.

Taken individually, no single one of these practices is necessarily a driving force affecting the desired goal; however, much of the work of the development manager and the senior manager is concerned with the proper alignment of the organization and the removal of obstacles. Implementing these practices within the enterprise helps smooth the path for management personnel. Constant vigilance and management talent are still required to bring a program to successful completion.

G. Follow-up system

With any program there is an enormous number of details associated with specific tasks, and these details require follow-up. These details are so numerous that it would be impossible to schedule all of them on the program's Gantt or PERT chart. In addition, there are subtasks associated with the management of the project that might need investigation. Items such as scenario development, "what if" investigations, logic, and schedule manipulations all require follow-up.

To best keep track of these details, implement an action item follow-up system. This follow-up system also can serve as a permanent project record for future use in demonstrating diligence or resolution of issues. With the use of computers and mechanized methods, any number of implementations can be used. Regardless of which one is used, be sure that the selected system includes the following components:

- Item number
- Description
- Action required
- Assignment date
- Responsibility
- Completion date
- Running status

An example of a format for a follow-up system is shown in Figure 6-10. The example system represents only the information to be recorded and maintained.

To be effective, this needs to be managed every day. As each item becomes complete, it should become part of the permanent record of the project. Follow-up dates should be adhered to strictly to prevent loose, unresolved details from bogging down the project.

2. Management Reporting

If the development team is doing a great job in executing a project and senior management doesn't know about it, is the team still doing a great job?

Senior management has a right and a need to know the status of a program and rates of progress. The development manager has the responsibility to keep senior management properly informed. This section will focus on setting up systems to accomplish just that.

A. Set up a good, accurate, easy-to-use reporting system

Program management is not about sophisticated management reporting systems; it's about getting results. Therefore, set up a management reporting system that is easy to use and effective at communicating essential information. Make it specific enough to be accurate and have depth, but make it general enough for a senior management overview. Have

Action item follow-up system								
Item	Description	Action required	Assignment date	Assigned individual	Completion date	Running status	Follow-up date	Comments
1								
2								
3								
4								
5								

Figure 6-10. Follow-up system.

audited data entered and make it as automated as possible. The manager's job is to manage the development, not to reinterpret data and present it to senior management.

Although many systems generate their own reports for status, this section will present a format that consists of the data in which most senior managers are interested.

B. Lay out the program in detail

Begin by laying out the project in total. Starting with Stage 0, include the following list in detail. By laying out all this data for each stage, the requirements for the program are readily apparent. Figure 6-11 represents an example of this process.

1. Funding

This is the stage's financing package. Detail the budgeted expenses and allocate them to tasks, performance, and deliverables. In this way you will have a good idea what the deliverables are costing the project.

2. Logic

Lay out the project in PERT format showing interconnected tasks and milestones. Take care to identify the connective logic, as this ensures diligence and quality. Haphazard redefinition of the logic is a sign of a program out of control.

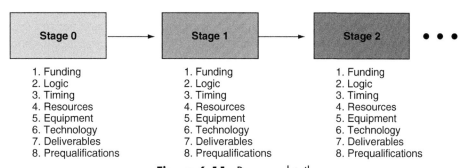

Figure 6-11. Program detail.

3. Timing

Lay in estimated times for each activity and task. The use of development personnel's estimates is the best choice here. Any other estimate won't be bought into by the team because the estimate is not theirs.

4. Resources

Identify the required resources and schedule them on the time line.

5. Equipment

In a similar manner, identify all of the capital and expense equipment required for the project. Specify dates on the time line when they will be required, and also factor in training time if it is required.

6. Technology

List and schedule the technological requirements for the project. Make sure to include collateral requirements to integrate the technology into the company. Also allow time for absorption of the technology by the team members. This technology transfer must take place for the project to be successful in the long run. This is because the licensing of technology is for one purpose—to get your company into the game initially. Staying in the game is the responsibility of the company.

7. Deliverables

Schedule and lay in the list of deliverables for the stage. This is the strict list of deliverables that must be completed in order to consider the task of the stage complete.

8. Prequalifications

State clearly on the schedule (now multidimensional) the prequalifications needed for the next stage that must be tended to now. An example of this type would be to start campaigning for funds needed for Stage 3 during Stage 2. Another example is the recruitment of talent required in the future so that personnel are in place and functional when they are required.

C. Time management and log

The next area for consideration is the development time and engineering log. This is a tool to manage the engineers' time and to ensure the project is getting the proper time allotted. Business development time is used in many ways, even when it is assigned to a specific program. As will be shown in more detail later, there is always danger of misallocation of time to nondevelopment activities, which delays programs and makes the group progress lethargic.

An example of a time reporting system is shown in Figure 6-12 (a truncated format for review; only Week 1 shown).

As shown in Figure 6-12, the use of an engineering time log for tracking development time spent generates the percentage of time used on various aspects of development activities. In this example, the actual development time represents only 47.4% of the hours expended by the engineer. Orders required 20.7% of the time.

Engineering time log								
		Mon	Tues	Wed	Thur	Fri	Sat	Week 1 Total	% Total
Orders									
		4	4	4	5	6	7	30	
		1	6	1	1	1	1	11	
		1	2	2	2	2	2	11	
		1	1	4	4	4	4	18	
		1	1	5	5	5	5	22	
Subtotal		8	14	16	17	18	19	92	20.77
Development projects									
		7	7	7	7	7	7	42	
		7	7	7	7	7	7	42	
		7	7	7	7	7	7	42	
		7	7	7	7	7	7	42	
		7	7	7	7	7	7	42	
Subtotal		35	35	35	35	35	35	210	47.40
Engineering Investigations									
		1.2	1.2	1.2	1.2	1.2	1.2	7.2	
		1.2	1.2	1.2	1.2	1.2	1.2	7.2	
		1.2	1.2	1.2	1.2	1.2	1.2	7.2	

Figure 6-12. Engineering time log.

Engineering changes, which have little contribution to value, represented 10.1% of the total. Through the tracking log, the nondevelopment time can be accounted for. If the manpower loading was allocated at 100% for development, it is apparent how the project will be significantly delayed with only 47.4% being expended. Moreover, this represents only one week. Left unchecked, one week dovetails into another, further reducing the effective development time.

D. Regular status meetings

The use of regular status review meetings is beneficial to maintaining the program on track. The status meeting accomplishes several things:

		1.2	1.2	1.2	1.2	1.2	1.2	7.2	
		1.2	1.2	1.2	1.2	1.2	1.2	7.2	
Subtotal		6	6	6	6	6	6	36	8.12
Engineering changes									
		1.5	1.5	1.5	1.5	1.5	1.5	9	
		1.5	1.5	1.5	1.5	1.5	1.5	9	
		1.5	1.5	1.5	1.5	1.5	1.5	9	
		1.5	1.5	1.5	1.5	1.5	1.5	9	
		1.5	1.5	1.5	1.5	1.5	1.5	9	
Subtotal									
Miscellaneous		7.5	7.5	7.5	7.5	7.5	7.5	45	10.16
		2	2	2	2	2	2	12	
		2	2	2	2	2	2	12	
		2	2	2	2	2	2	12	
		2	2	2	2	2	2	12	
		2	2	2	2	2	2	12	
Subtotal		10	10	10	10	10	10	60	13.54
Total		123	135	139	141	143	145	443	100

Figure 6-12 (Cont'd)

- It establishes consistency in progress assessment.
- It establishes accountability for the individual team members.
- It assists in identifying upcoming issues.
- It provides a forum for discussion on resolving problems.
- It focuses attention on critical path tasks.

Depending on the type of development project and the group's level of experience, the optimum time for these review sessions can range from every week to every two weeks. These are nuts-and-bolts sessions that are designed to assess progress and navigate around and through barriers. They are neither strategy meetings nor commiseration sessions. Their sole purpose is getting things done. There is another type of progress meeting that may be held every two months. This is a management review meeting to discuss overall progress issues with senior management. Specifically, the idea is to not get pulled into details on each project but, rather, to inform management properly on progress and to resolve linking objective issues.

Two-month intervals are not magical by any means, and the frequency can vary based, again, on the project; however, this series of meetings is designed to evaluate development operations and programs for effectiveness.

E. Product maintenance versus development

The issue of product maintenance versus development is an interesting one. Consider, for example, a case in which a fixed number of development people develop Product 1. As the product ages, there will be changes to the product driven by factors external to the organization. The support of these changes is referred to as product maintenance. As more products are added to the portfolio, the demand for development engineers' time begins to grow. Time reporting will show, as in the previous example, that development time for new products does not get expended on development. It gets parsed out to all of these other activities, which results in severe project delays that are not recoverable. Figure 6-12 illustrates how this can happen and how it appears on a time report.

Figure 6-13(A) shows three scenarios to illustrate the loss of development effectiveness because of the effect of dilution due to product maintenance. Each scenario is for a development program requiring 6000 hours of development time. The available hours are the total hours available for the group. The maintenance hours are the required hours needed to support the product previously introduced. The development hours are the hours that can be expended on the development project. The remaining hours are the hours left to complete the project after the month's development hours have been expended.

In Scenario 1, for example, the program requires 6000 hours. With no maintenance requirement, the 2000 hours available each month can be expended on the development program. It is complete in exactly three months. In Scenario 2, there are maintenance hours, which dilute the development effort. This results in a 4.28-month time frame for program execution. Scenario 3 consists of maintenance hours for two previously introduced products, which dilute the engineering effort even more. At some point in assembling a portfolio of products, maintenance can exceed development given a fixed number of people. The crossover is shown graphically in Figure 6-13(B).

F. Periodic management reports: A model

A must for the objective communication of product development progress is the use of a standardized management report. There are three sections that comprise the report, namely:

1. Product status report

The product status report is a short synopsis of the progress toward achieving functionality of the product and other specification items such as performance, size, form, and fit. The entire functional specification should be outlined in a tabular format, with check marks for parameter achievement. A gradient may also be used to grade achievability for each item of interest. Also in this section of the report, the original target manufacturing cost is compared with current estimated projections.

Product development / Maintenance trade-offs

Item	Month 1	Month 2	Month 3	Month 4	Month 5	Month 6	Month 7	Month 8	Month 9	Month 10	Month 11	Month 12
		Scenario 1			Scenario 2					Scenario 3		
Project hours required	6,000			6,000				6,000				
Available hours	2,000	2,000	2,000	2,000	2,000	2,000	2,000	2,000	2,000	2,000	2,000	2,000
Maintenance hours	0	0	0	600	600	600	600	1,160	1,160	1,160	1,160	1,160
Development hours	2,000	2,000	2,000	1,400	1,400	1,400	1,400	840	840	840	840	840
Remaining hours	4,000	2,000	0	4,600	3,200	1,800	400	5,160	4,320	3,480	2,640	1,800
		3 Months				4.28 Months				7.28 Months		

Figure 6-13. (A) Scenarios in development versus maintenance;

A

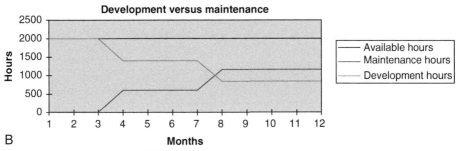

Figure 6-13. (B) Development/maintenance hours.

2. *Project status report*

The project status report summarizes the investment and exposure of the company in funding the program and compares the sunken investment to the total and remaining investments. It summarizes personnel investment, hours, and dollars.

3. *General business condition*

Finally, the general business condition report gives senior management an idea of the outlook of the business in terms of forecast sales, incoming order rate, shipments, and profitability.

A model for this format, which includes a place for narrative in each of the three sections, is shown in Figure 6-14.

G. Team push/management pull

As a final note to this section, one of the aims of management reports is the creation of a synergistic exchange between the business development team and the senior management group. The team pushes the project through the process by effort and deliverable results; the management pulls the project by setting high expectation levels and removing barriers for the team. When the two forces work together, they create a fireball of energy and enthusiasm for completing the project successfully.

INTELLECTUAL PROPERTY PROTOCOL

1. Cooperative Development with Other Firms

This is one of the most sensitive subjects in the business development process. It addresses the core of each company's competitive assets and must be handled very delicately. The protocol for their communication should be prescribed by the original agreement. If a patent is involved, each company should have a process by which to evaluate ideas and determine eligibility for applying for a patent. Prior to applying, each company should have a protocol for handling the technology so that they can advance the business development program and product development without divulging any information to third parties. If the one of the companies is overseas or has international facilities, even more diligent care taken mast be taken. This is because of the rules of patent law engagement for foreign filing. Since we are in a global economy, these issues can affect every company because of sourcing, end-use markets, and the proliferation of communications taking place.

Management report format					
				Percent achievement	
			0%–30%	31%–66%	67–100%
Product report	Specification	Description			
Narrative	Critical				
		Feature 1	X		
		Feature 2		X	
		Feature 3			X
		Feature 4			X
		Feature 5	X		
	Needs				
		Feature 1	X		
		Feature 2	X		
		Feature 3	X		
		Feature 4		X	
		Feature 5		X	
	Wants				
		Feature 1			X
		Feature 2			X
		Feature 3	X		
		Feature 4		X	

Figure 6-14. Management report format.

There is even a protocol with respect to timing. Today's partner can easily become tomorrow's competitor. This being the case, the manager should balance cooperation with future competitive status on these intellectual property issues.

			Feature 5	X		
Narrative	Costing					
				Variance		
	Target cost ($)		100			
	Estimated (Original) ($)		118.5	18.5		
	Estimated (Present) ($)		137	37		
Project progress report	Investment		Planned	Expended	Remaining	
Narrative						
	Number of personnel deployed		25	20	5	
	Number of hours		50,000	40,000	10,000	
	Appropriated funds ($)		2,500,000	2,000,000	500,000	
			Original		Estimated	
	Completion date		January XX		March XX	
General business condition report						
Narrative	Outlook					
	Incoming order rate			$ / Month		
	Shipments ($)			$ / Month		
	Profitability			$ / Month		
	90-Day backlog			$		
	Profit in backlog			$		
	90-Day sales forecast			$		

Figure 6-14 (Cont'd)

2. Near- and Long-Term Codevelopment Protocol

This brings up the issue of how to deal with a partner company for the near and long term. No corporate relationship will last forever. The process of change is a given in dynamic markets. So how do you codevelop technology to capture a business opportunity with a new product knowing that your partner today could be your market adversary tomorrow?

The answer lies in two areas: the original agreement's dissolution protocol and the integrity of the two companies involved. A well-written agreement should have a protocol for amicable dissolution of the venture written into it at the beginning. If the agreement does not have this protocol, there is a greater potential for trouble down the road.

The next area is integrity. Both companies must maintain integrity to the program and to their own value systems. The trouble can occur when the value systems of each company differ significantly. Such value systems can extend to how they approach the market, resolve customer problems, and mitigate liability.

If one of the players is a wildly irresponsible partner, when the going gets tough in the marketplace you may find yourself out on a limb with no support and few options. It is best to start slowly with a partner company, give out some safe information, and watch how the partner company treats it. This can give an early indication of future behavior.

COMMUNICATIONS PROTOCOL

1. Communications Records

Keeping records of communication in our litigious society is an absolute must. In addition, it simply is good business practice to do so. Establishing a record of communication that is retrievable and sequential helps in negotiations, separating fact from feelings, and can be invaluable in troubleshooting problems. In fact, there could be a case made for establishing a formal communications review during a program. Because everyone is so busy, we all process communications and copy everyone on a project "just to make sure" everyone has the information. In reality, there is very little respect of the protocol of communications anymore. Anyone can e-mail anyone. In fact, with e-mail programs you can get sensitive information to personnel without the main distribution list being aware of it. This blind copying can help overcome political obstacles, or it can create them.

At the beginning of a program it is essential to establish the protocol for communications and recordkeeping so that as each issue comes up, the announcement, assessment, and disposition can be retraced if need be.

There is also the recordkeeping required for liability mitigation. If something should happen during the course of the product life and the company needs to defend itself in a liability suit, the retention of good records helps in two ways. First, the actual issues and disposition (if handled properly) can assist in the company's defense or at least mitigate the damages. Second, the fact that the company was diligent in keeping the records in the first place goes a long way toward establishing credibility.

The manager should establish the protocol and differentiate it between the market issues, product development issues, and contractual issues.

2. Initiating and Protecting Communications

A. Set up communications

The communications system can determine the effectiveness of the project participants. A communications system is not necessarily medium-dependent. It is simply the unidirectional and bidirectional exchange of key information to enhance progress.

By definition, the communications system must effectively do the following:

- Transfer information without distortion.
- Acknowledge and confirm information.
- Maintain a certain amount of security.
- Allow a channel for meaningful feedback.
- Create understanding in the receiver intended by the transmitter.
- Act as a document and retrieval system.

These items require a fair amount of setup and security to work in a real development situation. In addition, they are multidimensional in that communications occur both internal and external to the company. The importance of a properly implemented communications system cannot be overstated. It will prove to be invaluable as the project progresses and as details become more prolific and harder to keep track of.

B. A model for communications

Communications within a company and outside of the company occur on all levels. There are communications within the company, through the channel, and to the customer. At all levels and in all directions, the transmission and receipt of messages, data, and decisions are opportunities for misunderstanding, inflated expectations, and inaccuracy.

We all have experimented with the game of "pass the message." This game illustrates how a message can change its meaning or emphasis as it passes from one person to another. Consider communications between companies in which the elements of time delay, personalities, and focus differ. These factors complicate the job of accurate receipt of transmitted data, as Figure 6-15 illustrates.

Depending on the stage of the transaction, there can be diagonal communications, such as the manufacturer's sales personnel communicating with the customer's engineering or finance personnel. Also consider how many individuals a given message passes through and the degree of interpretation that can take place as it is progressing the pathway from transmitter to receiver.

Another element to consider is the security of communications both within the company and outside of the company. The heavy black lines on the illustration indicate firewalls that prevent the breach of communications security. As we will see in the next section, the use of network communications has improved the speed of information transfer tremendously. It has also introduced a whole new set of security problems; such is the nature of progress.

C. LANs, WANs, intranet, and people

With respect to protecting intellectual property, trade secrets, and interests of the business in general, network communications cannot protect your company. There is no software, protocol, or arrangement of hardware that can make up for poor judgment on the part of team members. In order to use these tools effectively, each member of the company needs

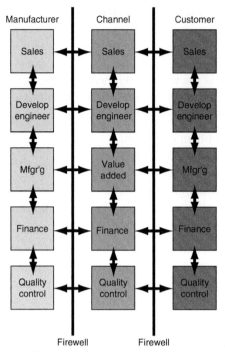

Figure 6-15. Communications.

to operate like a businessperson. This means being savvy enough to know what and what not to disclose to personnel outside of the company.

If the company personnel understand the business, the threats, and the opportunities and are truly interested in furthering the business, they will conduct themselves in your best interests. If they do not understand the issues, train them quickly or eliminate them as possible sources of information leaks. In Chapter 2 we discussed the spheres of influence that show how information could pass from one person in the company through people interfacing with them to others outside of the company.

Implement internal communications controls to be able to determine whether sensitive information is being made available to competitors or to the general market, where it can fall into the hands of competitors.

As a final note, secure people you can trust in every aspect of the company. There are no systems (electronic or otherwise) that can provide security as good as trustworthy people who have a vested interest in the business and are streetwise enough to perform their jobs and protect the company's interests.

CULTURE OF THE GROUP

1. Creating Corporate Culture

A. Set the pace—business development team

As discussed previously, not every team is fully functional at the outset. It is even less functional if the members never have worked together before. Consequently, the manager needs to establish the culture of the team. This means imbuing the group members with program values. It is best to establish this project culture as early as possible in the development cycle because time is at a premium, and time should be spent in execution rather than group alignment.

The pace of the program should also be set with the cooperation of both senior management and the group. In effect, the program manager needs to push the team for performance and also to urge senior management for funding and resources. Each faction needs to be urged by the other's commitment. For example, the team should work diligently because it is spending corporate resources. Senior management should take note of the commitment and performance of the team and release funds in a timely manner.

B. Actively confront that first obstacle

When a project is first starting out, actions speak volumes. Resolution of that first obstacle is crucial to setting the tone for subsequent activities. Raise the expectation level high and resolve the first obstacle as soon as is practicable. Do not address it with complacency because everyone will then view the group as ineffective and lacking commitment.

The group's test when encountering the first obstacle and every one thereafter is to overcome it without delay. An oft-quoted expression is applicable here: "An obstacle is what you see when you take your eyes off of the goal." In program management, time and complacency are the enemy.

C. Variable work effort applied to variable workload

What we have said up to this point does not imply that development should be slavery. Rather, the concept is to create a dynamic performance within the group that is variable. Because the needs of the program are dictated either by schedule or adversity, the demands of the team rise to meet the program's needs. This philosophy needs to become part of the operational characteristics of the team. Because the project is under some sort of project management control and the demand for personnel's time is identified initially, the demand variable should be left to overcoming the obstacles.

Another variable factor is time off, which can include vacations. Plan to work around personnel with critical skills during these times. As mentioned previously, the idea is to reduce the effects of uncertainty as best as possible.

2. Reinforcing Values and Value Systems

The true test of a manager is how he or she reacts when the program is going poorly. The project was started with a value system and, now that the progress is less than expected or an obstacle has severely backtracked the program, the manager must reinforce the value systems. This means relying on those values and getting the program on track. This demonstration of values goes a long way toward getting the entire team aligned again. Recovering from adversity is a very difficult thing to do.

When you are dealing with uncertainty and highly technical processes, and the people who are conversant with them, there are no rule books to follow. Any time you are in uncharted territory (just as in life), you must rely on the values you have. This is as important for the *team to understand* as the business development manager.

3. Incentives for Development Personnel

A. Incentives

This section does not delve into the means for compensation or incentive programs for employees, other than to endorse some means for encouraging the development group to bring the project to completion within the prescribed parameters. Members of the team must set aside their personal agendas to become an integral part of the team. Therefore, it is only fitting that the team be rewarded as a unit for their performance. Salary and benefits get the personnel in place and reporting for work each day. Incentive for group performance will drive them to achieve results on time and within budget.

B. Relate the incentive to time and quality of work

The incentive should address two aspects for development personnel: personal and professional. This means that one component should specifically reward the individual members of the group in a monetary fashion. The other component is to be spent on new discretionary equipment, training, or processes to assist the members in professional growth. This practice serves to tie performance to both personal gain and professional improvement.

C. Recognition of players

Special care needs to be exercised in rewarding key performers. In many cases, these are more valued members of the group in terms of contribution; however, care must be taken so as not to disenfranchise the other members. Fundamentally, most professional people realize the contribution capability of higher-level individuals and, to some extent, the incentive can be a driving force for their own improvement. The use of the second component has a leveling effect in that all of the personnel may have access to the discretionary equipment.

D. We win or lose together

Finally, the key to long-term effectiveness of the team is to eliminate prima donna attitudes among the top performers. People who know their value to a team can begin to become prima donnas, and these attitudes must be discouraged. They demoralize the balance of the group and, over the long term, will destroy any leverage that might be created.

The value that must become part of the team is that the group wins or loses together, regardless of how talented one or another member might or might not be. The group wins, or the group loses and some other group wins. This is the law of the product development jungle.

TIME BASE

1. Interface to Program Management

In any business development program it is important to understand the impact of time on potential success. In general, the development of a product or a business model is done under severe time constraints. The dynamics of the marketplace simply place time pressure on every aspect of the organization. Therefore, it is incumbent on the business development manager to demonstrate timeliness of progress to management while instilling a sense of urgency in the team and matrixed participants. Many times the problem occurs in getting the matrixed participants to keep to the same time base as the team. Accordingly, the team must direct forces, talent, and expertise to meet management's time base. The manager should assess progress and communicate all aspects of the progress to management. If contingency plans are initiated, inform management before you institute them.

2. Program Management Style and the Group Temperament

One final thought on time base relates to the program management profile and the temperament of the group. The two should generally match, with the manager leading the group to higher performance. If the manager's management profile is significantly different from the group's temperament, he or she may lose the support of the team. The team is what produces the results and acts as the ambassador to the corporation. This is especially true of a program involving two companies and a clash of cultures. In these cases the manager must be a driver and peacemaker at the same time.

3. The Program Kick-Off

A. Establish the start and end dates

As important a milestone as the end date of a project is, the start date is equally important. The beginning date establishes that point where planning, conversation, and forecasting diminish in relative importance to actual execution. To that end, the organization and the development group need to understand that development is in progress and the single-minded purpose now takes precedence: completing the program successfully.

Establishing the end date also sends the right message to the organization and the development group about the expectations for completion and general timeliness. Fundamentally, however, the handling of the first obstacle will set the tone. Setting the end date as relatively inviolate eliminates excuses for inaction in getting back on schedule. The dates and deliverables drive the activity.

B. Make use of training as part of the program

Each day brings new learning; so, too, with product development. Although there are high expectations for the development group and the company population in general, there is no need to expect that the team is at capacity in new learning. The manager needs to balance the needs of the project and the need for continued development of the team members.

If such training and new learning do not take place, the group will fall behind in capability with respect to the market. This, in turn, will reduce your competitive position for the next program. The operative lesson is to integrate training and new learning wherever possible.

C. One step at a time

Execute the program one step at a time. If the proverbial "journey of a thousand miles begins with a single step," then remember that the thousand miles is covered by countless steps, one after the other in relentless succession. This is also the case with product development. The program schedule might seem daunting at first, or even insurmountable, given all the tasks, subtasks, and activities. Take each task, however, and execute it one step at a time until all steps are complete.

D. Have fun!

To place the program in perspective, the new product development activity can be the hardest, most difficult work a person can do. But lighten up and have fun. This can be the hardest job you will ever love! The challenge of technology, its application, and the excitement of completing it within the allotted time frame can make the job truly enjoyable.

As with anything else in life, look two steps forward, and take the next step toward the goal. Plan the contingency and keep the group's eyes on the goal. If you feel that business development is like a race, you are probably accurate. The kick-off is the formal beginning of a long process. Mark the time, pull the trigger, and let's go ahead!

CORRECTIVE IMPROVEMENT

The term *corrective improvement* is combinational in that it describes an ongoing activity focused on making correction when something is less than optimum, and then looking for ways to improve the process so the event does not happen again. Many corrective actions systems fall short in that the problem is noticed and then addressed directly. This should be done to mitigate any further damage. However, there is another aspect, namely, how do we prevent the problems from happening again? How do we view the event and look upstream and downstream at contributing causes and improve the entire process? This is a very important tool in improving productivity. Without it, most corrective action systems generate inspection and labor-intensive steps that cost money, waste time, and add bureaucracy to the program. By taking some time to look for the improvement that negates the need for checks and approvals, you can begin to streamline and error-proof the process.

1. Disciplining the Team

Sometimes the corrective improvement is not for systems; it may need to be applied to personnel or the team. In the case where corrective improvement needs to be applied to the team, the manager needs to focus on the program goals and move the team to achieve these goals. The team needs to work until the task is completed and also work to catch up as best as possible. Often there is a tendency to redo or update the plan to accommodate the slippage, but in reality it is almost impossible to catch up after you have gone off track. If the plan was loaded properly to begin with, there is no float available for catch up. In this case, outline these facts for the team, catch up as best as you can, and make sure it doesn't happen again. Disciplining the team should never be punitive; it should focus on the goals and milestones of the project.

2. Recovering from Adverse Events

In the next chapter we will discuss in more detail some of the skills needed to recover from an adverse event.

3. Prevention of Future Events

There is only one thing good about a program setback: learning what was the cause, the effect, the overall impact, and the way to prevent it from happening again. In addition, when learning about the cause the manager needs to project how similar causes or phenomena could affect the program in different ways. A review of the event, its cause, and prevention is a good way to cement the philosophy in the team.

SUMMARY

This chapter is a transitional chapter. It describes the point in the development program at which planning yields to execution. Once the starting point is established, the balance of the effort is in producing tangible results.

In this chapter, we reviewed several of the elements that comprise the formulation of the development team and the tools for making them effective. Initially, the logistics of starting the program planning and the issue of assembling the group and transforming a loose collection of people into a functioning team was discussed. Next, the issue of organizational dynamics (both intercompany and intracompany) was discussed. This led to a discussion of organizational formats and how to match the format to the specific project under development.

Mentoring and apprenticeship are more important than ever before as we move forward in new product development. They can imbue new members of the team with the values of the company as well as with technical know-how. Management of the group was also discussed with respect to generating the results effectively. Tools for tracking and managing the time of the group were presented.

An important element in developing a product is to organize and lay out the project in total. This structuring of the project will yield big returns if the schedule should become compromised in any way. A clear listing and interconnection of the tasks make planning and execution more manageable.

With a fast-paced dynamic market, the intellectual property protocol and communications protocol were discussed. These two items, critical to the longevity of the business development initiative, are a sensitive issue, especially when two companies are collaborating on a program. Communications are a key element for successful programs; they can be a program's downfall or make it successful.

Also discussed was the culture of the development team and the company. The culture is developed over time but must be established as a baseline immediately. The team needs to be action-oriented and obstacles must be navigated around or eliminated. Develop a culture based on a sense of urgency. Integral with culture is the time base of the program and how to adhere to the time base under various conditions.

Finally, we discussed corrective improvement used in business development operations. The concept of corrective improvement goes further than loss mitigation and prevention to a systemic method for errorproofing.

With the completion of Chapter 6, the group is now ready to start the execution of the business development project with the information presented in Chapter 7.

EXECUTING THE PLAN

1. Program Management

A. Mechanics of product development

1. Flow chart development

As discussed in Chapter 6, the flow chart of the program is an important step in executing the development on time. There are many software programs available to support the task of documenting the logic, keeping track of the tasks at hand, and tracking the timing of the tasks. Regardless of which software program is used, discipline in enforcing the program tasks and schedule is of paramount importance. The software can only keep track and prompt. Managing a successfully executed program requires human interaction to stay on track.

2. Orchestrating timing, resources, and creativity to meet the objective

Only human energy can assess, revise, and navigate obstacles. This occurs by amassing horsepower to overcome or work around those obstacles. Timing, resources, and creativity must be orchestrated to meet the objective. Failure to employ these elements will cause endless delays and corrupt the time line of the program. The new product development manager must keep a watchful eye on all dimensions of the program. Figure 7-1 illustrates the philosophy of management.

As Figure 7-1 shows, the manager must look ahead in the project time line to anticipate future details as well as keep track of documentation and completing previous details. By looking ahead, the effect of uncertainty can be minimized. By mopping up details, nothing is left to chance and progress is truly in the forward direction, with little danger of past baggage haunting the program. Analogous to a symphony conductor, the development manager must coordinate the team resources with reference to the scheduled tasks.

3. Program management

The issue of program management is complex and pervasive, involving the entire company. One needs to think of program management as having two loops of control. The inner loop is dedicated to the specific project whereas the outer loop coordinates several

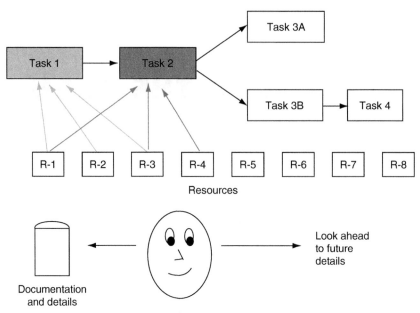

Figure 7-1. Orchestrating project management.

projects that contribute to the total activity within the corporation. This is represented in Figure 7-2.

The inner loop   The inner loop of project control is a classical feedback system in which the schedule (analogous to a command) is input to the development system. The development system executes action items to complete tasks of the program. The means for feedback to advise of status is analogous to the reporting system of a program. The status is compared with the schedule (command) in the assessment node and an action is directed to the individual development system.

The outer loop   The outer loop of control correlates the individual programs to the larger corporate strategy and operations. The overall corporate strategy is the command signal for the corporate business operations and development system. As depicted at the bottom of the illustration, the operations and development systems execute the business strategies and development. The output of the assessment node serves as the command for the individual programs. A management review evaluates the output performance of operations and feeds information back to the corporate assessment node, completing the outer loop. In addition, summary status reports taken from each individual project are input to the management review. This review serves as an entry point for feedback on all of these programs and extends to the higher-level corporate assessment node.

The corporate assessment node evaluates the difference between business strategy and operations performance and then generates appropriate action items and tasks to be executed by the operations and development system.

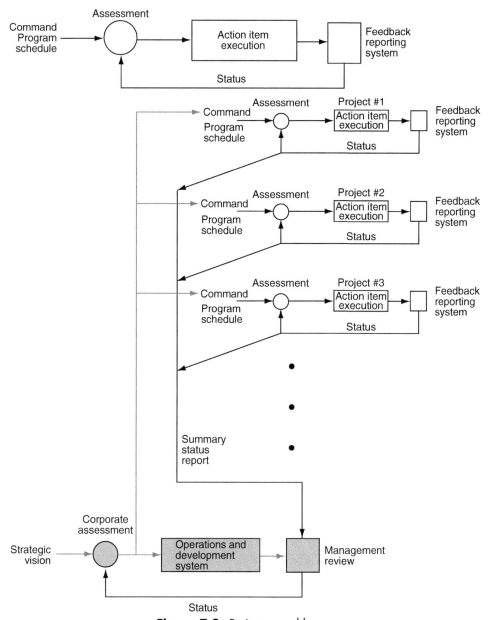

Figure 7-2. Project control loops.

2. Project Management: It's People Management

A. Team management

Forget the claims on the box covers of project management software. Software cannot manage a program. There is more to program management than keeping track of the details of logic, timing, and resources. Although a fully integrated, software-based project management system can be quite valuable in executing a project, once the details are tracked, project management is essentially people management. It involves the motivation and guidance of very talented personnel to accomplish a specified result in a scheduled time frame.

Therefore, development managers should become as conversant with people skills as they are with the requisite technical skills. Accomplishing things to exacting specifications is a difficult endeavor. Accomplishing them through other people elevates it to a higher level of difficulty, and accomplishing them through other people within a specified time frame elevates this to a unique challenge suited to only a few people.

B. The unpredictability factor with personnel

There is an unpredictability factor in development personnel that must be recognized. Progress might not be linear; therefore, the schedule must be tied to actual patterns of progress. For this reason, it is helpful for a team to have completed several projects together previously to refine the group's estimating and performance skills. Team members have both good days and nonproductive days. Progress is a pathway and the manager needs to guide the way.

C. Managing technological risk

Managing technological risk is sometimes an oversimplified practice. Often, managers look at risk and uncertainty, make an immediate assessment, and assume that overcoming it is part of the development. Risk is, however, a significant barrier to a program and, as in the case of resource allocation, simply redoubling efforts will not be sufficient to prevail. In fact, risk should be quantified in each stage of the project. Because the project may have a PERT chart to execute tasks, there should be a scaled uncertainty measure underlying each task. Uncertainty should be measured on a relative scale in which the most uncertain element represents 100%, as shown in Figure 7-3.

In this figure, uncertainty is charted along with the talent levels of resources assigned to the task. Although this is only a sample representation (not the actual form that might be used), it does serve to point out that uncertainty changes during the project and varies by task. The manager needs to amass the intellectual talent and bring it to bear on formidable tasks to resolve uncertainty in a timely manner and prevent delays. Simply stated, this means preparing for Task 3 while the team is still completing Task 1.

3. Executing the Product Development Plan

A. Any plan will work—the key is adherence

A plan is a pathway toward the completion of a program. The key to completing the program is adherence to the plan. There are several elements that must be coordinated as part of the product development plan:

- The plan itself with tasks and logic
- The schedule
- Chart of uncertainty
- Resources
- Work content

Impact of uncertainty							
Item	Task 1	Task 2	Task 3	Task 4	Task 5	Task 6	Task 7
Duration	5	7	35	23	12	16	25
Responsibility	MAA[1]	TSB	PCR	GHT	DFT	FTD	PQR
Level 1–10	4	3	8	6	7	3	7
Uncertainty Factor	10	15	100	35	40	23	80

[1]The 3 letter notations indicate responsible individuals initials as identifiers.

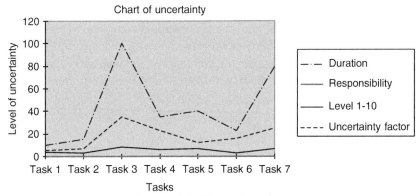

Figure 7-3. Uncertainty chart.

- Management reporting
- Measurement of performance

All of these elements must cohere throughout the development process. The manager's role is to give careful consideration to all of the elements and juggle them to serve the needs of the project in total. This might mean trading off elements for expediency; although this practice is not recommended, the manager might need to exercise such an action to move a program out of the doldrums.

B. A look at task mechanics

All tasks should be handled in the same manner. Standardization serves to establish a common denominator for scheduling a project. To further define the task execution, every task should include the following five integral aspects:

Assessment: Assessment is the analysis and breakdown of the tasks into their component parts. It results in identifying the deliverables that constitute completion. It can also be thought of as task scope.

Strategy: The strategy is the plan for implementing the task. It specifies the means by which the end occurs. It can be a detailed representation of the substeps needed to complete the task.

Execution: Execution involves the actual implementation of the substeps and components of the task. It is the core development activity. Many development people jump into it without taking the necessary initial steps.

Verification and metrics: Verification is another often-overlooked item. All too frequently, the verification step is waived in anticipation of the design review. There are a few problems with such a practice, however. The first is that verification is an integral part of the task process. It ensures that the start of a subsequent task activity is built on a solid foundation of the previous task. Second, precious time can be lost by not verifying (by starting the next step only to find that previous steps must first be retraced and corrected). Third, a design review is simply an overall review that is not intended to catch mistakes or serve as a comprehensive recheck of previous work. Metrics should be part of the verification step; test data prove that the verification has legitimacy and is tied to the original intent.

Documentation: "No job is done until the paperwork is complete." This age-old saying is no longer dependent on paper but it does speak with relevance to every task, subtask, and component of the project. Documenting the elements of the task shows diligence, charts progress, establishes points of reference with data, and serves as a reference library for future queries.

C. Task breakdown structure

Each task of the project must be developed within the context of a task breakdown structure. The task breakdown structure is an organized method for identifying the technical and administrative disciplines involved in that task. Each subtask component takes on the same format of assessment, strategy, execution, verification, and documentation. Time is allotted to each activity and talent codes are also assigned. The various disciplines engage the task and execute it. Figure 7-4 illustrates the task breakdown.

As Figure 7-4 shows, the assessment and strategy for each discipline should be started and completed at the same time to ensure that the task is initiated with all disciplines in communication and at a common starting point.

It is acceptable for the separate disciplines to conduct discrete investigations for optimizing a solution, so it is not absolutely necessary for the investigation start times to be concurrent. However, the activities should be concurrent. As shown in the example, the

Task breakdown structure							
Task 1	Discipline	Assess-ment	Strategy	Execution	Verifi-cation	Documen-tation	Total
Code							
M1	Mechanical	4	4	45	16	16	85
P1	Packaging	4	4	80	12	12	112
E1	Electrical	4	4	16	12	18	54
E2	Digital Electronics	4	4	60	40	40	148
S1	Integration	4	24	72	40	40	180
T1	Manufacturing Design	4	4	40	24	40	112
Totals		24	44	313	144	166	691

Figure 7-4. Task breakdown.

integration activity showing a strategy planned for 24 hours might cause a delay, depending on how other tasks interface. If this is the case, the program manager might elect to initiate this activity in advance so that execution can be simultaneous with the same phase of the task as performed by other disciplines.

With the basic metrics in place, it now becomes necessary to stage execution by the individual disciplines in a sequential fashion based on the interdependence of the requirements of a given task. The truncated Gantt chart shown in Figure 7-5 represents this staging; with the sequencing in place, all of the tasks and subtask activities are interlinked.

The resource requirement used in the execution of each task can be identified and the requirement for each period of time is known and scheduled (in weeks, in this example). This process can be made as specific as needed to manage the project effectively. Fundamentally, however, the Gantt chart emphasizes that, given the program timing, the task breakdown, and the logic, the required resources must be available and supplied. If this does not happen, timing and eventually the sequencing of the program will degrade, causing unrecoverable delay.

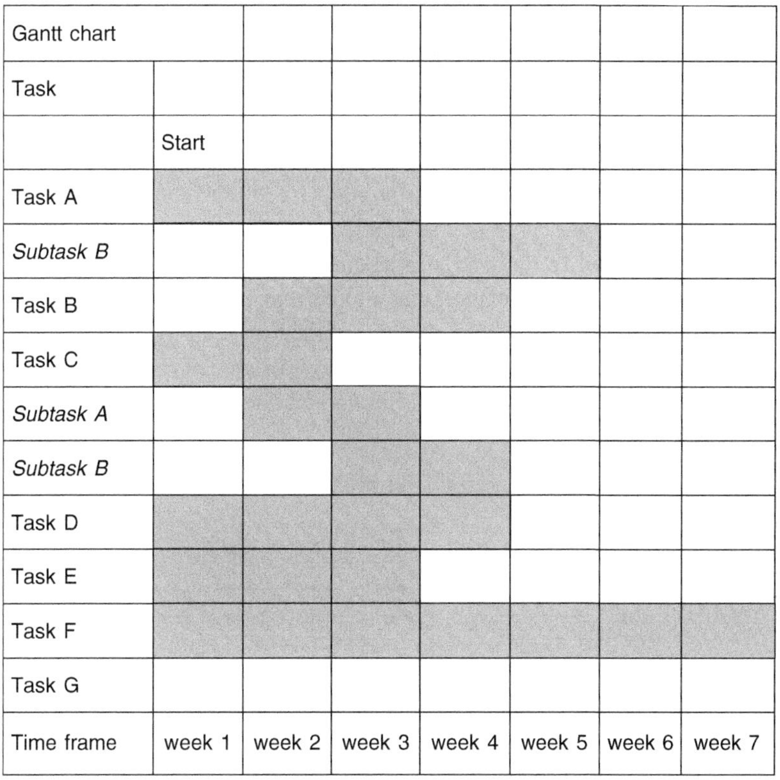

Gantt chart							
Task							
	Start						
Task A	▓		▓				
Subtask B			▓	▓	▓		
Task B		▓	▓				
Task C	▓						
Subtask A		▓					
Subtask B			▓				
Task D	▓						
Task E	▓						
Task F	▓	▓			▓	▓	▓
Task G							
Time frame	week 1	week 2	week 3	week 4	week 5	week 6	week 7

Figure 7-5. Gantt chart.

D. Logic versus timing trade-offs

There could occasionally be a need to trade off timing and logic. This practice is ill-advised, however, from the standpoint that sequential revision of a project plan only serves to confuse participants. Such a practice also tends to degrade the integrity of the logic and qualification, as subtasks will be missed or overlooked.

The other point to remember is that revising the sequencing disturbs the resource allocation, further confusing the scheduling. It is best to expend the energy required to be on time the first time.

E. Setting up a demand-pull system

To manage the development process effectively, construct a situation in which the market side of the business pulls development along. Focus on completion dates and sales lost when dates are missed. Focusing the group on dates creates a sense of urgency and participants in the subsequent task create the demand-pull on participants in the previous tasks. The manager's role is then to enforce quality and consistency in the completion of the tasks. A demand-pull system is shown in Figure 7-6.

To complete the presentation of demand flow, the personnel's first step as part of Task 2 is to verify the completion of Task 1. This process ensures that a program does not forge ahead with poor results from an isolated task. Building intermediate checks into the execution process also ensures a higher degree of quality for the program overall.

4. Dealing with Shifting Linking Objectives

A. Company culture

The linking objectives are very important to a program because no project can be started and completed in the vacuum of other departments. Given this, senior management should not expect a group to operate without the timely and effective participation of the other departments of the company. To that end, it is a senior management issue and they are solely responsible for ensuring these objectives are linked and prioritized. If the culture of the organization does not allow for this, precipitate clarification with management. If you are in a role to prevent these missing linking objectives, it will save a significant amount of effort later. Failure to address this will cause problems and these problems generally will not go away.

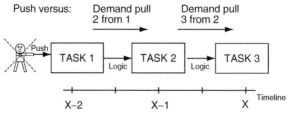

Figure 7-6. Demand-pulll system.

Some company cultures encourage competition for interdepartmental resources among program managers. If this is the case in your company, decide whether this is a situation you will put up with or is intolerable. If you decide it is tolerable, then become proficient at it in order to be effective in executing the program.

B. Project team form

To demonstrate the requirement for aligned and linking objectives, consider Figure 7-7. It shows several example projects to be executed within the company. The various departments need to contribute resources to complete the projects effectively. Project A is the highest-priority project, which should be on all departments' top priority lists, not just those of engineering and development. In this example, the balance of the departments does not consider Project A as a priority and, worse, manufacturing is focusing on Project D, the seemingly lowest-priority project. At the right of the chart is a concurrence assessment. In this exaggerated example, there is no concurrence with any of the projects or departments, which indicates a severe senior management problem.

The business of running a company must have a balanced focus for all of the departments. The objectives require both execution of the normal business activities (as shown at the bottom of the chart) and the project work associated with advancing the business through new product development.

Linking corporate objectives						
Project Priority		Sales and market-ing	Engi-neering and develop-ment	Manufacturing, material procurement, methods, manufacturing systems	Adminis-tration, finance, account-ing, and order entry	Concurrence of linking objectives?
Highest Priority	Project A	B	A	D	C	No
High Priority	Project B	C	B	C	B	No
Low Priority	Project C	D	C	B	A	No
Lowest Priority	Project D	A	D	A	D	No
Non-New Product Development Vocational Activities		Normal sales and marketing activities with existing accounts and products	Product mainten-ance and order processing	Scheduling, manpower loading, and shipments	Treasury functions, cash flows, financial statements, business analysis	

Figure 7-7. Linking objectives.

C. Coordinating all company programs

The object lesson is that all company programs need to be organized and resources applied in a productive fashion in order for any of them to progress. A haphazard approach using freedom to shift resources might appear feasible on paper but, in reality, this shifting causes loss of time on reassignment, start-up, and shut-down.

In program management the one thing a company needs to do to maximize the development output for each dollar spent by establishing and nurturing momentum. Constant reassignment depletes any momentum that has been established.

D. Executing without resources

There are occasions when a program must be executed in the wake of resource limitations and reordering of priorities and personnel shifts. Given the corporate hierarchical form, development managers are left with little choice but to consistently inform senior management of the resulting impacts. Often, managers establish a pattern of bringing issues to brinkmanship in some meeting. The more effective way might be to create small brinkmanship episodes, but a more frequent approach is to widen the scope of personnel involvement and knowledge. This tactic gets the word out in several different perspectives and also informs the rest of the organization about the impact of company decisions. Consider the approach as incremental brinksmanship: low-key but frequent, with an ever-widening audience.

E. Is senior management aware of the situation?

If this is a severe issue, the development management needs to determine whether senior management is aware of the situation. If they are but do nothing to resolve it or even to acknowledge it, the development team has a much greater problem. There must be a covenant between the business development management and senior management. If there is no corrective action, there is no covenant and the team is potentially wasting its time. In these situations, when senior management fails to address these issues the team must make a decision whether to continue under these circumstances or to precipitate a means for resolving it.

5. Documentation System Design

The importance of documentation systems in program and business development cannot be overemphasized. With rapidly changing dynamics in the corporation, the need for record-keeping is paramount. It provides the continuity in a discontinuous world. The documentation system should have several attributes, namely, information retrieval, product liability recordkeeping, test data, logistics, and minutes of discussions and results of decisions made in meetings.

A. Information availability and retrieval

As important as the documentation is, it is necessary to create a format and a vehicle to retrieve the information readily. The documentation should track the project chronologically and by discipline. As shown in Figure 7-8, this matrix represents a progression through the various phases of a project and also segregates them by discipline. For your own company,

Documentations systems							
Item	Sales	Marketing	Engineering	Manufacturing	Quality	Purch	Treas
Program stages							
Program stages							
Program stages							
Program stages							
Program stages							

Figure 7-8. Documentation systems.

you may want to organize them differently; however, this matrix may make it easier for the team to track the project from a chronological perspective.

B. Product liability documentation

Equally important is to keep documentation on the product development for future liability defense and loss mitigation. Each decision made and each compromise and trade-off should be documented and filed. It shows diligence on the team's part.

C. Test data

Validation: This is the test data that validates the design specification and the actual product performance envelope. After the development is complete, the actual performance of the product should meet or exceed the design specification as well as the published specification.

Verification: This is a check on the marketing requirements specification and how the developed product measures up to the original market specification. Between the two measurements there can be a degree of assurance that the product design will function according to spec and is pretty close to what the customer wanted.

D. Minutes of meetings

Each meeting should have a protocol for calling the meeting, assembling the members, recording the attendance, listing of discussion points and topics issues, and the disposition of these items.

E. Documentation of discussions

Many times there are impromptu discussions about the business and the product. It is just as important to document these discussions and their outcome as any other meeting

F. Decision-making processes

There are three elements in decision making: researching alternatives and data; actually making the decision; and managing the decision after it is made. Each of these is important to document. In many cases, future reexamination of a decision becomes necessary, so by documenting the process you retain the methodology, assumptions, and facts in existence at the time of the decision.

G. Standards

Finally, it is important to keep a record of the standards used in the development of the product. The prevailing standards at the time of development are good reference documents to keep with the file.

CORE COMPETENCIES

1. Assessment

Skill assessment is an introspective exercise that sometimes may not be very comfortable. Every organization has certain skill sets among its personnel. As a normal course of business, personnel absorb new knowledge and develop skills. However, certain developments may require additional skills not resident within the corporation. As the project is being considered, it is important to assess what additional skills may be needed to execute the development and factor them in accordingly.

The following questions are useful for performing an initial skills assessment:

- How do we view our ability to employ technology?
- How do we view our ability in application development?
- How do we view our ability in setting pricing?
- How do we view our ability to absorb uncertainty?
- How do we view our ability to derive the optimized solution?
- How do we view our ability to profit from an opportunity?

Honest answers to these questions will go a long way toward getting your team on the right path in a timely manner.

2. Improvement

Given that no company necessarily possesses all of the skills to execute a program at the beginning, the ability to improve skills is the true measure. This is very important because once a shortfall in skills is realized, the ability to improve is quite helpful in the long run. It prevents navigating around the weakness or understating the need for the skill, or some other natural corporate defense mechanism.

The overall goal is to get to the point where the firm is able to insert new technology and absorb the uncertainty associated with it along a time line and within budget. This requires management and personnel development.

CORE TECHNOLOGY DEVELOPMENT

1. Internally Generated

A firm's core technology should be cultivated consistently and regularly. This is required to stay current with the market pace and also to provide a sharpened skill set for the corporation to assimilate technology and be able to apply it. It is very difficult for an organization to do this from scratch but, as a normal course of business, these skill sets can become sharp enough to allow accurate forecasting of project tasks and completion dates.

There are two important terms when discussing core technology: *assimilation efficiency* and *diffusion index*. The *assimilation efficiency* is the ability to assimilate, understand, and apply new technology with respect to how much effort time and money it would take. The *diffusion index* is the measure of the organization's ability to diffuse the information from a central expert throughout the corporation. It is a measure of the internal transfer of technology. When deciding to incorporate new technology, the manager must work within the parameters of the organization or work to improve those parameters.

2. Externally Generated

When the firm does not regularly keep up with the core technology in their field, it becomes a problem to catch up. The ante in this game is often too high from a personnel as well as an expense perspective. In these cases the team or management may opt for getting the expertise externally. This is fine, but there is another critical component to this: that is the eventual assimilation of the technology into the organization. This is especially true when the firm wants to maintain absolute control of their product destiny. Again, the assimilation efficiency and diffusion index are critical parameters.

OUTSOURCING DEVELOPMENT

1. Pushing Core Competencies

When an actual partner is involved in the core technology profile, it becomes still more complicated because two different companies have different agendas, viewpoints, and definitions of completeness. In these cases, the frequent objective is to use the expertise of the donating company, learn as much as possible from them, and eventually integrate that expertise into your firm. If this is the case, the team must be ready to absorb it and nurture it.

We previously discussed navigating the relationship and how to address the eventual separation of the two companies. However, for this discussion, the focus must be on absorbing the know-how as fast as possible so you can become self-sufficient.

2. Shedding Non-Core-Competency Work

As part of the assimilation process in these cases, we need to mitigate the impact of the product maintenance function on the development personnel. This frees them up for working with the new material. An alternative within the organization must be cultivated to absorb this maintenance function and still provide the rest of the organization with the service it needs to continue the business.

CORE PRODUCT DEVELOPMENT

1. Product Development Phases

This section addresses the core mechanics of the product development phases. Whereas the previous section discussed program charting and sequencing, this section focuses on the stages of the program and the accompanying deliverables.

A. The six stages of product development

There are six distinct stages of product development. These stages outline the maturity of a product, from the inception of an idea all the way through to an optimized product. These are the **B**usiness **D**evelopment **P**rocesses:

BDP 0:   Market assessment and internal assessment
BDP 1:   Prototyping
BDP 2:   Core development and testing
BDP 3:   Industrialization
BDP 4:   Commercialization
BDP 5:   Optimization

Let us now examine each stage and outline the deliverables. The presentation of these deliverables represents only a general order for executing the tasks required for the deliverable. The actual sequencing of the tasks and deliverables could be different in an actual project because some activities will be done in parallel.

- *BDP 0: Market assessment and internal assessment*
  The first stage is the assessment phase. This is an internal introspection of the company and its direction, and an external look at product opportunities. The goal of this stage is to obtain a clear idea of how the product opportunity fits with company goals and objectives. The following is a list of deliverables for this stage:
  - *Market opportunity analysis:* The market opportunity analysis is a summary of the opportunity for a new product or product line in the marketplace. It is the germ of a business idea where the product concept starts. This study is cursory in nature but must result in identification of a clear opportunity.
  - *Competitive analysis:* The competitive analysis is a summary of the current activity being conducted by the competition, who could be traditional or nontraditional competitors. The analysis should be specific enough and accurate enough to outline the current products available because the product your firm will design must leapfrog competitors' products with features advanced enough that, when it becomes available

for sale, the offering is still competitive. This analysis should also consider the life histories of competing products and technologies.

- *Product concept:* The product concept is generated from analyses of the opportunity and the competitive position and from a projection into the future. It outlines the product to be developed. It should use whatever medium is appropriate to describe what the product is and how it will be positioned in the marketplace to capture market share.
- *Feasibility study:* Given the product concept and the company capabilities, this study answers the question of whether the intended pathway is feasible for the company to follow. Although some products are significant opportunities, their execution might not be feasible for the company. Time is the other important factor to take into consideration at this point in the evaluation. Is it feasible for the company to develop this product concept in the allotted time frame?
- *Company strategy:* How does the product concept mesh with the overall company strategy? This position statement can describe a strategic initiative satisfied by the product opportunity, or can describe how the product will allow the company to further its long-range goals. The fitability study (next deliverable) is a refinement of the strategy study. If the strategy summary is a broad, overview look at the pathway provided by the product, then this is a detailed analysis of how the product fits into the existing scheme of products and offerings from the company.
- *Fitability study:* The fitability study is a refinement of the strategy study. If the strategy summary is a broad, overview look at the pathway provided by the product, then this is a detailed analysis of how the product fits into the existing scheme of products and offerings from the company.
- *Timing:* All things being equal, products are opportunities for a limited time and, as such, they need to be developed and marketed within that time frame or the opportunity is lost. This section of the summary outlines the overall time frame for the program. It identifies the program in terms of priority, funding, and resources.
- *Market dynamics assessment:* What are the market dynamics like? Is the market rapidly changing and difficult to read, or is it more stable and predictable? Is this market similar to others that you have seen to the extent that you can fit a pattern to it and predict the next trend? All of these considerations contribute to this assessment of the general market activity.
- *Go or no-go decision:* Given all the data in the summary and the experience base of the people involved, document the go or no-go decision and the circumstances and assumptions surrounding it. By documenting the decision, analysis for future programs can be better informed.
- *Market requirements specification:* For a go decision, the market requirements specification must be prepared. This is the document that connects the customer's and the market's needs for the product specific to the company developing it. The specification represents the customer's voice on what they want to have in the product and serves as the base specification for the product upon which all later versions are built.

- **BDP 1: Prototyping**
The next stage is the prototyping stage. This is the physical embodiment of a product concept as a starting point—the stage at which the product concept evolves into a piece of hardware that is functional. It is desirable to deliver this prototype to the customer as

rapidly as is practical to secure early feedback, while still maintaining the integrity of the intellectual property. The deliverables for this stage are as follows:

- *Preliminary design specification:* The preliminary design specification is development's response to the functional specification, which represents the voice of the customer. The design specification is the best assessment of what can be accomplished.

- *Intellectual property review:* At this point, it is a wise idea to prepare a patent review to determine whether there are any patent issues outstanding or whether there is a potential conflict with existing patents. This review should be performed as early as possible to allow alternate approaches if necessary. This deliverable should identify opportunities for securing patents, trade secrets, copyrights, and trademarks.

- *Design strategy:* The design strategy is the development approach to inventing the product. It deals with issues such as selection of technology, degree of risk, product reliability needs, and cost. It also has implications for development time (to the extent that one approach might be more expedient than another) and for possible allocation of resources. This issue can have far-reaching effects if not considered properly, so it is best to take care in this initial evaluation.

- *Test strategy:* The test strategy follows the design strategy and is part of the overall production strategy. It is a plan for how the new product will be tested. It correlates closely with the design strategy because both are technology-dependent.

- *Manufacturing strategy:* The manufacturing strategy is a plan that outlines the philosophy of manufacturing—for example, three circuit boards fastened to a frame with collateral components bolted down, versus three-dimensional circuitry molded into the product itself. Is the product to be a throwaway item or is it repairable? Answers to these questions constitute the manufacturing strategy.

- *Quality plan:* The quality plan is the master plan for the quality claims of the product throughout its life cycle. To differentiate the issue of quality from that of reliability, the quality of a product is strictly a measure of how accurate and repeatable the final product is when it comes off of the production line, and how accurately it matches the drawings and documentation used to describe its manufacture.

- *Reliability requirements:* Reliability is a measure of how long and how accurately a product will perform its intended function. A description of product reliability has its own mathematical relationships. Most consumers are interested in this measurement, although they often refer to it as quality. Reliability is a measure of the value in use that the consumer will obtain from the product.

- *Safety requirements:* The safety review is a comprehensive analysis of the issues of safety and how they affect both the product and the user under various scenarios of use and misuse. This review needs to be a structured and somewhat exhaustive effort to demonstrate diligence on the part of the manufacturer. It has been demonstrated in recent years that this effort needs to affect product design—building prevention of misuse directly into the design rather than relying on the use of warnings.

- *Cost analysis with sensitivity:* As part of the initial product planning exercise, the product's cost needs to be paramount. From this point onward, most development and manufacturing activities tend to add cost. Therefore, the initial cost estimate must be low enough to support all of the incremental costs associated with developing the product.

- *Statutory and regulatory approvals:* Most, if not all, industries have a means for regulating that industry. Products involving electrical circuits, for example, could come under the regulatory control of Underwriters Laboratories, Inc., the Canadian Standard Association,

The European CE Compliance Directive, The German VDE Certification Institute, and many other organizations. The need at this stage of the program is to articulate the plethora of standards the product needs to meet in order to be accepted by the marketplace.

- *Value engineering study:* This study is the means by which the needs, wants, and desires of the marketplace are organized, evaluated, and compared with implementation costs. These implementation costs affect manufacturing costs, development costs, and time.
- *Aftermarket plans:* Consideration of the aftermarket for the product must be factored into the overall product plan. If the product is disposable, the issue is minimal. If, however, the product is repairable or needs to be maintained in some way, an entire business infrastructure will be required to support the aftermarket plans. In addition, the eventual disposal of the product should be planned by the manufacturer and communicated to the consumer in accordance with environmental regulations.
- *Preliminary return on investment:* To obtain management approval or to determine whether the program is even worth doing, a preliminary return on investment (ROI) calculation needs to be made. As simple as this calculation is, many companies fail to perform it and consequently forfeit a dispassionate means to evaluate one opportunity against another. Companies might then embark on what might be the least expedient pathway toward their strategic goals.
- *Design review results:* Along with all of the previously mentioned studies related to product and technology, there should be a cross-functional design review to ensure that the pathway toward the new product is correct. The design review should check that all of the studies have been performed and that they were performed accurately and against the appropriate standard(s), not just glossed-over.
- *Packaging optimization:* If the product is packaging-dependent, it is a good idea to attempt to optimize the package at this point. Packaging is difficult to optimize when the team is immersed in the technical challenges and functional requirements of a program. Hence, by constraining the packaging needs early in the program and enforcing them throughout the program, the packaging requirement can be met without undue strain.
- *Capital requirements plan:* The capital requirements plan needs to be outlined at this point in the program from the perspectives of the manufacturing approach and the design strategy. By doing the capital requirements plan at this stage, misunderstandings about future expenditures and potential refocusing on a different design approach can be avoided.
- *Technology integration study:* As products become more complex and several technologies are used in the implementation of a product, it becomes necessary to plan how the development team will integrate the technologies into the product. This plan is necessary because test plans and design verification will require it.
- *Final design specification:* With all of the fact-gathering and assumptions complete, it is time to prepare the final design specification. This will be the permanent plan of record regarding what the new product has been determined to be at this stage of the development. There can be subsequent changes; however, this is the plan of record that can be used as a measure of the extent of future changes.
- *Alpha unit developed:* The final deliverable for this stage is the production of an alpha prototype. Alpha is the designation for the engineering prototype. It may be made in a lab, without tooling or specialized parts. It could have a form that will not be the final

form. It might not have the aesthetics of the final unit, but it will demonstrate the functionality and the performance of the final unit. It is suitable for showing to a prospective customer to obtain feedback.

- *BDP 2: Core Development and Testing*
  In this stage, the embodiment of the development is tested by design verification and design qualification. At this stage the product matures from the physical embodiment of an idea into a product that can be produced with predictable performance. Its list of deliverables is as follows:
  - *Criticality analysis:* The criticality analysis is an exercise in determining the degree to which one failure in the product could prove critical. It draws a relationship between the failure mode and the effect within the application. This can also serve as a basis for further examination during the safety review. A design failure modes and effects analysis (DFMEA) may be included here.
  - *Safety review:* The safety review is a summary of the general effects of the safety issues on operation of the device and on personnel. It must show due diligence on the part of the manufacturer such that each issue was identified and accounted for in the design so as to mitigate potential damage.
  - *Preliminary forecast:* The preliminary forecast is important from two perspectives: that of the ROI calculation and that of commitment. The marketing function must commit itself and its salesforce to deliver the forecasted numbers. Requiring this forecast as part of the project plan helps ensure commitment.
  - *Beta unit:* The beta unit is, for all practical purposes, as close to a production unit as is possible without the manufacturing setup being complete. In fact, there can be circumstances in which the beta unit will be made from production tooling.
  - *Change control system:* Once the product has matured from the laboratory stage, an engineering change control system is required to enforce some means of product configuration control. This is needed to prevent changes from creeping in without the team agreeing to them. The engineering change control system is often the only means for documenting, tracking, and enforcing product configuration control.
  - *Design verification test results:* The design verification results show that the product is designed to meet the requirements of the design specification. These tests document where the product meets, exceeds, and might fall short of the design specification. The evaluation is conducted with metrics.
  - *Design validation results:* The design validation test results show that the product meets the intended use by the customer as defined in the functional specification. It documents where the product will meet, exceed, or fall short of the functional specification.
  - *Standardization tests:* As part of the qualification process for meeting certain standards, the test results must be documented. Test results could show that changes are required for the product in order to meet the standards. This deliverable is helpful in managing the transition from a nonqualified unit to a qualified unit.
  - *Pilot run plan:* The pilot run is the first practice production run of the product. It tests drives, so to speak, the manufacturing documents, the design documentation, and the processes. The logistics and the details of these are found in the pilot run plan. A process failure modes and effects analysis (PFMEA) may be used here.
  - *Life test results:* In order to substantiate the claims made for a new product, life testing must be initiated and completed. This life testing will reveal product weaknesses and

allow the development group to make specific changes. If a claim is made in the marketplace, then value in use must be integral to the product.

- *Design review results:* This is a formal design review process in which all of the plans and certification efforts undergo a peer review. This is a healthy exercise in that it allows dispassionate review of a product and also extends knowledge about the product, approach, and technology to others outside of the development group.
- *Significant performance and application features (SPAF) document:* If the functional specification is the voice of the user or consumer and the design specification is the voice of the development group, then the SPAF document is the voice of nature. This is a development-initiated document that defines the absolute boundaries of product operation.
- *Pilot run results:* The results of the pilot run must be documented and any anomalies resolved. A development environment is much more forgiving than is a manufacturing environment; problems could surface in the pilot run that will require attention.
- *Final production strategy*: The final production strategy is created from the results of the pilot run. It incorporates the solutions, changes, or both that arise as a result of the pilot run.
- *Final test strategy:* In the same manner, the final test strategy is driven by any changes from the final production strategy. These are specific test issues that might need to be added or deleted to accommodate any changes in the pilot run.
- *Final quality plan:* The final quality plan is also included in the list of changes of a production nature.

- *BDP 3: Industrialization*
  - *Capital equipment and setup:* Next is the industrialization stage in which the product is made manufacturable and the manufacturing processes are developed and put in place. At this stage, the product has matured from a development environment to a manufacturing environment. Development involvement takes place primarily from a support and integration function. Changes and enhancements are limited.
  - *Manufacturing setup*: Concurrent with the capital equipment setup, the balance of the manufacturing line must be set up to accept the new product. This includes all documentation systems and procedures.
  - *Tooling and equipment qualification:* If any tooling and equipment are required for the product, this is the time to set them up and also eliminate any variability associated with them.
  - *Final pilot run:* Sometimes an additional pilot run is necessary to test any changes resulting from the first run or any subsequent changes to the product. The deliverable for this phase is final pilot run certification.
  - *Product certification:* At this point in the program it becomes necessary to certify the product. This ensures that the product has been designed, configured, manufactured, and tested according to specifications and standards with the appropriate safety review input.
  - *Field test data and feedback:* The beta test program is required to test the entire customer and user side of the product development. This means customer validation of the usability and completeness of user manuals, part numbering systems, the order entry process, commissioning, and start-up (if required). The results of this testing program verify acceptance. The feedback can also be used to modify the marketing plan used at the commercialization stage.

- *Design freeze:* There comes a point in the product development where the design simply must be frozen and any improvements must be scheduled for the subsequent release of product enhancements or versions. The product does not change for non-safety-related issues beyond this point. This is a difficult deliverable to achieve.
- *Change control system active:* As part of the frozen product design status, the product needs to be under strict change control. All product design and process changes occur under this control, with a protocol for the release of the change.
- *Procure production materials:* In preparation for the production run, and with the design frozen, this deliverable confirms that forecasted product quantities are ordered from suppliers and that items with long lead times will not impact availability for sale dates.
- *Quality database integration:* The quality database needs to be an integrated database complete enough to feed back field problems and be detailed enough to effect solutions. If, for example, root cause analysis is required at the component level and a demonstrable corrective action must be initiated and completed, then the "hooks" for the data need to be implemented in the database.
- *Final cost evaluation:* At this point in the development and manufacturing implementation, the final cost data need to be prepared. They will be used for comparative analysis with original estimates as well as for preparing an assessment of the gross margin and market-level pricing. Occurring just prior to product introduction, this evaluation will ensure that the program is launched with the correct pricing and that there is little misunderstanding regarding margins.
- *Freeze manufacturing, testing, and quality plans:* With the design frozen, the manufacturing, testing, and quality plans also need to be frozen. These plans must likewise come under change control.
- *Finalize marketing strategy:* With the field feedback in place and the cost of the product known, the marketing strategy can be finalized. This must occur to ensure that the marketing effort is appropriate and directed to produce early results, rather than stumbling around trying to discover what works and what doesn't in the marketplace.
- *Prepare launch plans:* With the marketing strategy in place, the launch plans can be finalized. Action items, responsibilities, and completion dates will summarize the required effort.
- *Initial stock in place:* Depending on the route to market for the new product, there might need to be a significant amount of initial stock required to launch the product. Manufacturing orders must be initiated to build the stock. Marketing effort must be expended to place the stock where it can have the greatest impact on generating sales.

- ***BDP 4: Commercialization***
  The fifth stage is the commercialization stage, when the product is rolled out to the marketplace and units are placed in the hands of consumers. This is the pivotal test, towering in importance over all the others because it demonstrates the final acceptance or rejection of the product by the consumer.
  - *Final competitive assessment:* To assist the marketing personnel in their activities, a final competitive analysis that takes into account all changes made to date needs to be created. This analysis shows how the product compares to competing products. It also serves as a basis point for the product evolution flow chart and for the timing of enhancements and improvements.

- *Final literature:* If any material changes were made as a result of the industrialization phase and the field test program, the literature (in whatever medium) needs to be updated so that all documentation (both internal and external to the company) is consistent.

- *Forecast for the product:* Upon initial commercialization of the product, the marketing function, in conjunction with the sales function and with input from the sales channel, generates a usable forecast. This will be used to correct any accuracy problems in the previous forecast.

- *Sales channel initiation and training:* At the time of product introduction, it is critical to spread word of the new product as quickly and uniformly as possible. To that end, the sales channel needs to be trained and "turned on" to the new product and focused on selling it.

- *Customer visit schedule:* It is also a good idea to have the sales channel document its customer visit schedule as a deliverable. For use as a planning tool, this item also becomes a check on the ability of the sales channel to be effective in promoting the product.

- *New product release:* Finally, the deliverable of new product release is the formal release of the new product to the marketplace. This is achieved when a fully tested product is in stock at the right price and is marketed through the most effective route to the user. In other words, no excuses—the product is ready in every sense of the program.

- **BDP 5: Optimization**
The final stage is the optimization stage in which feedback and improvements are incorporated. The product has matured beyond its initial release and the traditional learning curve cost reductions occur due to this effort. As the marketing effort is scaled up, the cost is reduced and the enhancements are incorporated. This is the stage at which gross profit is generally maximized for the corporation.

  - *Field feedback plan:* With the product having some longevity in the marketplace, it is necessary to compile and analyze field feedback. Such feedback will be used to generate the required enhancements in functionality and also to identify the areas in which cost reductions are necessary.

  - *Manufacturing feedback:* In the same manner, there needs to be manufacturing feedback to determine what product changes are necessary to facilitate improved or less costly manufacturing methods.

  - *Expanded production plans:* Depending on the ramp-up in sales volume, the use of additional manufacturing capacity—either through capital equipment or time-shifting of manufacturing cycles—might need to be considered. In either case, the marketplace demand must warrant the ramp-up in capacity. This deliverable should be staged in advance of the market need but not so far ahead that it would be underutilized.

  - *Value management:* During the initial product planning stage, when the requirements specification was being prepared, value engineering was introduced. Value engineering determines the value the user places on certain features, and is the basis of the pricing/functionality matrix. Now that the product is on the market and the marketing and sales forces have experience with it, value management should be applied to recheck where the consumer places value in the product and what features are worth what portion of the price.

- *Product evolution analysis:* Product experience will allow an informed product evolution flow chart to be updated. Given that the product exists and that the business will lead the company in some direction, using the product evolution flow chart is a useful deliverable in ascertaining the next form of the product.
- *Cost reduction:* The value management exercise and the cumulative experience with marketplace preferences and manufacturing capabilities should enable a cost reduction in the product.
- *Product obsolescence*: As part of the product evolution flow chart, the product eventually must be retired. This deliverable will define the manner in which the retirement will take place. If it is a repairable product, how long will parts for it be on hand? If it is not repairable, what is to be offered in its place? These issues are part of the product obsolescence plan.

B. First-time-through versus existing system

There is an important perspective to be aware of in executing business and product development: the fundamental difference in the human energy level required to implement a development within the scope of an established system of procedures versus starting from scratch. The job is quite complicated if an organization must institute procedures and put them in place; it then becomes even more difficult to institute a system and make it prevail throughout the organization.

If an entirely new system needs to be established concurrent with the product development, you must factor in more time to execute the program. If the organization resists the implementation, the chances of completing the program on time and getting procedures in place that will prevail are quite slim. Extra effort at the outset for the purpose of establishing authority will have far-reaching benefits for the program and the development group.

C. Putting procedures in place and making them stick

Regardless of the circumstances or the expectations, these procedures and processes must be put in place in an effective manner. They must also be used as part of the program to cement them in place in the overall organization. Failure to do this will result, at best, in a program with a one-time success (in spite of the lack of procedures) and little future leveraging of talent, or, at worst, failure of the program and dismantling of the development group.

D. The manager as minesweeper

With the team in place and functioning properly, the manager needs to act as a minesweeper for potential problems that may affect the program. The manager must lead the group through the potential conflicts and resolve the issues along the way. Figure 7-9 illustrates this point.

As shown in Figure 7-9, the manager needs to clear the pathway for the team by precipitating issues that can adversely affect the group's performance and then resolving them. Eventually, the group will become proficient at doing this and do it as second nature.

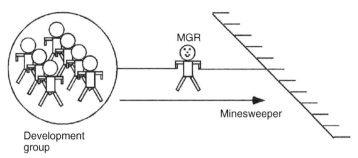

Figure 7-9. Manager as mine sweeper.

2. Requirements for Testing and Qualification

This section discusses the suite of tests required as part of the development process. This testing includes misuse testing.

Several tests need to be done as the product is being developed to certify that the product meets both the original specification and the customers' needs. The suite of tests can be broken down into seven groupings: design validation testing, design verification testing, safety testing, third-party certification testing, customer evaluation testing, beta testing, and production testing.

A. Design validation

As discussed previously, the design validation test results show that the product meets the intended use by the customer as defined in the functional specification. It documents where the product will meet, exceed, or fall short of the functional specification.

B. Performance and design verification

The design verification results show that the product is designed to meet the requirements of the design specification. These tests document where the product meets, exceeds, and might fall short of the design specification. The evaluation is conducted with metrics.

The prototype testing program is part of the design verification testing. This testing compares the operational characteristics of the alpha prototype model (designed and fabricated by engineering) to the original design specification. In other words: Here is the list of items we want the product to do; does the initial prototype meet this functionality? It is a starting-point test to ensure that the prototype is in compliance with the design intention.

Also included in this test regimen are the initial safety and reliability tests. These can only be initial tests because the design, form, and fit are not finalized.

C. Safety

Testing the product from the perspective of product liability defense is an important step in the actual defense against a potential lawsuit. This testing can be thought of as an extension of the test program for design verification and validation. Its objective is to uncover potential problems in the course of the normal engagement and use of the product. It also extends to the potential misuse of the product due to either lack of knowledge or flagrant misuse.

Figure 7-10 shows a pathway for the analysis of product misuse. The testing program starts with the design use evaluation. In this analysis, the scope of use for the product is defined and agreed-to. This analysis is performed on paper before the product is developed, as the product is still being conceptualized.

The second step in the evolution of the liability testing model involves the anticipated use of the product. This step defines the practical and anticipated use boundaries of the product. The product features, benefits, and specifications define these boundaries, which are identified upon completion of product development.

Next, the testing moves toward an analysis based on unintended use and misuse of the product due to user unfamiliarity. This is a projection of common user misinterpretations of product use and tries to accommodate them by warnings and instructions. At this stage, the company can also design error proofing features to prevent accidental misuse (e.g., parts not intended to fit together will not fit together; or connections that are keyed with mating connections to prevent electrical shorts).

The final phase of testing is for the flagrant misuse of the product. The progression to each succeeding stage of testing shows an increasing degree of user failure rather than manufacturer's judgment. To the degree to which it is practical, the design of the product should try to accommodate potential misuse by prevention through form, fit, and function.

An example highlighting the various stages of product misuse analysis would be the design of a fictitious gasoline fill nozzle. The design use is conceptualized as a simple fill nozzle. It has specifications for safety and performance and can be used by individuals not normally trained to fill fuel cans. The safety issue in the example is that noncertified fuel cans are not to be filled with gasoline. An unintentional misuse of the product might be filling a noncertified can with fuel.

To prevent this from happening, a manufacturer might create a discriminating interface between the nozzle and the fuel can so that only a certified can could be filled with gasoline.

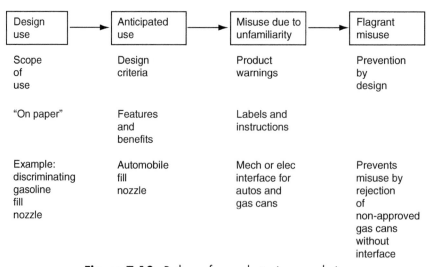

Figure 7-10. Pathway for product misuse analysis.

This is the same interface used in automobiles. Accidental or flagrant misuse is discouraged by this error proofing scheme because the valve in the nozzle must have confirmation of a certified interface before it allows flow of fuel.

D. Third-party certifications

The use of third-party testing to gain certification is a requirement in almost any endeavor. Safety, compatibility, and fitness for use are now prerequisites to selling the product. In many markets you are actually precluded from selling certain products unless you have such certifications. Third-party certifications yield an unbiased assessment of the product's performance and adherence to standards. It provides a good package of values that accompanies the product and can aid in promotion as well as customer acceptance.

E. Returns from customer evaluation

This suite of tests is important for the long-term viability of the product line. When a unit is returned from the field, it contains a wealth of information about the application of the product in an actual environment. These units should be mined for potential weaknesses in the product. In addition, the failure incidents should be investigated to determine the root cause of the problems. With accurate root cause information, trends can be established and corrective action can be taken to reduce the likelihood of additional field failures. At this point the database is loaded with fresh field information.

The field returns should also be analyzed for any similar issues that might arise. Often, nature reveals issues not even considered in the laboratory or protected test environment. The analysis of field returns can also be helpful in determining misuse or misapplication, and this data can then be used to alter marketing plans.

F. Beta testing

The beta unit stage test program consists of reconfirming the design verification and functionality testing. The details of such testing can vary, depending on the degree of change in the form, fit, and functionality from the alpha unit.

External design validation testing also should be performed at this stage. The answer to the question, "To what degree does this product meet the customer's expectation?" is required to proceed further. Beta testing also encompasses product testing at the customer's location in an actual application. It verifies design, requirements, manufacturing processes, and the effectiveness of the literature and instruction. Extended life and reliability testing should also be included.

In this stage of testing, there also should be misuse testing. The misuse testing is needed for potential future liability defense and also to obtain feedback for improving the product's ease of use.

The last segment in this suite of tests includes the standards and third-party certification testing for sale of the product. Third-party testing gives a nonprejudicial assessment of the product and also provides governing agency approvals prior to sale. Confirming tests of the initial manufacturer-initiated tests, such as safety-related tests, might need to be rerun at this time under witness of the governing agencies. The third-party tests ensure industry validation and go a long way toward making a product liability defense.

G. Pilot and production

The pilot run stage of testing consists of several tests. The pilot run is a test of the manufacturability of the product. Cost data and timed labor data will have to be verified. The manufacturer's process capability will be validated to ensure that there are sufficient process experience and controls to manufacture the product. At this time, the database used for managing the product will be defined and put in place. Acceptable levels of variability for the product and the processes will also be established.

Suites of tests are also required for manufacture of the product. This suite of production tests must verify that the units are built according to the design and must test all critical components comprising the product. Experience and statistical sampling methods have dictated what determines how many components represents "all components," without creating too much non-value-added production testing.

3. Designing the Quality System and Database

A. A model for quality management system

The International Standards Organization (ISO) has developed a homogeneous set of standards for ensuring that an organization produces quality products and services. Although these standards do not represent equivalency of every agency approval or quality requirement, a study of the initial standard provides a solid basis for constructing a quality system.

These standards provide for several levels of certification depending on the type of business. We will consider the initial version of (ISO 9000), which encompasses sales, engineering, development, manufacturing, and aftermarket support and service. Our consideration is not from the standpoint of certification, but rather as a *framework for a quality management system*. Updates to the ISO Standards have added specifics. However, this is presented to outline the spirit of the system.

There are several distinct components to consider in the certification. They span the transaction from initial customer engagement and order negotiation through aftermarket service. The following represents a basic framework of the ISO standards components. They are presented in the operational order in which a contract would be executed in the company, as follows:

1. Quality system requirements
2. Management responsibility
   a. Quality policy
      The quality policy clearly states the organization's goals and the needs and expectations of its customers. It is to be published throughout the organization. There should be demonstrable management and employee commitment to this policy.
   b. Organization
      This is a chart showing how the company is organized; it identifies the departments and the key managers.
   c. Responsibility and authority
      Key managers with executive authority to make decisions and control all elements of the quality policy should be identified. Setting policy, taking corrective action, setting objectives, and monitoring achievement are all part of this activity.

d. Resources

The manufacturer needs to provide adequate resources to implement the quality policy, satisfy customer expectations within the framework of the policy, and achieve its objectives.

e. Management representative

The company needs to provide a management representative with delegated authority for arranging, coordinating, and overseeing the quality system in progress.

f. Management review

The company also needs to provide for a periodic review of effectiveness of the quality system. It must ensure adequate staffing, conformity to standards, compliance with quality policy, and feedback and resolution both internal and external to the organization.

3. Quality system

a. General

The quality system is the dynamic mechanism that implements the quality policy. Often embodied in the quality manual, the quality system is free-functioning and operation-oriented.

b. Quality system procedures

These procedures are required for the applicable requirements of the standard and should be consistent with company policy. The procedure should specify the who, what, when, where, and how of the activities to be executed.

c. Quality planning

The manufacturer must demonstrate how the quality system elements are implemented and maintained.

4. Contract review

a. General

Contract review is the primary interface with customers. It should include what the customer requirements are and how they will be reviewed and communicated within the organization. This occurs prior to accepting an order.

b. Review

This process includes review of the requirement itself, agreement with the organization, and resolution of differences.

c. Amendment to contract

When changes occur in the contract, there should be subsequent review of the contract to confirm reasonableness and agreement. Acknowledgments both internal to the affected parties in the organization and external to the customer must also be part of the process.

d. Records

As with all activities where a potential for misunderstanding exists, there should be recordkeeping.

5. Design control

a. General

The design control is essential to ensure all of the quality aspects, such as safety, performance, and reliability of the product. This encompasses all phases of design and design process.

b. Design and development planning

The manufacturer should have established procedures which include scope and objectives, work schedules, timing and frequency of design verification and validation, safety, reliability and performance, methods of measurement, test and acceptance criteria, and assignment of responsibilities.

c. Organizational and technical interfaces

The organization should define, document, coordinate, and control all aspects and interrelationships and interfaces within the company as they relate to development, manufacturing, and aftermarket activities.

d. Design input

The design input is the translation of customer requirements into a definition of the performance, function, environmental, and safety and regulatory agency requirements. Typically in the form of a product requirements specification, the design input is the formal starting point for development.

e. Design output

Design outputs are the requirements for purchasing, production, installation, inspection, testing, and service. A release procedure should be in place to control the design output to the balance of the organization. Design output needs to be documented to verify and validate against design input and customer requirements.

f. Design review

The design review is a formal peer review of the development results. The degree of competence of the participants should be commensurate with the technology and process under development. A review must be objective and detailed to be of value.

g. Design verification

Design verification is necessary to ensure that the design output matches the design input. Test data and demonstrations are the principal and objective means to effect the verification. Any verification should be conducted in accordance with relevant standards, practices, and predetermined acceptance criteria.

h. Design validation

The design validation checks the completed effort of the development process and compares it with the customer's original requirements. The results of all of these tests should be part of the design records.

i. Design changes

Changes to the product are required under a formal design change system. The following can prompt these changes:

1. Omissions or errors during design
2. Manufacturing, installation, or servicing difficulties discovered after design
3. Customer- or contractor-requested changes
4. Functional or performance improvements
5. Changes due to safety or regulatory requirements
6. Change due to design review, verification, and validation
7. Corrective or preventive actions

6. Document and data control

a. General

Document and data control consists of the information pertinent to design, procurement, process, quality standards, inspection of materials, and the quality system itself. It should reflect the present documentation, present data, and their historical pathways.

   b. Document and data approval and issue

The manufacturer should have clear and precise control of procedures and responsibility for approval, distribution, and administration of internal and external data and documentation. This includes removal of obsolete information to prevent misuse or misunderstanding.

   c. Document and data changes

The same level of change control that applies to the product must also apply to documentation. Procedures should be established and implemented for controlling all changes.

7. Purchasing

   a. General

The control of procurement must ensure that purchased products that become part of or that affect the quality of the products, as well as the statutory or regulatory requirements, are adequate. Standards cover evaluation and selection, procurement specifications, performance and verification, and receiving inspection procedures.

   b. Evaluation of subcontractors

The manufacturer should have a means for evaluating the capability of subcontractors and ensuring conformity of procured product.

   c. Purchasing data

The manufacturer should have objective data to measure the subcontractor's product fabrication and shipment performance. This is to be kept up-to-date as with other documentation.

   d. Verification of purchased product

When specified, the manufacturer may engage in supplier verification at the subcontractor's premises. In addition, there may be customer verification of subcontracted product.

8. Control of customer-supplied product

Depending on how the manufacturer and end-customer construct the contract, the customer may supply product that will be incorporated into their own end-product. This product supplied to the manufacturer must have controls similar to those required for any other supplier.

9. Product identification and traceability

Where required, the manufacturer needs to provide a means for product identification. Similar-looking parts with differing functions might need to be marked differently. Traceability is required when parts must be traced back to their origin or source in the event of nonconformity and is often used to determine the affected batch.

10. Process control

The bases of process control are to invoke procedures and methods to ensure conformity in the completed product rather than to inspect for nonconformity. Process control should include procedural control, maintenance, and essential material control.

11. Inspection and testing

   a. General

The issue of inspection spans the organization from receiving through delivery and service. All aspects of this verification system must be under control because the material formulates the product.

b. Receiving inspection and testing

Inspection and testing procedures in the receiving department ensure conformance of supplied product. It also is the first point for verification of a conforming product. Failure of this verification could result in a nonconforming product.

c. In-process inspection and testing

In a similar manner, in-process inspection allows early recognition and disposition of nonconformities.

d. Final inspection and testing

Final inspection involves examination, inspection, measurement, or testing to ensure conformance prior to shipment of product.

e. Inspection and test records

As always, records of activities involving the product's manufacture and control points should be maintained for future use in solving or mitigating problems.

12. Control of inspection, measuring, and test equipment

a. General

Simply stated, the effectiveness of the product configuration and conformance to specification rely on the accuracy of inspection, measurement, and test equipment. Although this aspect of internal controls is often overlooked, it is mandatory for any type of certification.

b. Control procedure

Consistency is also required in maintaining control of the equipment.

13. Inspection and test status

The manufacturer should have means for identifying the status of tests and inspections performed on the product as it progresses through the manufacturing system. It should denote conformance, acceptance, rejection, or a hold, pending disposition.

14. Control of nonconforming product

a. General

b. Review and disposition of nonconforming product

15. Corrective and preventive action

a. Corrective action—How are Corrective actions conducted and problems contained?

b. Preventive action—How are future problems avoided by changing procedures and processes?

16. Handling, storage, packaging, preservation, and delivery

a. General

1. Handling—How are parts handled to avoid damage?

2. Storage—How are parts stored to avoid damage and rotate inventory?

3. Packaging—How are products packaged for shipment to protect contents?

4. Preservation—How are at risk products preserved for shipment?

5. Delivery—How are products designed for delivery processes?

17. Control of quality records

The manufacturer's records should show evidence of the quality system elements and components that fall under jurisdiction of ISO requirements. They should be identified, prepared, and controlled by authorized personnel, and steps should be taken to prevent unauthorized tampering with records. Some examples of these types of records are management review records, contract review records, inspection and test records, and internal quality audit records.

18. Internal quality audits

    To ensure that an organization has a vital, effective quality system, it must conduct introspection on operations periodically. This means self-audits on the elements of the organization covered by the ISO system. Follow-up activities on audits ensure results that show an effective system.

19. Training

    The needs of the marketplace dictate not only change but also requirements for training of personnel. The organization must take steps to ensure adequate training and documentation, and ensure that these pervade all levels of the organization. To achieve the quality objective, each and every employee should be trained in procedures and processes that affect them and that they in turn affect.

20. Servicing

    If servicing is required to maintain the products sold in normal operation and use, then the servicing organization and activities fall under ISO control. This control extends beyond repair procedures. It involves identifying responsibilities, tools, processes, equipment, documentation scope, and technical backup.

21. Statistical techniques

    a. Identification of need

       Statistical methods can be beneficial to the manufacturer to demonstrate conformance, variation, process control, and the materiality of each variance. The manufacturer should select statistical techniques appropriate to the item desired to be under control.

    b. Procedure

       To ensure uniformity of methods, the procedures for the treatment of statistical data should be under document control. Collected properly, the data obtained and analyzed under these methods can be effective in demonstrating conformity.

2. Implementing a quality management system

The implementation of a quality management system should happen in stages. Installing all of the recommended procedures at once in an enterprise is simply too overwhelming. In addition, an oppressive level of procedures will much more likely be questioned for validity if they are not implemented gradually. In a similar vein, the implementation of such a system is more difficult if it is happening concurrently with a development project. It could take the run of several projects to implement an entire system.

If a phased approach is used, it is important to implement a limited portion of every clause of the quality management system and add detail in subsequent phases, rather than implement one clause at a time in its entirety. This way, it is easier to see how the clauses interrelate.

C. How business development is affected

As can be seen with the design control element of the quality management system, development activities are greatly affected. Each step of the development program must be executed and documented according to procedure. This process is often considered oppressive but actually will assist in the long run.

The design control process matures the development activities from a "Let's see whether it can be developed" proof of concept to development being an integral part of the customer–manufacturer interface and contract.

The quality management system can also be thought of as a means for the team members to establish ownership of each element of the development program. The ownership value is diffused throughout the organization among personnel who are not necessarily identified as participating in the project. This is an important point because the quality management system can establish a unifying function across several programs and departments.

D. It's integral, not an add-on

A final thought on the quality management system is that it is integral to the organization and the project, not an add-on. It must be part of the core development function and be a contributing factor in all elements and all phases of the project.

CRITICAL PROGRAM SKILLS

1. Model for Skill Sets Development

The following model defines three levels of capability in program participants. Each level builds on the previous and requires increased commitment from the participant.

- *Practitioner level:* An individual at the Practitioner level expands their knowledge in actual training and use of the concepts, processes, and tools used in implementing a development program. The typical student embarking on the Practitioner level is part of a development team and already has some vocational knowledge and background in their own area of expertise, but lacks understanding and working knowledge of the other members' duties and tasks. The purpose of such training is to provide uniform training to all members of a team so that they can work more effectively with team members in a fast-paced, single-pass development program.
- *Lead level:* People at the Lead level are oriented above the Practitioner level; their basic skill sets are further developed to the point where they can synthesize plans and interpret and analyze events beyond the surface level. This skill set can be developed through on-the-job training and seminars. At this level the skill sets are honed such that contingency planning is inherent and the associated wisdom is developed to create workable plans and backup plans for projects.

The Lead level person will refine specific planning and analysis tools in each of the disciplines and will be able to actively manage a team in the planning, analysis and control of a program. They will also be more adept at absorbing additional responsibility and will control more of the financial aspects with increasing financial exposure being in their charge. The philosophy of the Lead level differs from the Participant level in that the individuals are to be adept at anticipating problems and navigate the solutions before they materially impact a program.

- *Master level:* The Master level provides an opportunity for the participant to direct a program within their own company on an actual project. They will be afforded the opportunity to take what they have learned and use the anticipatory thinking they have developed in the Lead regimen to apply their skill set to an actual program.

The Master level differs from the Lead program in that actual programs are executed and real funds are expended. The philosophy of the Master level is that the individuals are now expected to demonstrate executive leadership actions that result in actual program success. The participant can organize and manage a program with senior management responsibility and measured results. They will have personnel authority and profit and loss (P&L) responsibility.

2. Problem Solving

A. Problem-solving categories

The issue of effective problem solving is one of the most important in new product development. The ability to employ effective problem-solving skills is a major requirement for the group. Fundamentally, little can be accomplished without this skill and almost anything can be overcome with this skill. Consequently, this section focuses on the problem-solving process in a generic fashion so it can be applied in cases specific to your projects.

In applying the principles of problem solving, the first step is to determine what the present situation is. This is referred to as *issues review*. The objective of this review is to determine whether there is a problem to resolve and what type of problem it is. Most often, the problems encountered fall into one of three categories:

- Cause assessment
- Decision management
- Planning architecture

1. Issues review

Let's examine the situation first. There are four basic areas of consideration involved in this review. These questions are sequential and are arranged to uncover what type of problem is being considered. The points for consideration are as follows:

- Examine the process and observe the broad or complex concerns. List the concerns for examination.
- Separate the concerns into distinct issues.
  - How does the collection of concerns relate to each other and to the higher-level issue?
  - Is there more than one issue involved?
  - Will one action resolve or cause explain the issue?
  - What evidence surrounds the issue(s)?
  - How can you be more specific?
- Categorize and break down larger issues into root issues.
  - What is the relative importance of each single issue?
  - What are the impact, seriousness, and urgency of the effects of each issue?
  - How is each issue changing as time progresses?
  - Based on impact, urgency, and trend, what is the hierarchical order of the concerns?
- Determine a starting point for analysis.
  - Is there a change from the standard, or is one expected?
  - What is the extent of the change?
  - Under what conditions did it occur?
  - Is the cause known?
  - Is the cause relevant? (If the answer to these last two questions is yes, then you need to determine the cause.)

- Is there a choice to be made between two alternatives? (If yes, a decision needs to be made. Use the decision management perspectives described later.)
- Is there a project or decision to be implemented? (If yes, use the planning architecture described later.)

By using this series of questions, an issue can be analyzed to determine what type of problem is occurring and how to proceed. It will be most helpful when the issues and data are written down and considered within a framework. Assuming that the initial issues review is complete, the following three analyses will be used:

2. Cause assessment

The cause assessment is used when development managers need to identify root causes of problems to effect corrections and place the program back on track. This next series of questions can assist managers to determine the root cause systematically based on examination of facts and deductive reasoning.

There are several basic points for consideration in determining root cause of an observed problem. They are:

- Specifically state the issues.
  - Specifically, what is it?
  - Specifically, what is wrong?
  - Specifically, what standard exists?
  - How is that a change from the expected?
- Describe the problem. List the examined facts.
  - In what context is the issue observed?
  - What exactly is wrong?
  - State the location of the issue in the observed context.
  - Where in the object or context does the issue occur?
  - Time- and date-stamp the examined phenomenon.
  - When in the life cycle of the context was the first issue observed?
  - State any correlations in the examined issue.
  - Quantify how much of the context is defective.
  - Quantify how many issues are occurring.
  - What are the trends?
- List the relational facts.
  - In what similar circumstance might we expect to see this issue, but do not?
  - What similar issues might we expect to see this issue, but do not?
  - At what other locations might we expect to see this issue, but do not?
  - Where else on the object might we expect to see this issue, but do not?
  - When else could we have observed the phenomenon, but did not?
  - When else in the object's or context's life cycle could the issue have occurred, but did not?
  - Where might you expect correlations, but do not see them?
  - How much of the object or context might you expect to be affected by the issue, but is not?
  - With how many objects or contexts might we expect to have issues, but do not?
  - What might the trend be, but is not?
- Identify differences between the examined and relational observations.
  - What is different or unique about (observation) compared to (matching relational fact)? Repeat for each set of observations.

- List relevant recent changes.
- Generate likely causes.
- Test likely causes.
  - If (state a likely cause) is the real cause, does it explain why it is (state an observed fact) and not (state the corresponding comparative fact)?
  - How does this likely cause explain each set of observations and analysis?
  - Can this likely cause be eliminated because it fails to explain the observations and analysis?
- Select and verify the cause most likely to be suspect.
  - How can we prove that this is the most likely cause?

3. Decision management

The next area of analysis is the decision management process. Making a decision is relatively easy in light of alternatives given or choices to be made. Making an informed decision that has basis in and is supportable with options for correction and recovery is another matter. This section focuses on decision making with the objective of managing the process for results. The decision-making process is broken down into several elements for consideration. They are:

- What is t he purpose of the decision?
- What are the criteria for the decision?
  - What factors should be considered?
  - What are the critical elements that should be included in the criteria?
- Categorize the criteria.
  - Are there compromises in the criteria?
  - Are the criteria clear and measurable?
  - Will we be able or willing to accept the alternatives if these criteria are not met as stated?
- Generate the alternatives.
  - What are the possible courses of action?
  - What options are available or can be engineered?
- Compare the alternatives.
  - How well does each alternative perform against the criteria?
- Identify risks and assess risks.
  - Engineer recovery from adverse risks.

4. Decide and act

Taking into account the alternatives and factoring the risk assessment into the analysis, select the best, most balanced alternative. There are times when all of the facts are present and other times when there are not enough facts to make a decision. In these cases, consider the model and contrast shown in Figure 7-11. In the case shown on the right, you gather as many facts as you can and make an initial decision, monitor progress, and continue to assimilate new information coming in. Then evaluate the direction and progress and feedback to the initial decision point.

5. Planning architecture

The planning process is a structured method with which to analyze and plan. It consists of several major steps in the planning process. They are:

- State the deliverable of the plan.
  - What is the desired outcome? Who, what, where?

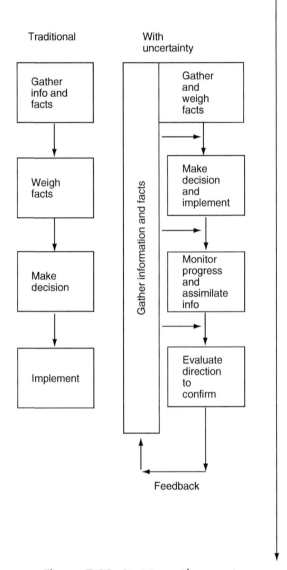

Figure 7-11. Decisions with uncertainty.

- Categorize and prioritize action items.
  - Identify prerequisites and dependent activities.
- Identify critical pathways, elements, and personnel.
- Identify and list potential problems.
  - Consider probabilities.
- Identify causes of critical potential problems.
  - What specific influences or events could cause each potential problem?

- Develop preventive and alternate action(s).
  - Identify new, alternate critical pathways based on circumstances.
- Design contingent action(s).
- Identify and list contingency trigger points.
  - How will we know when it is time to put the contingency into action?

As with any of these guidelines for consideration, they are simply guidelines for a thought process. They are not the magical means to solve all problems but they can assist analysis and, most important, they can enhance the completeness of the analysis in terms of thoroughly researching the issues and alternatives prior to drawing a conclusion. So often, companies rush to conclusions only to discover that further research and fact-finding are required.

B. Innovation as a problem-solving tool

Most progress is gained by hard work, effort, and innovation. Innovation is a key component in problem solving. It also can be a means for generating unique approaches to marketing or capitalizing on an idea. Given a fixed resource budget, a fixed number of people, and a competitive arena, innovation might be the only mechanism for competing effectively. It also can be the means to level the playing field against an opponent with limitless resources. There are several basic steps to the innovation process. They are:

- Develop the goal statement.
  - To (fill in specific goal).
  - Lead with an action verb (e.g., improve, enhance, reduce).
  - Structure the statement so that the choice is open-ended.
  - Describe the end result desired in positive terms.
- Develop the design attributes.
  - Develop single-issue attributes stated positively with the end result.
  - Select the key design attributes relating to the goal statement.
- Expand the ideas from each key design attribute.
- Select and combine ideas.
- Evaluate ideas in relation to the goal statement.
- Modify and develop solutions.

C. Root cause analysis

Determining the root cause of a problem or issue is critical to resolving it successfully. Often, companies immerse themselves in the details without determining the root cause. This tactic results in wasted time and effort. In addition, the participants end up trying to make a situation fit observed facts rather than mining the observed facts for the data and patterns of the root cause.

D. Fishbone, Pareto, et al.

The classic fishbone chart is a means for determining root cause by breaking down issues into component causes. It ensures a complete analysis and allows multiple pathways for multiple issues. The term *fishbone* derives from the appearance of the chart, in which each secondary issue or tertiary issue fans out from the higher-level issue that precedes it. Figure 7-12 illustrates a generic fishbone analysis.

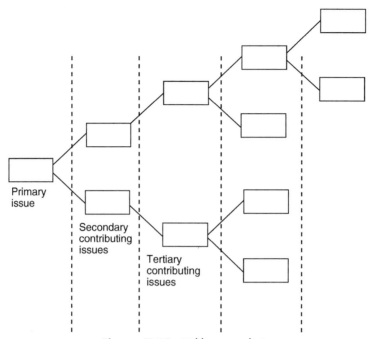

Figure 7-12. Fishbone analysis.

As shown, each cause is broken down into secondary or subcategory contributing causes. These secondary or subcategory contributing causes are then further broken down into tertiary causes, which contribute to the secondary causes. For the analysis to be valid, it is important to break down each possible issue into all of its contributing causes. Failure to miss one can mean a breakdown of the process.

The next diagramming option is the Pareto chart. The Pareto chart is a means for analyzing loose, disjointed occurrences or incidents or data and assigning them to similar categories. These are then totaled by incident type and are numerically presented in a bar-chart format. The highest number of incidents is presented first, followed by the balance in decreasing order. This ordering allows the loose, disjointed data to be presented in an organized fashion for problem resolution. In the example in Figure 7-13, there are nine pieces of data assigned to four categories. Specifically, incidents numbered 3, 7, and 9 fit the conditions in Category 1. Incidents numbered 1 and 2 fit the conditions for Category 2, and so on.

By creating a Pareto chart of the categories, we see that the Category 1 incidents represent a more serious problem than the other categories, but not by much. There is much to be learned from the chart. There are the sheer number of incidents per category, the distribution of incidents among the categories, and the number of categories. In this way, a group can focus on the most severe problems first and also ensure that all problems are addressed in some way.

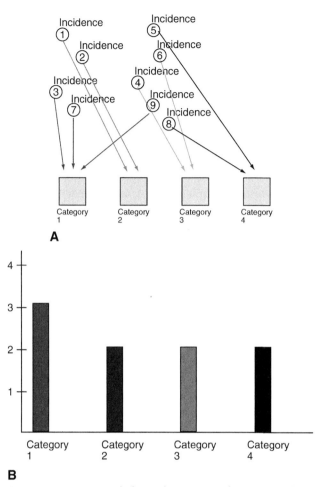

Figure 7-13. (A) Underlying data; (B) Resulting Pareto chart.

E. Attacking the specific obstacle

There is one basic principle to guide you through problem solving—the use of overwhelming force to overcome obstacles. The disruption that an obstacle can cause is irrecoverable in many cases. Therefore, the objective is to minimize the time spent in overcoming it. Consequently, bring overwhelming force and persistence to bear on the obstacle.

F. Pushing the right and watching the left

Some problems seem to defy resolution. They simply are too complex and intertwined with other problems and therefore are difficult to understand. In these cases, it is a good idea to introduce causation of some known type and observe the reaction. The idea is to profile some type of transfer function or characteristic to understand the dynamics of the problem. By causing input and observing the reaction, it might be possible to reduce the problem to understandable components for quicker resolution.

G. Solving people problems

The issue of personnel problem resolution can be more involved than a static or dynamic problem. This is because people are human and have feelings, egos, and emotions. They are part of your team for several reasons, the sum of which might not even resemble the reasons you may have selected them. In spite of all this, the manager has a job to do and a program to deliver to senior management. The following represent several basic tenets in dealing with personnel issues:

- *Speed:* Do not delay; act decisively and swiftly, assuming the facts are known.
- *Accuracy:* Ensure that the facts used to make a decision are accurate; avoid guesswork.
- *Quality, fair play:* Administer discipline, advice, and appreciation in a fair, equitable, and consistent manner.
- *Changing players:* If players must be changed, determine actions and replacements and act swiftly and decisively. If termination is required, act decisively with the individual and the group. Do not get persuaded into second chances if the best choice for the program is known.
- *Dealing with nonconcurrence and disagreement:* Encourage some disagreement and constructive confrontation. It is a healthy means for decision making and alternative evaluation. Management can overrule certain objections where disagreement might occur or preferences might be voiced; however, if a nonconcurrence is voiced by a member of the group, take it seriously. This person is, in effect, stopping the production line because he or she feels so strongly about an issue. Personnel generally do not behave this cavalierly, so respect their nonconcurrence.

3. Contingency Planning

A. Think ahead and plan contingencies

As the program plan is developed and implemented, thought should be given to contingency planning. At critical points in the program the team might want to think through contingency pathways. Figure 7-14 shows a contingency plan developed for a path on which

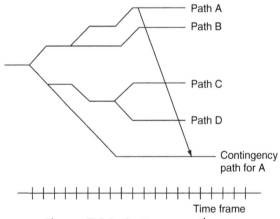

Figure 7-14. Contingency pathways.

a milestone cannot be reached. In this example, the pathway follows the line to the alternate pathway. These contingency paths ensure mitigation of the time damage to a program and allow for a more seamless transition to an alternate path. The team should plan these paths and crossover points in addition to the basic program flow chart. These paths can be thought of as bridges to an alternative flow chart.

B. Examining direction after a setback

The true measure of a team's character and commitment is how it reacts to a setback. A setback in a program is one of the most difficult things to overcome. It generally follows a period when a tremendous amount of human energy has been expended in the attempt to overcome an obstacle. At this time, the team is already stretched and must endure the disappointment of the setback. This is the point, however, at which the maximum amount of energy and tenacity must be applied to overcome the problem.

Rather than redoubling the group's efforts, now would be an opportune time to reexamine the tasks and milestones currently in progress and refocus in the perspective of the overall program. This process is illustrated in Figure 7-15.

The field of view shown in Figure 7-15 is the manager's wide-range view of events and task requirements, involving the fact-gathering that must take place prior to resetting the group's effort and getting the program back on track. The idea is to prevent another setback immediately following the first, which could demoralize the team.

C. Resetting the team after a setback

If the program does suffer a setback, a management action is needed to reset the group's effort. This is important because the group's momentum has been compromised. Resetting

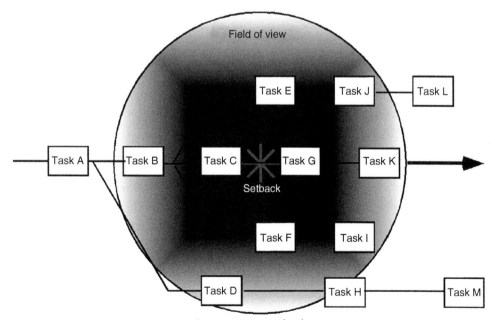

Figure 7-15. Setback.

the effort has two components: morale issues and vocational aspects. The morale issue must be resolved by the manager's leadership. The vocational aspect results from the manager's sweep of the field of view, reinitiation of group activities, and driving progress.

4. Recovery Skills

A. Don't let adversity slow the program down

If it is a sure bet that adversity is present in an uncharted field of endeavor, then recovery skills are an imperative tool in the manager's portfolio. How fast a group can recover is an indication of how timely the program will be completed. In this section, we focus on commonsense recovery moves to minimize the damage caused by a setback.

B. Increase the pace to offset losses

To demonstrate the importance of keeping on schedule, the manager must increase the pace of the development effort to make up for lost time. At best, this tactic will keep the program from losing additional ground. The increased pace sends a clear message to the team that delays will not be tolerated. In addition, team members will tend to operate much more diligently in the future to avoid further setbacks, knowing that the result will be an increased pace.

C. Next time, team will overcome by itself, naturally

An important value to instill into the team is that the way to prevent setbacks is by operating in a manner that will avoid pitfalls. By extension, this means that the team will begin to look ahead for possible future problems and to plan contingencies—all of the skills we previously discussed. In this way, the group leverages the manager's knowledge base.

D. "Failure is not an option"

A line from the film *Apollo 13* can serve as a motto for business development groups. At the darkest point in the effort to return the astronauts to Earth, the comment was made that "Failure is not an option." This slogan can be used to inspire a team to perform. Each individual is part of something larger than himself or herself. Team members are interdependent; each member relies the performance of the others.

E. The goal: Focus on near term and long term simultaneously

The development manager can be effective only if there is focus on both short-term and long-term goals of the company. The short-term goals are oriented toward the direct management of the tasks of the project. The long-term goals of the company are the various milestones and product introductions for the company. This requires various perspectives and the ability to operate on various levels of detail, as shown in Figure 7-16.

5. Communications Breakdowns

There is one thought here: Fix communications breakdowns, and fast. Communications are the fuel of a product development effort. If there are problems with personnel communicating, the program grinds to a halt. This happens most often when someone is waiting for

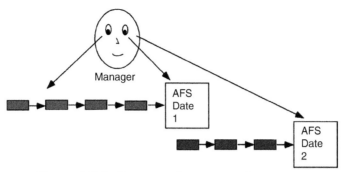

Figure 7-16. Short-versus long-term goal orientation.

an answer. The team needs to be shown that time is of the essence and that no program can wait for a decision. The program must be actively worked on by everyone, and everyone must understand the importance of accuracy and timeliness in communications.

6. Obstacle Removal

A. Become at ease with problem solving

Obstacles are a fact of new product development. They occur at every turn, and it is the group's and manager's responsibility to navigate these obstacles and bring the program to successful completion. Such navigation should become second nature to the group. Team members need to be at ease with problem solving and overcoming obstacles. Furthermore, they need to become at ease with resolving these issues within a time frame.

B. Don't just solve the problem for the team

There is a saying that goes something like this: Give a man food, and he will eat for a day. Teach a man to farm, and he will eat for a lifetime. The same holds true for the management of development groups. Use the overcoming of obstacles as an object lesson to teach the group how to navigate for itself—how to evaluate and act.

C. Multiple pathways to the goal

The manager needs to develop a philosophy within the group that there are several ways to achieve the objective. Alternative means must always be available to get the program on track or to keep it there. The group needs to develop contingent strategies around these obstacles and exercise them without delay.

D. Strategies for contingencies

The timing for coping with contingencies is crucial to minimizing lost time. All too often, the manager and group wait until they actually run into an obstacle before reacting to it. This approach wastes time and limits options. It is critical to recognize potential obstacles looming in the future and to take effective action earlier. Consider Figure 7-17, which shows two scenarios for engaging an obstacle. The first is the method generally used

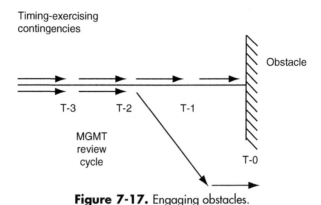

Figure 7-17. Engaging obstacles.

by default. The project rolls along until it encounters the obstacle. Management then scrambles to recover from the lost time and, in the process, uses precious time to determine a way out of the situation.

In the second scenario, management looks into the near future and tries to recognize potentials for obstacles or patterns that might contribute to them. Generally there are signs of impending obstacles that can be recognized well in advance of the actual impediment. The key is early recognition and reaction to mitigate the damage of time lost. In the example shown in Figure 7-17, the recognition and the action take place at T-2, so that an agreed-to approach is in place and workable at T-0. If a problem is real and inevitable (or even likely), do not wait to exercise the option of working around the problem early.

7. Tracking Performance

There is an important distinction between tracking a project and freeing up personnel for optimum performance. The idea behind tracking performance on a project is to understand the progress and effect corrective action if delays occur. To be truly effective in a program, the manager must clear the pathway for maintaining steady and predictable progress to keep the program on schedule. Looking at it in a different perspective, tracking performance should confirm an on-time schedule.

Operationally, project tracking is like a control loop. The project plan is the command signal; the summing node is the management review; and the development team makes progress. A feedback reporting system (project tracking) summarizes progress for comparison to the plan at the summing mode or management review. Management intervention acts on the development group to effect changes in the progress. Ideally, as stated previously, proactive management should relegate the role of project tracking to that of confirming an on-schedule situation. This is illustrated in Figure 7-18.

A. The plan dictates the type of tracking

There are different types of plans and different means for tracking progress of the plan. The tracking mechanism must be commensurate with the plan type.

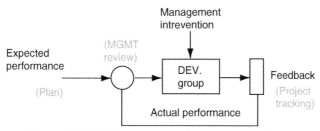

Figure 7-18. Management role in project tracking.

B. All systems work with discipline

As stated previously, most systems will be effective as long as they have discipline. Discipline is the operative word. The type of discipline needed for development is not the "bull of the woods" type but, rather, steady progress and follow-up to ensure results within an allotted time frame. Because it has been discussed that time lost cannot be made up, the unwavering need is to keep on schedule at almost any cost. Staying on schedule will occur through the steady disciplinary process of following up on action items, responsibilities, and completion dates.

C. With interdisciplinary projects, use critical path methods

One of the most important factors in program management is staying true to the logic of the project. The logic is the interconnection of activities and the order in which they are initiated and completed. In addition, the interdependency of the activities must be preserved to maintain integrity of the overall program. To that end, it is advisable to make use of the critical path method of program control. This method shows the interdependencies as well as the longest or critical path through the program. With these two bases of information, the logic integrity will be preserved and the critical path is the focus.

D. Management of multiple programs

This is all well and good for a single project run by a single manager of a single group, but what about the management of multiple programs? Herein lie the mysteries of effective management, as multiple projects have multiple priorities planned along multiple schedules. Coordinating these requires a level of management that is in a league of its own. The projects must be broken down into their component parts, tasks, and subtasks, and talent must be assigned to the activities called for in the specific time periods. Schedules cannot be missed under any circumstances, or the flow of progress will unravel like music played out of rhythm.

Given the proper resources, such tracking ensures timely completion of all the programs. The key, however, is resources. Rarely are there enough resources to spread over the several projects called for in a senior manager's plan. Program management then becomes a juggling act to place resources with the tasks at hand, with eventual sloppiness about completing details. This process becomes self-defeating over the long term. Be sure not to overload the development personnel or the team will become so overloaded as not to be able to complete anything.

E. Wildcatting development

Not every company is interested in implementing a structured product development system. Some are interested in a pure opportunistic approach where the employees are directed to work on the "product opportunity of the day." Typically in smaller companies, senior management fails to reinforce a system for generating new products and fails to use discipline in executing a program in the wake of changing market conditions, lost orders, and competitive thrusts.

Unfortunately, this type of company exhibits a high cost of development engineering on its income statement, with little completed to show for the effort. There is also a significant chasm between the sales effort and the development effort. Materials are a high portion of manufacturing cost because design-to-cost efforts are erratic. Rework is high because development is rarely completed on a product before it is released for production.

Despite all of the negatives, there is still a way to manage a development group immersed in this type of environment. The use of *management by queues* seems to work well in such situations, although it offers no predictability about when a program will be completed. The queue is the list of projects assigned to an individual development resource. These projects are waiting for the resource to be applied to them. When the market condition changes, the priorities of the queues are reordered. The manager in such a scenario also must manage the lack of continuity caused by the changing conditions.

This environment requires a special type of product development professional. He or she must personally provide the continuity that is lacking in the system. Management by queues also requires the development personnel to have experience in the development arena and to have the initiative to stop, start, and retain the pertinent data for a program.

F. The role of action items

In Chapter 6 we discussed the use of action items in a program management venue. For effective program management, the action item is only one part of the equation; it is one link in the chain of hierarchical control. The other essential links are the assignment or responsibility of action items, and the completion date. As shown in Figure 7-19, the action item supports the subtask. The subtask supports the tasks, and the tasks support the stages of the project. The stages then support the project itself.

G. Follow-up systems

To establish and maintain continuity in tracking performance, the use of follow-up meetings is essential. The follow-up meeting yields the appropriate performance feedback, keeps all parties aware of project issues, and creates expectations and accountability among members of the group.

There are many mechanized systems for following up on details. The important point to remember is that all of the details must be followed up on—each and every one of them, all of the time. If the manager is cavalier about details, the group eventually will become cavalier about the program. Using mechanized, software-based systems can be helpful and can create automatic follow-up based on dates and times. These systems are only effective, however, when the manager obtains action with the follow-up on a timely basis.

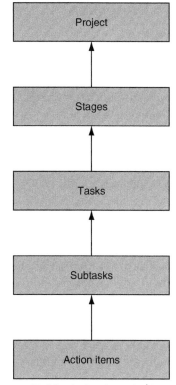

Figure 7-19. Action items in hierarchy.

LEGAL ISSUES

This section addresses how legal issues such as intellectual property protection affect and enhance business and product development activities. This discussion will answer some of the following questions:

- Patents
    - When do we apply, prosecute, and secure a patent, and how they can benefit a company (large or small)?
    - How do we obtain a patent and, more important, why should we even consider one?
    - How do we use patents for additional revenue and how do we minimize the expense in securing patents for small and medium-sized companies?
- Trade secret protection
    - How do we use trade secrets to enhance patent positions and to protect small-company technology with a minimum of legal costs?
- Product liability
    - How do we avoid product liability issues and prepare for potential issues, and how can we mitigate exposure for small and medium-sized companies?
- Cooperative development
    - How can we position our company to protect ourselves when conducting joint cooperative development of technology and products with peer companies and larger companies?

1. Intellectual Property Protection

A. Background of patent protection

The issue of patent protection is important to the overall success of the development program. There are several advantages for the company in obtaining a patent. Generally, if a product is not patented, it is, as a practical matter, available for copying by the competition. In addition, any agreement or venture entered into by two or more companies is fertile ground for disagreement over origin and split of profits. The use of assigned patents clears up any potential misunderstanding.

The intellectual property field encompasses more than patents, as we will discuss shortly. It includes trademarks, copyrights, and trade secrets. *The information presented in the following pages is designed to serve as background information only, not as legal advice. Readers are encouraged to seek counsel for a considered opinion and direction for their own circumstances.*

Historical perspectives   Historically, the speed of information transfer was slower than in recent times. Increased connectivity among product developers, manufacturers, suppliers and employees has accelerated the pace of information-sharing.

Because of the accelerated activity of product development and the rapid turnover of technology, the importance of obtaining patents is even more critical today than in the past. More firms are developing technologies and applying them in products, therefore requiring protection. Their competitors are developing improvements and patenting these improvements, thereby generating additional intellectual property activity.

Importance of patent protection   The patent protects a company's investment in research and development. It protects the invention by granting the applicant certain rights, thereby slowing or stopping the progress of competitors toward those same rights.

A patent demonstrates expertise in the technological aspect of the invention. It has commercial advantage with federal government agencies when contracts are negotiated and let. Finally, a patent is a demonstrable asset that can aid in securing capital financing from investors and commercial institutions. With all of these advantages, it makes good business sense to secure developer's rights by obtaining a patent.

Impact on the development program   The intellectual property issue is increasingly affecting product development. Trade secret information and technological know-how, applications, and processes are being developed for competitive advantage and are being protected. Planning for the protection of certain aspects of new products must begin at the conceptual stage, and security is more and more difficult to maintain. The activity level of intellectual property development and protection is growing, with the result that it is becoming a vital component of new product development.

B. The basic protection mechanisms

Patent protection   A patent is basically an agreement between the inventor and the federal government. The inventor "teaches" the public how to use the invention in enough detail to enable a person to make and use the invention without undue experimentation. The patent is designed for the sole purpose of encouraging the development and use of new technologies and methods. In return for this national benefit, the government then grants the inventor the sole rights to make, use, or sell the invention for a period of time. Historically, a patent grant lasts for 20 years from the date of application.

There are two basic types of patents: the utility patent and the design patent. The *utility patent* protects inventions that are functional and have utilitarian value. The *design patent* protects ornamental designs. The utility patent is available for inventions such as new and useful processes, machines, manufacturing methods, or composition of matter. Recently, software that is also nonobvious or anticipatory can be patented. The following fare beyond the scope of patentability: mere printed matter, methods of doing business, things that are unaltered from their natural state, or abstract scientific principles.

The execution of a patent is adversarial in that there are statutory bars to obtaining a patent, and the applicant must document timing, novelty, and potential usage and benefit. These statutory bars include the following:

- The applicant is not entitled to a patent on an invention that was publicly known before the inventor conceives it. This sounds like an obvious fact, but it does come into play in cases in which timing and contention might be problematic.
- Patents that have previously been issued are demonstrable proof of public knowledge.
- The applicant for a patent can lose rights by premature disclosure, sale, or use in public view.
- Another way to lose rights to a patent is through abandonment. The idea behind the patent system is to move ideas for the betterment of humankind into the public use. If someone sits on an invention for some period of time and shows no demonstrable intent to use it, he or she might effectively lose rights to it.
- Simultaneity of patent applications, both domestically and in concert with foreign application, might prevent the inventor from certain rights.

The inventors must submit the application under their own names and must be the ones to have first made the invention in the United States. The invention must also be nonobvious to a person of ordinary skill in the art. This restriction is designed to ensure actual novelty of the invention. The merit of the patent application is judged based on the following factors:

- The scope and content of prior patents and publications.
- The level of ordinary skill in the art.
- The differences between the invention as claimed and the prior art.
- Whether the invention provides unexpected results, fulfills a long-felt need, is commercially significant, or all three.

Copyright protection   Copyright protection is primarily intended for protecting an author's works. These can include booklets, brochures, artistic designs, maps, and architectural blueprints. In the argument of form versus substance, a copyright is oriented toward protecting the form of ideas expressed, not necessarily the ideas themselves. Publication is not necessarily required for copyright protection but it is still an important concept.

The term of a copyright is generally considered to be the author's life plus an additional 50 years beyond the author's death. If the work is a work for hire, the term is generally considered 75 years from date of publication or 100 years from date of creation, whichever is shorter. A copyright is obtained by filling out the appropriate application and submitting it with a fee for registration.

Trademark protection   A trademark is product protection oriented toward the source or origin of a product. It is designed to distinguish goods and services of one company versus another. A trademark is used to capitalize the goodwill and reputation of a company through its products. Therefore, it is illegal for one company to infringe on another by use of its trademark. Prior to considering a trademark, the company must perform a trademark search through the U.S. Patent and Trademark Office (USPTO).

The protection provided by the trademark is directly related to its distinctiveness and nongeneric nature as it relates to the product. For example, Skater™ would not necessarily be considered a trademark for roller skates because the term Skater™ does not uniquely identify the origin. At issue is the relationship between the mark and the product or market segment it is serving. The term Skater™ might be quite acceptable for an unrelated market.

There are four basic types of trademarks with varying degrees of protection, as follows:

- *Generic mark:* This is a trademark that uses a generic term related to the product family. An example of this is the earlier example of Skater™ applied to roller skates. It simply is not unique and therefore has little value.
- *Descriptive mark*: This is a mark that uses a description of the technology or part of the product to try to identify its origin. As in the previous example, Ball Bearing™ would be the trademark—not very unique as to identity.
- *Suggestive mark*: This mark requires the consumer's imagination to infer the nature of the product from perception and thought. An example of this would be Lightening™ brand roller skates. This mark evokes thoughts of extremely fast roller skates.
- *Fanciful mark:* This mark is the most valued because it is a word created expressly for the product and the trademark. Another option is the use of an arbitrary word for the product trademark. An example of a fanciful mark would be Whoosh!™. An example of an arbitrary word applied could be Warp 10™.

The basic requirement for trademarking a product is its use. Barring any other previous use in the industry or product segment, the commercial use of the trademark is generally sufficient. Trademarks are placed in use by registering them with the USPTO. The term of the trademark is potentially infinite, but the registration must be renewed periodically.

Trade secret protection   This type of protection is generally considered for proprietary information not in the public domain. A trade secret is a technology, methodology, or process which, by being kept secret from the competition, prevents competitors from being able to copy it, thereby inuring benefit to the owner(s) of the trade secret.

A trade secret is a fragile entity because its value lies in secrecy. Once a trade secret becomes known, the protection is lost. This applies no matter how the secret is disclosed. The primary disadvantage of this type of protection is that it offers no protection against someone else developing the same technology independently. Moreover, the second party could seek patent protection on improvements, thereby foreclosing on further development and use of the base technology. Certain technologies embodied in a product that can be reverse-engineered are not suitable for trade secret status because copying is relatively straightforward.

A trade secret is maintained by secure disclosure, limited access, and clear definitions of its proprietary nature on all detail drawings, printouts, flow charts, schematics, assembly drawings, technical data, and test results. Any disclosure requires confidentiality agreements duly executed prior to disclosure.

Trade secrets can be exploited through explicit licensing agreements; however, an agreement will govern only to the extent that the second party honors it. If the party to which the trade secret has been disclosed does not honor the agreement, the trade secret is essentially lost.

Comparison of protection methods Figure 7-20 is a chart summarizing the different methods of intellectual property protection.

C. International perspectives of patents

The issue of patent protection does not end at the boundaries of the United States. The entire world is developing and introducing technology to the international marketplace. Consequently, there needs to be a broader perspective in patent protection, both domestically and internationally.

Comparison of protection methods			
Protection mechanism	**Term**	**Scope of protection**	**Compatible concurrent forms of protection**
Trade Secret	Potentially infinite	No protection against independent development	Trademark
Trademark	Potentially infinite	Protects against others trading on trademark owner's reputation	Trade secret, design patent, utility patent, copyright
Design Patent	14 years from date of filing	Protects ornamental features only	Trademark, utility patent, copyright
Utility Patent	20 years from date of filing	Protects the concept of invention	Trademark, design patent, copyright
Copyright	Life of last surviving author or, in the case of work for hire, the shorter of 75 years from publication or 100 years from creation	Protects only the form of expression, not substance or content	Trademark, design patent, copyright, utility patent

Figure 7-20. Summary of intellectual property protection.

For example, there is a protocol for protection of a patent filed in the United States and a certain amount of time required to file in other countries. If a patent is granted in the United States but is not filed and executed in foreign countries, there could be a possibility that the technology is embodied overseas in a product imported into the United States and sold here in competition with your patented product. There is a strategy for patent applications and use domestically and internationally designed to provide maximum benefit to the inventor. Consequently, protection of an invention should be arranged in light of various international issues.

The U.S. philosophy on patents is that the first inventor (as defined by U.S. law) is entitled to obtain a U.S. patent on the invention. Subsequent inventors may patent improvements but not the basic invention. In other countries, the criterion for holding a patent is the first party to file, not necessarily the first to invent. This approach is in conflict with the U.S. system. The first party to invent is proved by the existence of lab notebooks, records, and related chronology and witnesses' documentation corroborating the testimony of the inventor.

Paris Convention for the Protection of Industrial Property   The Paris Convention for the Protection of Industrial Property is a treaty relating to patents, which is adhered to by 168 countries, including the United States. It provides that each country guarantees to the citizens of the other countries the same rights in patent and trademark matters that it gives to its own citizens. The treaty also provides for the right of priority in the case of patents, trademarks, and industrial designs (design patents). This right means that, on the basis of a regular first application filed in one of the member countries, the applicant may, within a certain period of time, apply for protection in all the other member countries.

Patent Cooperation Treaty   The Patent Cooperation Treaty was drafted in June 1970, and came into force on January 24, 1978. It is presently (as of December 14, 2004) adhered to by more than 124 countries, including the United States. The treaty facilitates the filing of applications for patent on the same invention in member countries by providing for, among other things, centralized filing procedures and a standardized application format.

D. Procedures for executing a patent

There is a protocol for obtaining a patent in the United States. It is procedurally legal and volleys between the patent attorney and the patent examiner. The basic deliverables for the application of a U.S. patent are as follows: (*Note that this is an abbreviated list and is summarized for illustrative purposes. Actual documentation will have additional categories for completeness.*)

Title of the invention   The title is the formal statement of what the product basically is. It is a generic description of the product so that it is recognizable and comparable to other products. The title should be as brief as possible.

Cross-references to related applications   This is a listing of U.S. applications that are prior to the one in question. They are for reference only and are selected as those related to the current application. This listing is also used in the definition of scope and further defines the specifics of the invention as related to other applications.

Background of the invention The background is a brief description of the field of art of the invention. It is a definition and description of the problems solved by the invention and references applicable prior art.

Summary of invention The summary describes the essential elements of the invention that are particular to the claims of the applicant.

Brief description of drawings A patent application generally includes drawings used to describe the invention. This section of the application describes the details, drawings, and figures of the invention.

Detailed description of the preferred embodiment This is a summary showing the public how to use the invention. Every element of the invention is described, along with the drawing referencing it. Numerically ordered and referenced, the explanation is complete enough so the average person "skilled in the art" can understand the application of the invention. A number of embodiments of the invention may be presented here; however, the best one must be included at filing.

The claims The claims are a definition of the specific rights inured to the inventor by the patent. They define the boundaries of intellectual property. The claims are the value of the patent in that they define the scope of protection. Drafting them properly is a job for an experienced intellectual property attorney.

The claims of the application vary in scope, starting with the most general and moving on to the most specific. This progression gives advantage to the applicant if the patent examiner challenges the initial claims. The general claims may be invalidated by relevant prior art, but the further definition of claims may not be invalidated. This progression therefore allows for highly specific protection on a particular product.

The other point to remember is that claims that are too specific allow the patent to be designed around more easily by competitors.

Abstract of the disclosure The abstract is a short, albeit complete, description of the invention itself.

Declaration and power of attorney The declaration identifies and documents the name, address, and citizen status of the applicant(s). It also states that the applicant(s) have reviewed the application and understand the contents and that the applicant(s) believe they were the first and original inventors of the invention claimed. The law is strict with respect to the declaration, in that no changes may be made after the inventor has signed it.

Letter of transmittal This is the formal communication to the USPTO in submission of the application.

Check for application fee The application fee accompanies the application.

Small or large entity statement　Fees may vary depending on whether the company or individual is a small concern or a large concern. Accordingly, a statement to this effect should accompany the application.

The following is a list of some of the critical documents and related instructions used in submitting an application.

- Introduction
- Non-Provisional Utility Patent Application Requirements or a Utility Patent Application Transmittal Form or Transmittal Letter Fee Transmittal Form and Appropriate Fee
- Application Data Sheet
- Specification
- Title of Invention
- Cross-Reference to Related Applications
- Statement Regarding Federally Sponsored Research or Development
- Reference to a Sequence Listing, a Table, or a Computer Program Listing Compact Disc
- Background of the Invention
- Brief Summary of the Invention
- Brief Description of the Drawings
- Detailed Description of the Invention
- Claims
- Abstract of the Disclosure
- Drawings
- Oath or Declaration
- Sequence Listing (If Necessary)
- Identification of Drawings

The basic procedure is as follows: A completed application is submitted to the USPTO. The application then gets assigned to a patent examiner who is technically capable in the field of the invention. The examiner then does an investigation to determine prior patents and the prior art of the submitted invention. When this process is completed, the patent examiner issues an Office Action. This is the formal response to the application and deals with the specifics of the application claims. Equally important, the examiner also cites references upon which the claims will be evaluated. The First Office Action is valuable to the submitter's patent attorney in terms of working out a strategy for executing the application to a successful completion.

The First Office Action then gets mailed to the patent attorney, (if used) who then returns it to the inventor for review, formulation of strategy, and eventual response. The inventor provides detail to the attorney, and together they draft a response to the Office Action by arguing for the stated claims. This stage in the process is a negotiation between the examiner and the patent attorney until the exact scope of the claims is agreed-to. Once there is agreement, the Patent Office initiates the process to issue as a U.S. patent.

E. Security involved in obtaining a patent

The key element of patents (as well as of product development) is the combination of diligence and secrecy. As a normal course of development, certain aspects that might become patentable should be treated with trade secret status.

Once the team feels that it has a product concept that might be patentable, it needs to conduct a thorough state-of-the-art search of the technology and products. This can be done by canvassing the patent files already in existence and conducting an extensive literature search. The idea is to determine whether there is infringement on some previous patent. It is beneficial to conduct this search at this time because an alteration in the product concept at this stage can avert an infringement and save considerable time and money. In this way, the patent system can be used to research available technologies. This activity can also make research and development more efficient. By knowing what is currently available, precious time and money need not be spent on something that has already been developed.

Next is the determination of patentability. The product should be examined for items that have specific novelty by themselves or in combination with other items. The patent attorney can assist in the patentability assessment. Finally, a full disclosure should be made to the patent attorney in preparing the patent.

F. Integrating the patent application with the product strategy

To the extent possible, a new product development is best served by integrating the patent strategy with the technical and commercial aspects of the product. Merely seeking a patent on a product after it is developed, with no plan on how to capitalize on it or how to use the timing and commercial aspects to advantage, is a waste of development funds.

An additional consideration is that a new product development and patent program must gain momentum as products are developed. The idea is to constantly raise the barrier to entry for competition. Your company can raise that barrier two ways: by increasing the level of embodied technology and by patent protection on the embodiment. With a well-considered program, the industry and your competitors will find it very difficult to displace your position as a premium and entrenched supplier of the product. Securing a patent will enhance your technological edge over the competition.

2. Product Liability

In this section, we examine the effect of product liability on the development process. Historically, there has been some connection between the design process and potential liability primarily focused on implementing good design practice. In actuality, however, there should be a direct connection—a means for corrective action and a documentation system in place as the product is developed.

The planning for product liability must be both proactive and defensive at the same time. It must be proactive from the perspective of attempting, through design, to minimize the chances of injury or property damage through use of the product. It must be defensive from the perspective of accumulating the documentation that indicates due diligence in the design and manufacturing process. Such documentation will be required in the unfortunate event of litigation.

A. Corporate protection

Given the conditions in which manufacturers must conduct business operations, how can manufacturers protect the corporation? The answer lies in the methods and procedures used in the development process. The protection of a company consists of two basic elements—preventive and defensive.

The preventive actions take place long before an unfortunate accident and impending lawsuit. They occur through good design practices, a thorough understanding of how the product is used, the diligence in understanding the potential misuse of the product, and high-quality manufacturing processes.

The defensive posture and positioning take place after the accident and prior to or upon the initiation of the lawsuit. In such a case, records will be taken, fact-finding takes place, and structuring of the corporate position occurs in a loss mitigation phase. The corporation is drawn into litigation either in or out of court, and energy and funds are expended to fight off the suit.

There are three basic tenets in the prevention and defense of a product liability action, as follows:

Plan ahead   Adopt a realistic attitude toward product liability by clearly developing an understanding of the hazards associated with the product. Create and disseminate efficient and effective warning systems. Establish a solid information base about product hazards, misuse, etc.; make recommendations; and document the dissemination of safe use and handling of information.

In addition, seek a third-party assessment of the product design to double-check the company's design operations and provide an unbiased opinion of the design practice in the event of future litigation.

Demonstrate concern for safety in design operations   When conducting design operations, make user safety of paramount importance. In a potential litigation, the manufacturer's interest can be served by a demonstrable concern for safety. As issues surface, create a system for addressing them, resolving them, and documenting the agreed-to approach to market. The manufacturer must demonstrate a comprehensive understanding of the product's uses and limitations and take reasonable steps to advise and protect users. Juries tend to punish manufacturers for flagrant disregard for safety, but they also have a tendency to be understanding in cases where the manufacturer truly has made an effort to design safety into the product.

Exercise the duty to warn   The onus is on the manufacturer to continuously exercise the duty to warn users of potential harm through the use or misuse of the product. Appropriateness of warning labels and instruction will be evaluated by potential juries for reasonableness. If a lack of information makes the product unsafe, it is the manufacturer's requirement to warn. Design engineers should eliminate hazards wherever possible and specify collateral safety devices where appropriate. In cases where these measures are not effective enough, warnings and instructions are required.

C. Personal protection

The steps to protect the corporation have been taken as part of the product development process, but what about individual participants in the process? Are they protected under the corporate form and under corporate policies? In theory, the individuals responsible for the design and control of the product are somewhat vulnerable to lawsuits; however, the corporation—not individuals—has the sufficient funds to satisfy personal injury lawsuits.

Moreover, a group or team under the umbrella of company operations was responsible for the design and control of the product, and the company is a much more likely target of potential litigation.

There are insurance policies that protect both the organization and its employees from losses in the event of a lawsuit. These insurance products can be all-inclusive or only catastrophic in scope. The officers of the company remain liable, but they can be covered under some type of professional liability insurance. The issue of personal protection should fall within the company's scope of normal operations insurance. Individual team members should have some degree of protection under this policy.

D. Project phase documentation

The need for project documentation does not have to be oppressive to the program. It should be a normal part of the filing and categorizing of issues related to the phases of the project. Issues should be resolved, closed out, and readily accessible. This task is made easier by the use of mechanization in the form of e-mail and electronic file transfer. For example, as issues are discussed via electronic means, they can be answered and the decision communicated to all parties involved and filed in an appropriate manner. Retrieval of issues then becomes easier and communications are referenced to the program stage in which they occurred. Document issues as if you must defend your actions in court; some day you might need to.

E. Loss mitigation

As the decisions concerning a product line are made and compromises are discussed, a strategy for the defense of possible lawsuits should also be considered. In the event of an actual lawsuit, the organization can take steps to minimize its financial impact. The following steps can serve as recommendations to take in the event of a suit:

- Notify key personnel.
- Assign an internal liaison team to facilitate counsel's requests for information and assistance.

The defense occurs over three phases: investigation, discovery, and trial. At each phase and at each decisive point, the company must be well-fortified with information and documentation to mitigate the damages. Unfortunately, failure to be prepared could result in a loss that will invite similar claims.

SUMMARY

This chapter presented the core development phase of a development program and discussed a wide spectrum of material. Initially, we started with a section on the mechanics of development. This section discussed the control theory surrounding the project and the people involved.

Next we reviewed the company's core competency capability through assessment and improvement. This was followed by a review of the core technology development viewed from both an internally generated and externally generated perspective. Outsourcing development was reviewed next, paving the way for core product development.

The basic execution of the plan was reviewed, along with tasks and project management systems. Each of the product development phases was discussed and the specific deliverables were outlined for each stage to ensure completeness. Perspectives on tracking performance, removing obstacles, and backup plans were also reviewed. Dealing with the program objectives within the larger framework of corporate objectives was discussed as well.

Tools were presented for various methods for problem solving. Included were tools for issues review, cause assessment, decision making, and management and planning. A section was included on the critical program skills needed in an organization, and decision management and recovery skills in the wake of problems or failure were presented.

A quality management system was presented and a model for a standard was discussed for an effective system.

A section on legal issues, in both patent and intellectual property protection and product liability protection, was included to stress its importance and the many deliverables involved. A section on intellectual property protection was presented for consideration as an integral part of the development process, and the issue of product liability was presented in light of the need for a quality management system and the need for documentation.

With the execution of the development well underway, we are now ready to focus on the manufacturing development process, which is an integral extension to the development process.

MANUFACTURING DEVELOPMENT

MANUFACTURING PROGRAM MANAGEMENT

1. Concurrency of Development Efforts

A. Manufacturing—the forgotten development?

Manufacturing development is an integral part of the new product development process. Often, a development occurs in a manufacturing vacuum, with the result that the new product might suffer lower than expected gross margin or, worse yet, might entirely fail to capitalize on the market opportunity.

An enterprise cannot add on the manufacturing process, factor it in at a later date, or ignore it and still reap the rewards of the product investment. Manufacturing must be an integral process developed as part of the product. When factoring design to cost, the manufacturing element represents a significant portion of the direct and indirect costs associated with the product. Failure to address the issue properly results in a nonoptimized situation.

B. Manufacturing is part of the product development process

The importance of integrating the manufacturing process into the product development process cannot be overstated. There are trade-offs between design and manufacturing that must be reconciled as part of the development project rather than in production, where cost usually increases. The manufacturing process must be designed while the product is being designed, and both design phases must be concurrent.

Manufacturing cannot drive the product design solely, nor can development drive the design in a vacuum of other processes. All processes must work concurrently to optimize the solution offered to the marketplace.

C. Outline the deliverables

Design engineering has deliverables to mark the completion of certain steps in the design process; the manufacturing system should be developed in a similar manner. One can think of the correlation between these two elements as illustrated in Figure 8-1.

Design / Development

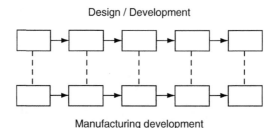

Manufacturing development

Figure 8-1. Correlating design development with manufacturing developement.

Each subelement of the product design should be a factor in the manufacturing system setup. In this way, a manufacturing system can be designed and laid out to manufacture effectively, with synchronization and gathering appropriate supporting data for management reporting. As we will see later in the chapter, the manufacturing system involves more than putting assemblies together. It is a critical feedback loop in the process of bringing the product to the customer. Along with the feedback loop is the need for objective data to evaluate performance and effect corrective action.

D. Laying out the manufacturing line

As the product is being developed, the layout of the product line manufacturing should be completed. If the product is composed of subassemblies, then each subassembly feeder line should be laid out and optimized. The manufacturing line should function in a synchronized fashion with all elements operating on a common time base.

E. Integrating the supplier element

The supplier element is a key factor in facilitating the manufacturing process. The supplier is a partner in the process. As the product and the manufacturing process are designed, the third step is to bring in the suppliers as an integral part of manufacturing.

The organization can leverage off of the work the suppliers do in being at the forefront of their technologies. The suppliers can assist in the development effort by introducing new manufacturing concepts to the organization. They have the benefit of involvement with other manufacturers and can be a key link to state-of-the-art ideas.

2. Integration of Multiple Disciplines into the Process

A. Diagram flow of multiple disciplines

The development of a new product is an interdisciplinary process. All aspects of the organization should be represented in the program. These include development, quality, purchasing, sales and marketing, finance, and manufacturing. Each group contributes its portion vocationally and serves as a check against other areas where vocational conflicts may naturally occur. For example, sales and marketing keep development in check against cost overruns. Manufacturing keeps development changes in check. Accounting and finance hold the purse strings tight on expenditures. Quality keeps manufacturing processes in balance. Figure 8-2 presents these interrelationships.

B. Give and take between manufacturing and development

Part of the integration process between the design engineering group and the manufacturing group lies in the give and take between them. Neither group can secure 100% of the desired program elements program to satisfy its parochial goals. Rather, the success of the new product lies in the cooperation that must take place between the two groups.

Significant progress toward cost reduction and improved quality can be made by design engineering assuming responsibility for the reduction of manufacturing problems by virtue of design. If, however, the design is conducted without reference to manufacturing and the rest of the company, little is gained because the design group is often unaware of the potential problems of manufacturing.

Some companies resolve this issue of cooperation by periodically assigning development engineers to work in manufacturing and requiring them to produce the same results expected of manufacturing personnel. This practice can be healthy in terms of highlighting manufacturing problems to the development people directly.

C. The partnership

The interface between the departments therefore should be one of partnership in overcoming the dynamics of the competitive marketplace for the financial benefit of the company. Often, the relationship can degenerate into competition between departments; however, it is the program manager's responsibility to galvanize the departments to focus on accomplishing the greater goal. If compromise is required, there should be an accompanying action plan to optimize it at a later date.

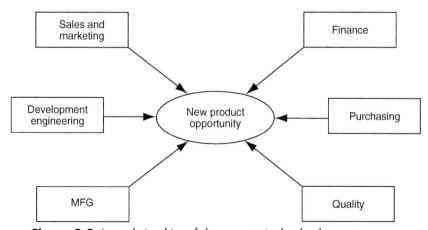

Figure 8-2. Interrelationships of departments in the development process.

D. Manufacturability: It starts and ends with design

Up until this point, the focus of new product development has been from the perspectives of design, functionality, and cost. The focus must now be broadened to include manufacturability. The ease of manufacturing a product does not originate in manufacturing, nor can it be accomplished by the addition of capital equipment after the design process has been completed. Ease of manufacturing must be integral to the program. Furthermore, there are several basic elements that comprise a manufacturable product. These will be reviewed in more detail in the next section.

The issue of manufacturability goes beyond margin improvement and the organization's ability to harvest the new product development investment effectively. It goes directly to the issue of produceability. What if your sales channel brings in significant amounts of business—can your manufacturing organization deliver? If it is ill-equipped to produce, the entire effort suffers at the last stage. Keep in mind the lessons of lost opportunities, bloated inventories, incorrect inventory parts forecasts, poor throughput, and problems due to routing issues or manufacturing gridlock.

E. Manufacturing can't fix what development doesn't provide for

In many cases, the organization has high expectations of manufacturing. However, manufacturing cannot repair or create what development has not provided for in the new product design. Development ultimately must have the vision of how the product is configured and assembled. To be effective, the development people must have a working knowledge of manufacturing.

The team concept is somewhat helpful in accomplishing this. In a team atmosphere, however, development has more often than not designed the product by the time the first meeting with manufacturing occurs. The manufacturing development personnel merely react as best as they can to the product concept. A better approach is to require several product concepts to be produced by engineering and manufacturing together, and then select the most appropriate platform. It is even healthy to summarily reject a product concept that came solely from development, and to require interdisciplinary involvement in producing the concepts.

F. Considering the diverse and fast-changing workforce

"I am not so sure about these new kids..." How often have you heard this statement? It or something like it has probably referred to each successive generation. Fundamentally, however, it underscores the basic fact that the workforce is ever-changing and is composed of personnel with divergent viewpoints and value systems. Most likely, gone are the days of the uniform "nose to the grindstone" work ethic among an entire workforce. Individualism will reign and the manager's responsibility is to accommodate and make use of the strengths of individual personnel to accomplish the corporate goals.

For manufacturing, this means continually dealing with training, tardiness, attitudinal, and work quality issues. In addition, economics and shifts in focus will cause periodic spot shortages and surpluses in trained personnel. To a certain extent, design can accommodate the work quality issues by designing for one-way fits, elimination of errors, and product configuration. To leverage profits from a workforce, however, uniform and consistent effort must be directed to the manufacturing tasks at hand in spite of the diversity in the human resources applied.

G. Manufacturing as a competitive weapon

With a product diligently integrated into manufacturing, and an effective sales channel, manufacturing can be a formidable competitive weapon. Consider the power an organization has when it can produce a product desired by the market and can deliver it in significant quantity relative to the market segment, and has the potential to generate significant amounts of profit. This can be a prescription for immediate success and the building of long-term momentum for the corporation.

Certain market segments can be defended to the point of locking out competition forever. The strength your company gains will cause your competitors to become weaker not only in a relative but also in an absolute sense, as they are losing sales with nothing else to fill in the shortfall.

DESIGN FOR MANUFACTURING

1. Manufacturing Needs

At this point in the discussion it is a good idea to review the basic tenets of design for manufacturability. The following section outlines a list of desirable attributes of a product design to facilitate manufacturing. They are mostly common-sense items but they are soon compromised as the development team begins to get into the project. In fact, a great deal of the compromise originates in the mind of the development engineer even before the first team meeting. Often, development engineers fixate on their concept and defend it to the detriment of manufacturability. The challenge for management is to refocus on these basic design tenets throughout the development process—especially at the beginning, when compromise is irreversible.

A. Philosophy

The basic philosophy behind design for manufacturability is to design, arrange, and position elements to go together in a preferentially compatible manner. This means that parts go together with minimal effort, force, fitting, and fastening. It takes advantage of natural forces such as gravity and accommodates the use of fixturing and tooling to speed up the process while ensuring an acceptable level of quality.

B. Minimizing the number of parts

One of the easiest ways to improve a design for manufacturability is to reduce the raw number of parts. This is accomplished by integrating functionality within a more highly engineered and tooled part. By combining functionality and features in these parts, their sheer number can be reduced. A reduction in parts count helps reduce inventory, space, work in progress, and potential scrap. Also, it reduces the per-unit consumption of materials and supplies.

C. Consider modular design

A modular design reduces manufacturing costs by allowing the use of subassemblies. It allows feeder lines to be set up in manufacturing and adds quality checks and controllability earlier in the assembly process. The use of known goods and certified subassemblies in final production results in a faster final assembly as well as improved first-pass yield.

D. Take advantage of gravity

As the assembly process is laid out, parts and subassemblies should drop into place and be secured by integral fasteners wherever possible. Take advantage of gravity, or at least do not fight it. In this manner, you can reduce the time to manufacture by reducing the amount of human intervention and repositioning. It also can pave the way toward automated assembly.

E. Reduce variability

The design of the product should aim for reduced part variability. This means that if two parts or subassemblies are nearly alike in different versions of the product, the designer should consider reducing the variability by designing one part or subassembly that will accomplish both functions. This tactic reduces inventory levels and also improves throughput by eliminating "wrong parts only" in inventory. A reduction in highly similar parts also improves quality and eventually contributes to reduced cost because quantities of identical parts are increased.

F. Provide easy access

Easy access to critical or important components is both a serviceability issue and an assembly issue. Access to such components increases the availability of equipment by reducing the mean time to repair (MTTR). It also reduces manufacturing time by allowing free movement within the subassembly.

Providing for easy access does not necessarily cause an increased amount of product real estate if the design is handled correctly. Creative and frugal use of product real estate can ease these access issues.

G. Eliminate fasteners

Fasteners are an age-old means for assembling parts and assemblies. Although they are convenient, relatively low in cost, and allow a certain amount of flexibility, they add labor cost, contribute to repetitive motion injuries, and increase the likelihood of assembly mistakes. Their use generally requires additional procedures and tooling to ensure proper component torque. A part or assembly that can have a built-in fastener results in reduced mistakes, improvement in quality, and reduced hazards within the workforce.

H. Increase part symmetry

Although not necessarily a requirement for manufacturability, the use of symmetry can reduce certain costs when secondary operations are required. For example, machining a certain assembly can be made easier by the use of symmetry—reducing setup time and fixture placement on subsequent operations. Symmetry also has certain advantages where material handling is concerned. Fixtures for handling can be consolidated and additional flexibility can be created.

I. Ease of part handling

The issues of material handling and part placement do not need to be limited to the consideration of symmetry alone. They can extend to all parts and assemblies in a product. Ease of part handling contributes to lower cost of material movement tooling, better quality due to fewer mishaps, and a more laminar process flow. The parts should be designed for ease of alignment and ease in maintaining location. They should fit into their locations easily without the use of alignment, locating, and tightening operations.

J. Design parts with vendors

Become proactive in seeking vendor assistance in designing product parts. They have the knowledge of the state of the art in part design, materials, processes, and trade secrets. The proprietary nature of the part can be preserved while still taking advantage of vendors' expertise.

K. Create processes to eliminate errors

Reduce dependence on the workforce to operate as craftsmen by designing the parts and the processes of assembling those parts to eliminate errors in assembly. It is futile to attempt to inspect quality into a product after the manufacturing operation is completed. Splitting up the inspection process and placing it at the level of individual operations only mitigates the amount of waste; it does not prevent errors completely. The prevention of errors and subsequent waste must be systemic, not subject to human diligence.

L. Refine processes to reduce per-unit resource consumption

Design and manufacturing processes must strive to reduce material consumption for each part utilized. For example, if adhesive is used to join parts, can the parts be designed to snap together and hold rather than relying on adhesive? This would eliminate concerns about the temperature of the adhesive at application, the cleanliness of the parts, the receipt of the correct adhesive in the receiving department, and the human craftsmanship required to apply the adhesive and join the parts. Clearly, there are many cost savings and quality improvements that can be gained from this philosophy of design and manufacturing.

M. Create a process to build and ship rapidly

Executive to materials manager: "Make sure we always have enough parts and materials to build the products." Although the goal of meeting shipments and customer commitments is a noble one, the focus should be not on laying inventory in reserve but, rather, on designing manufacturing process that are rapid in terms of throughput. By focusing on throughput, inventory is reduced, work in process can be reduced, and the labor dollars sitting in inventory waiting to be consumed are eliminated. The clear directive here: Focus on rapid manufacturing throughput, not exclusively on inventory.

N. Create a mindset of minimization through the manufacturing pipeline

The product team should develop a philosophy of having manufacturing endeavor to pull the product through the process frugally (from the perspective of materials and labor). This can be accomplished not by degrading quality but, rather, by striving for the minimum material requirements while still achieving the quality measurable and customer satisfaction. Often, the product progresses through manufacturing with labor being thrown at it, accumulating costs and variability.

2. Manufacturing and Process Layout

A. The solution depends on the product

There is no single optimized manufacturing system for all product lines. Each product line within the organization might require its own optimized solution. A general layout for a product composed of discrete parts is represented in Figure 8-3.

The example in Figure 8-3 consists of five stations of different operations. These are denoted as Operation 1 through Operation 5. Parts bins feed each station and the personnel at each station assemble the parts into assemblies. The personnel are also responsible for inputting manufacturing data into the networked computers supporting the information system. Separate feeder lines are staffed by personnel who feed parts to the major line. All of these items must be coordinated and synchronized. With proper synchronization and layout, this short line can produce product and keep track of manufacturing data.

B. Review industry practices

When setting up a manufacturing line for a product, it is often helpful to gather data about other products that might be similar in terms of both configuration and assembly methods. Do the necessary detective work to get detailed information about the following:

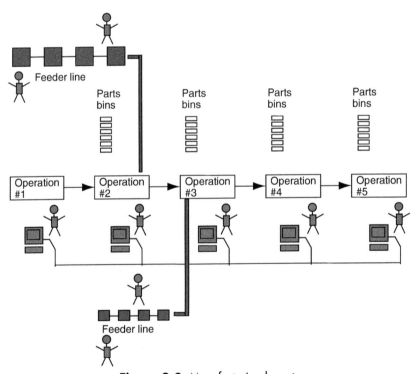

Figure 8-3. Manufacturing layout.

1. Processes

Determine the type of process capability the competitors have. How do they control their processes and what are the measurements for each process used? Then design a manufacturing system that can exceed these numbers, if possible.

2. Yields

What are the expected and controllable yields used in industry today? Are these numbers achievable in your environment? Determine why or why not, and implement corrective action.

3. Cost structures

Find out what the typical cost structures are for the type of assembly and construction techniques your company uses or anticipates. What is a typical labor percentage for this type of construction? Can a similar percentage be achieved at your company? If the assembly technology limits the margin by virtue of the cost structure, there could be minimal payoff in trying to optimize the existing approach. It might be preferable to focus on a newer or more radical approach to optimize profit.

4. Manufacturing throughput

Determine the manufacturing throughput available for similar processes and products in your industry today. Is this the best that can be hoped for and planned for in your situation? Are you in a position to increase capacity easily? Is there a plan to increase capacity incrementally?

5. Indirect support issues

What is the indirect support required to produce the volumes anticipated? Is engineering going to be spending all of its time supporting manufacturing because of engineering circumstances? How much of the engineering talent pool will be absorbed by the need to provide indirect support?

6. Product life cycle

How rapidly will the product life cycle turn from version to version? Will an initial product platform offering have a two-year run before an enhancement becomes necessary? When will market needs demand a completely new platform? What is the window of opportunity to secure planned profit? Will there ever be a pure revenue harvest mode of operation?

7. Redesign frequency

Do products of this type require constant redesign attention as they are manufactured? Do they have a high sensitivity to purchasing, supplier, and variability issues that require constant product maintenance, or are they stable in nature? Each of these elements requires careful consideration and planning to understand the market expectations, the organization's objectives, and the need to exceed the competitors' ability to produce product for delivery to the marketplace.

C. Realistic objectives and numbers

As stated previously, there is a give and take in setting up manufacturing. The best design that calls for the best tooling and capital equipment might not be warranted if the projected lifetime product volume is too low; the return on investment would be poor in this case. Rather, the investment and tooling expenses must be commensurate with the anticipated volume, profit potential, and financial return on the product.

D. How to set up for competitive battle

Is your manufacturing organization set up for competitive battle? Do you have the ability to fulfill large volumes of orders? Can you respond to opportunities and deliver goods and services to meet customers' expectations? These and other questions require careful consideration prior to organizing the manufacturing aspect of the new product.

If you are engaged in a high-volume market opportunity, secure arrangements with suppliers with volume commitments to protect margins and allow response to competitive price pressure. If the product is highly engineered, secure and retain engineering talent to pursue opportunities in a timely manner. If the product is to be purchased as a commodity, ensure that you are set up to be a low-cost producer while minimizing sensitivity to cost changes.

E. Perspectives on inventory

Inventory, by its very nature, serves a temporary holding function for materials and assemblies; its primary purpose is to match imbalances in unsynchronized manufacturing processes. If the output of Process 1 cannot go directly in a balanced fashion to Process 2, there needs to be some provision for inventory so that one process in the chain does not run dry as the line is run.

In a similar manner, using inventory is a popular method for delivering orders rapidly. This goal is accomplished by having subassemblies on hand for quick final assembly. The rapidity with which a product proliferates different versions through changes in hardware determines the extent to which an organization will most likely have the wrong inventory on hand to fill an order.

In practice, inventory seems to have an opposite effect to the one desired. It seems to slow down an organization's ability to react nimbly to the changing marketplace. There is a cycle time associated with completing assemblies, which carries a cost in materials and labor added to it. Stock jobs (in which a run of assemblies is completed to store in inventory for future use) seem to absorb labor, rarely get completed, and linger within the manufacturing organization. These are signs of an organization burdened with inventory. The goal is to synchronize manufacturing to reduce the level of inventory required to execute a given level of business. It is desirable to run the business with the minimum amount of inventory by organizing suppliers, controlling product configuration, and establishing an effective mechanism for forecasting. With these strategies in place, the correct inventory is delivered for secondary operations and the product is completed at the proper time for shipment to the customer. Stock room space is reduced and material handling (which becomes overbearing with bloated inventories) is controlled.

F. Changing parts in a fast-paced environment

The consequences of changing a part in an assembly reach farther than most people realize. As will be discussed in the next section, the configuration of a product is almost a sacred trust between the design function, procurement, and the manufacturing function. When a part changes, it affects the product integrity, performance, certification, and liability.

Therefore, changing a part must have a strict protocol, regardless of whether the impetus for the change is internal or vendor-driven. To understand the significance of changing a part, consider a simple manufacturing model into which a replacement part is introduced, and its effect on the overall system. Figure 8-4 illustrates the issue.

Parts substitution relates to form, fit, and function. If any of these elements of a part changes, there will be an impact on the final product. Consider, for example, the manufacturing line shown in Figure 8-4. Its five operations are fed by a feeder line where subassembly is done and by parts bins located at the line.

If a part in the parts bins changes in form, fit, or function, several aspects must be considered. A decision must be made whether to substitute the new part immediately or to consume the existing parts first. In the figure, the dashed line going around the existing part denotes an immediate substitution. In either case, the following elements are affected and must be verified:

- Design change review
- Testing and verification
- Assessment of manufacturability
- Process changes
- Work in process, field recall, or both

These issues affect the final product and must be investigated. If the substitution is immediate, there is little time to investigate. If the substitution does not take place immediately, there might be more time available while the existing parts are being consumed. Scrapping existing parts can be expensive, so there might be a tendency to consume existing parts first;

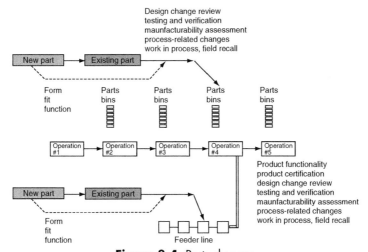

Figure 8-4. Parts changes.

however, a safety-related parts change could mandate an immediate substitution regardless of the impact on cost.

Similarly, a parts substitution at the feeder line might require investigating additional elements. Because the feeder line has its own manufacturability issues that dovetail with those of the main product manufacturing, the following issues must also be considered:

- Product functionality and performance
- Product certification

If the parts change, the product might have to be recertified, depending on the degree of difference between the existing part and the new part. Consequently, these changes need to be managed and grouped together or the product will have little stability in terms of design, cost, or manufacturing.

G. Designing the checks up-front

To minimize the disruption of parts substitution when the product is in manufacturing, certain steps can be taken at the part selection and specification stages. Several issues contribute to a part profile and need to be considered:

1. Price and price stability

Price and price stability are measures of the cost contribution to the product. They have several elements that range from initial cost, long-term price stability with reference to outside driving forces, and total installed cost, including handling and inventory arrangements.

2. Availability

Availability is a measure of the organization's ability and energy required to secure the part for use in the product. Is the part readily available for general use? Does it satisfy export compliance directives in all international manufacturing venues? The best specified part is of little value in generating profit for the company if it is not readily available.

3. Performance

Part specifications can play an important role in its overall use. If the part parameters are right on the tolerance edge, there could be additional qualification requirements in production to ensure functionality. If the part has specified parameters that exceed the minimum requirements, however, the tolerance/performance gap might be favorable enough to eliminate the need for additional production measures.

4. Quality

Quality is the measure of how repeatably the organization can produce parts or product that meet the requirements specified in the drawings and specifications. Clearly, the degree to which these match will alleviate long-term problems in using a part. If the production quality varies widely, there could be problems.

5. Reliability

Reliability is a measure of how long the part will meet it specifications. This must be correlated to the company's claims for product reliability and life.

6. Commonality

This is a measure of how specialized the part is. If it is a highly engineered part, you will be subject to supplier limitations in quality and delivery. If common parts are used, there is more flexibility in the product line over the longer term and in the organization's ability to support it.

7. Ease of use

Does the part require special handling or procedures that the organization must put in place? Can unskilled or semiskilled personnel handle the part? These types of issues determine the part's ease-of-use factor. Try to specify parts that don't require special handling, as these affect manufacturing, service, and repair, especially when these latter functions are off-site.

8. Integration with manufacturing processes

How does the part fit into the present manufacturing system? Can the present or planned manufacturing process and process control handle the part? Be sure to specify parts that are correlated with capital equipment plans and capabilities.

9. Longevity of run

If applicable, determine how long the vendor plans to run the part. This will determine the redesign cycle for the product. Common parts seem to have longer runs than do highly specialized parts. In addition, highly specialized parts generally increase in cost as time progresses because the supplier has no other customers to sell them to and learning-curve cost reduction doesn't take place.

10. Last-time buy arrangements

Determine the supplier's policy for last-time part buys. Also determine the advance warning time and the general time allowed for last-time buys. This will assist the company in obsolescence plans and redesign/replacement efforts.

11. Replacement factor in the product design

It is important to understand clearly and document the product's dependence on the part from a performance perspective. Is the part easily replaceable or is it designed into the product in such a way as to make it very difficult to replace? Knowing the answer will save time and development energy in the future if you don't design yourself into a corner at the outset.

12. Industry shortages and sensitivity to volume

Finally, what industry dynamics can affect part pricing and availability? Project the usage forward and try to determine the actions of the parts market (possibly your company's competitor) in the future; factor these into your plans accordingly.

3. Product Configuration

Careful consideration must be given to how the product line is conceptualized and structured. In this section we review the product configuration, which is the protocol for the various versions to be offered as well as how the versions are organized from a design and manufacturing point of view. The product configuration can contribute to the success or failure of the product. Good design practice and experience at the design level can alleviate future problems in manufacturing.

A. Design stability

Progress in manufacturing efficiency cannot be achieved without design stability. In the quest for a new product release to the marketplace, companies often release product to manufacturing before the design is complete. In these cases, further development changes to the product cause multiple problems in design and manufacturing because the design is still in transition. Ensure that the design is fundamentally frozen prior to a full-scale manufacturing run.

B. Manufacturing stability

With the design frozen, the team can concentrate on manufacturing process capability. This is a measure of manufacturing performance in terms of quality, repeatability, cost-effectiveness, and throughput. Process capability assumes that the design is stable and considers only manufacturing variability. Manufacturing variability can be controlled by implementing documented procedures and work instructions. These are medium-independent but necessary, along with employee training and refresher courses to ensure consistency in the processes.

C. Executing a change

Executing a change in manufacturing can have a serious impact on the product integrity. Previously, we discussed the coordination requirements involved in substituting a part. A change to the product is even more significant in that the entire battery of product certification testing might have to be redone. Given the seriousness of a non-part-substitution change, companies should establish a procedure that must be followed to ensure preservation of the product configuration and integrity as a result of the change.

Product changes can have effects in the future, depending on the compatibility of the change with future product. Two basic terms describe the issue:

- Forward compatibility
- Backward compatibility

Forward compatibility ensures that the product line can support change in the future by designing a product that allows a future improvement, enhancement, or change to be compatible with the host product being shipped now. Today's product is designed to be repaired with future parts in mind.

Backward compatibility ensures that previously designed and manufactured units are compatible with new parts. Yesterday's units can accept today's parts. The current product is backward-compatible with the old version (parts). Both are similar. Whether you are looking to accommodate the future version or past version is more of a time reference.

Compatibility is one of the most important aspects of a product change to handle correctly. Failure to do so can cause huge headaches in repair, service, and spares. One can imagine a compatibility problem when a host unit in the field needs repair and the new replacement part is incompatible. This circumstance would force replacement with a totally new unit at a significant cost penalty to either the customer or the company itself. An alternative would be to establish a new part number for the replacement (incompatible) unit and to keep a stock of the old style. The problem is that many times a part is obsolete due to inability to buy subcomponent parts for it. It is worth the extra design effort to design for forward and backward compatibility wherever possible.

D. Formal process for requests and changes

In order to execute a change correctly within the organization and on the product line, an effective and formal change system should be in place. Informal, "by-the-seat-of-the-pants," undocumented changes are not acceptable in today's manufacturing environment. Because the development process is interdisciplinary, so should be the change process. In addition, executing a change is more difficult than the original development in terms of managing actual cutover and tracking for future reference. Contrary to the beliefs of some, the process of formalized change involves more than adding or substituting on a bill of material or changing tooling.

As shown in Figure 8-5, the formal change process must be marked by the entry of a change request that is considered by a cognizant party, an investigation and impact assessment, planning the actual change, and only then the actual implementation. Implementation is followed by an audit of the change—a process that is frequently overlooked or forgotten.

Important elements of the change process are the initial request and the evaluation, whereby the seriousness and timing of the change are determined. These steps determine how the change will be engaged and executed by the organization.

E. Where the process leaves off

It cannot be overstated that the process of changing the product does not stop at the implementation. It stops only after the audit of the change. The audit of the change considers the dynamics of the change, the specifics of manufacturing implementation, and, equally important, the response from the field in terms of effectiveness and whether any new issues or problems were introduced.

F. Speed of execution: Documentation and approvals

What is the price to be paid for all of the diligence and approvals involved with a formal change system? Undisciplined people profess that diligence will cause significant delays in any improvements. In fact, the diligence required might prevent mistakes in the so-called improvements. Certainly, the organization can execute the change in an informed manner. Figure 8-6 illustrates time frames that vary with the level of diligence and bureaucracy involved with the changes.

The straight line illustrates the difficulty of change with the addition of diligence in executing a change. The graph does not attempt to depict the time required for a change; if there is a

Figure 8-5. Formal change process.

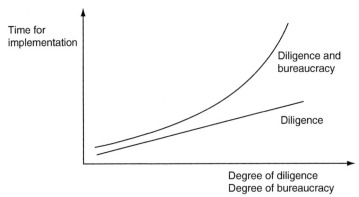

Figure 8-6. Change difficulty.

mistake, that will have to be corrected later (it will extend the x-axis of the characteristic). Referring to the curved line, the more checks and balances and approvals required, the longer the time required to change the product. If the organization is lethargic and laden with a certain amount of bureaucracy, the combination of approvals and bureaucracy will cause gridlock. This is unhealthy and can cause the product line to fail. The solution is to reduce the bureaucracy, not eliminate diligence and approvals. Fundamentally, a balance needs to be struck between the need for diligence and the need to execute a product change in timely manner.

G. Suppliers

What about supplier-driven changes? The product development team must be careful when coping with supplier-driven obsolescence. Parts obsolescence causes disruption in the product flow, consumes development resources for product maintenance, and creates unpredictability. The organization must be fully aware of the suppliers' obsolescence plans, and last-time buys and phasing of replacement parts must be coordinated in manageable "block" changes to the product. Otherwise, the product is completely at the mercy of the vendors' plans, not yours.

H. Cost incrementalism versus cost reduction

It happens all too frequently: A parts substitution for a soon-to-be-obsolete part is available, but it costs slightly more. It is a fact that last-time parts buys generally carry a heavier price tag. The dynamics of the change cause the product team to weigh convenience against increased product cost. In and of itself, a simple increase in a single part will not cause product failure; however, this trend over a range of parts will eventually affect profitability.

4. Changes to the Product Line: Development Versus Production

A. Format for a change system

An engineering change system consists of two basic procedures:

- Engineering change request
- Engineering change

The *engineering change request* is a formal request for consideration to the personnel who are cognizant of the product configuration. It is merely a request that initiates investigative action; it does not initiate change. It is accepted or rejected or is placed in a pending state to be addressed later. A good engineering change system also provides a means for ensuring that each request is addressed and followed-up on. In addition, to be effective, the system should provide for communication back to the initiator regarding status and disposition.

An *engineering change* is the pathway and the record of the initiation, execution, and resolution of a change to the product. It also includes provisions to audit the change at a later date. The engineering change initiates action to change the product configuration. A protocol is required to execute the change so that the appropriate approvals, investigative work, and documentation are completed. Failure to enforce these procedures results in unauthorized, incomplete changes and loss of product configuration control.

The change in the product configuration should involve all affected departments and secure their approval. Those departments not directly affected should, at minimum, be notified as part of the process.

An engineering change system is not a substitute for a product development system. It governs the change and the recertification of the product only as it relates to the change. It is not a means for generating and certifying a new product. Major changes and new product programs should be organized under a new product development process. Although each company may create their own engineering change system, the following sections describe one that could be implemented. It includes many items and decisions that must be made that impact inventory and phasing-in the change.

B. Development versus production

Depending on the size of the organization, the scope of the change, and the complexity of the product, it could be necessary to track product changes before production occurs. Tracking these changes assists in securing the correct parts from the bills of material, should they change. Tracking also can measure the stability of the development effort and can be used to forecast whether development will complete its activities on time. It may be a good idea to institute this system for the development group if it has been newly assembled within the organization. For the development group, a tracking system can minimize development efforts that constantly redirect the organization with product changes, which would make manufacturing development all the more difficult. A tracking system is mandatory for production, however.

C. Documentation

The documentation for a product change should be a small-scale set of documentation for the initial product offering. It needs to track the change from initiation through to the audit. If a change request is entered and rejected, the documentation should reflect the dates, the circumstances, the assumptions, and the decision makers. If the change is implemented, the documentation should chart the progress, the results, and the cutover date. The need for procedures and documentation might seem overbearing and bureaucratic, but they could be required to track down a field problem or defend against litigation.

D. Communication

The most helpful partner to engineering change is a company-wide network of communication. Effective and accurate communication can overcome a lack of a system for executing a change, but the lack of communications cannot be made up for with a change system. It is crucial to keep communications open during the various phases of the change process, from request through audit. Here, too, there should be a protocol for communication between the various company departments and for communication to external parties as well.

E. Approvals

An approval protocol for any changes to the product should be in place. This protocol should be strict, as anything less will eventually degrade the change system altogether. This protocol should have backup in terms of personnel in case the primary personnel become unavailable. Danger generally surfaces when the primary personnel are unavailable for approval and the backup personnel must fill in. It is important that they follow the same checklist for approval as do the primary personnel. When protocol is broken, even once, the system begins to decay and eventually the product configuration is compromised. Disregarding the protocol also exposes the organization to liability if there is an incident with a product. Therefore, it is essential to establish the protocol and adhere to it.

F. Coordination

The coordination involved with a change is generally outlined by the procedures of the change system. Care should be taken to clearly spell out the procedures, the functional areas, and the levels of approvals required. Do not leave the coordination to chance or to communication. Establish a procedure and adhere to it.

To reinforce the importance of coordination in a component change, one can imagine the collisions and voids caused by two parts arriving for assembly—one being the old version and the other being the new version. Similarly, one can envision the interruption in production caused by the new part not being available and the old part depleted out of inventory. Furthermore, one can imagine the rework associated with a production run started without knowledge of an engineering change that requires an immediate part substitution.

Fundamentally, effective coordination in product changes should be enforced or there will be wasted time, effort, labor, and materials, and the product integrity will again be compromised.

PROCESS COMPETENCIES

1. Manufacturing Process Control

A. Manufacturing process documentation

There is always an information void between the development-generated drawings and bills of material and the complete set of documentation required to do an effective manufacturing job. This void is filled by the manufacturing process documentation—a methods and procedures document showing how the product goes together. Such documentation is also the gauge by which manufacturing will be measured and provides the continuity needed in manufacturing. Keep in mind that manufacturing deals with people-related variables, and establishing a uniform skill level in the personnel requires effective process documentation.

B. Methods and process documentation

Process documentation generally consists of two types of documentation:

- Methods
- Process

The *methods documentation* covers specific procedural approaches to be used by the manufacturing personnel. It describes how assemblies are to be put together. This documentation can take the form of drawings, illustrations, and exploded views that serve as procedural reminders to the personnel. They include both pictorial elements and written instructions.

The *process documentation* describes how the process-oriented elements of manufacturing are to be conducted and controlled. Its purpose is to ensure that a repeatable result is achieved for a given set of conditions with given equipment. These include ambient condition data points for machinery involved in the processes.

C. Perspectives on documentation systems

Documentation systems tend to be medium-independent. It might be wise, however, to consider a computer-based set of documentation whereby a networked system can be accessed from server stations and, most important, wherein the content is controlled via the computer, not on paper. Such a system renders all paper copies uncontrolled; the computer system has the only official content. The electronic method allows the information be to be updated and controlled in real time, and revisions can be time- and date-stamped.

D. Operating in a fast-paced, high-volume arena

Making process changes in a fast-paced environment can be challenging for the manufacturer. Fundamentally, the process changes must be understood, controlled, and predictable before cutover to the change is made. Failure to follow these principles can introduce new unknowns into the process and complicate resolution to an acceptable yield. Consider Figure 8-7, which illustrates this issue.

In this example, Operations 1 through 4 represent the standard manufacturing processes. If a process change occurs and results in a nonconformance, the parts from Operation 1 must be segregated somewhere off-line; otherwise, the production line must be stopped. This example applies to a process situation, not to a discrete parts manufacturing situation.

The pathway labeled Contingency Flow is simply the pathway for the incorrectly processed parts to take. This pathway must return to Operation 2 somehow. This situation is especially likely to arise with very-high-volume manufacturing. One can see that the inconvenience of this route can disrupt operations and wreak havoc on the quality of the end product. In addition, it is only a temporary holding area that will soon fill up and eventually stop the line anyway.

A similar contingency flow pathway is also shown occurring between Operations 3 and 4. The manufacturing time base signal depicted at the bottom of the illustration sets the pace

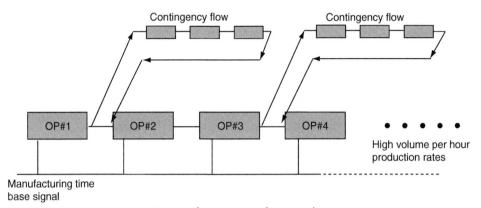

Figure 8-7. Manufacturing in a fast-paced environment.

for production. Detouring to the contingency pathway disrupts the timing. Therefore, it is best to certify the process change off-line or while the line is shut down before taking a chance on phasing-in the process change and risking scrap and schedule problems.

E. Process control systems for the skilled and unskilled workforce

Differences in skill levels within the workforce dictate differences in the way the process and manufacturing control systems are handled. Training is therefore a crucial part of the new product manufacturing development. Figure 8-8 illustrates differences in the training of skilled and unskilled members of the workforce.

As the figure shows, for skilled manufacturing personnel the training is oriented toward product-specific processes. The personnel are given process instruction and are certified to a specific level. Subsequently, they are given method sheets, which serve as reminders of the training and certification previously given.

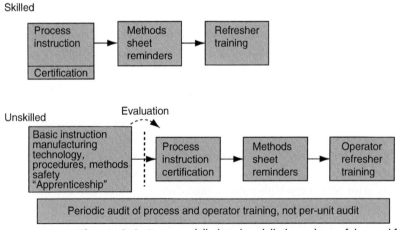

Figure 8-8. Training skilled and unskilled members of the workforce.

In the case of unskilled workers, initial training consists of basic instruction in manufacturing technology, general procedures and methods, and safety in the workplace. Workers are then evaluated to determine whether they come up to a minimum level of proficiency. Assuming this is the case, they progress to the actual product-specific training and process instruction and certification. Subsequent methods sheets serve as reminders of the training, and operator refresher training is also used. Throughout the entire process of orientation to manufacturing technology and the product-specific training, an audit function double-checks proficiency and the effectiveness of training. This type of audit is not associated with the audit steps used as part of the manufacturing process.

F. Measurement systems for process control

As part of the manufacturing development, there should be a systemic method for performance data entry and analysis. This method is used to determine yield data and analyze trends. As the products are being produced, the results of the various operations are catalogued and posted to a report. The report is analyzed periodically to identify problems slated for correction and trends indicating a degradation of the manufacturing process.

Figure 8-9 is an example of yield data in a manufacturing analysis. It represents a four-week window of data. This chart of data can be expanded to include any pertinent number

Manufacturing reporting system									
Operation	Volume for week 1	Number of reported errors	Yield data (ppm)	Volume for week 2	Number of reported errors	Yield data	Volume for week 3	Number of reported errors	Yield data
OP 1	12,250	3	245	15,000	2	133	17,500	1	57
OP 2	12,250	1	82	15,000	0	0	17,500	2	114
OP 3	12,250	5	408	15,000	4	267	17,500	3	171
OP 4	12,250	2	163	15,000	1	67	17,500	1	57
OP 5	12,250	1	82	15,000	1	67	17,500	1	57
OP 6	12,250	1	82	15,000	1	67	17,500	1	57
Cumulative Yield Data	73,500	13	177	90,000	9	100	105,000	9	86

Volume for week 4	Number of reported errors	Yield data	Total monthly volume	Number of reported errors per month	Yield data per month
20,000	4	200	64,750	10	154
20,000	2	100	64,750	5	77
20,000	1	50	64,750	13	201
20,000	1	50	64,750	5	77
20,000	2	100	64,750	5	77
20,000	1	50	64,750	4	62
120,000	11	92	388500	42	108

A

Figure 8-9. (A) Manufacturing errors.

Operation	Week 1	Week 2	Week 3	Week 4	Cumulative average by operation no.
OP 1	245	133	57	200	154
OP 2	82	0	114	100	77
OP 3	408	267	171	50	201
OP 4	163	67	57	50	77
OP 5	82	67	57	100	77
OP 6	82	67	57	50	62
Cumulative average by week	177	100	86	92	108

B

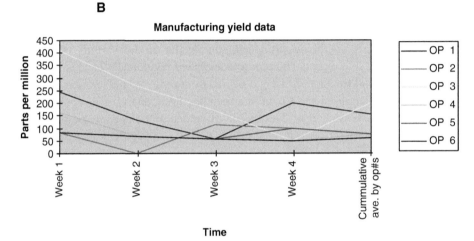

C

Figure 8-9. (Cont'd) (B) Cumulative data by operation; (C) Manufacturing yield data.

of interval data. The important factor in scaling the chart is to gather and process data at meaningful intervals to facilitate determining trends and taking corrective action.

Figure 8-9(A) is organized to record data on six operations labeled OP 1 through OP 6. In the first week, a total of 12,250 pieces of product were manufactured. The number of reported errors occurring at each operation is recorded and scaled to reflect the parts per million (PPM). In the second week, 15,000 pieces were manufactured, with the associated errors posted to that week's production and calculated PPM. At the bottom of the chart is the total number of operations (OP 1 through OP 6 at 12,250 each, or 73,500 total). This figure represents the total number of opportunities for errors. The number of errors are then totaled and the PPM is calculated. This represents the cumulative yield information on the entire assembly operation by week.

The figure also shows the total production unit volume, the total number of errors, and the calculated PPM for each operation. At the bottom of the chart are the cumulative data for all parts produced, with total errors and calculated total PPM for the month. Figure 8-9(B) gives the summary for the month, showing totals of the data.

Figure 8-9(C) illustrates the trends by operation for the four-week period. Each operation in Figure 8-9(C) is graphed and shows the trends of the operations in terms of PPM errors.

To the right of the chart, the intersection of the data line and the y-axis represents the cumulative average of the errors by operation number.

Through the use of manufacturing yield data, actual yield performance can be measured and recorded for the sake of manufacturing process improvement.

G. Verifying the process and process changes

The importance of verification and certification of the manufacturing process must be underscored. Product quality can be repeatable only if the process by which the product is made is predictable. This means that controlling the inputs and controlling the environment will yield predetermined results given that the subassembly itself is stable. Figure 8-10 illustrates this point.

As shown in Figure 8-10, the discrete parts and/or subassemblies are inputs to the process, which is influenced by the environment and the input process variables. These input process variables can be temperature, pressure, humidity, and time, for example. With the process under control, the parts exiting are predetermined. Furthermore, the process can be repeated by using the same "recipe" at a later date or in a different location. In a similar manner, any change to that process must also be verified and certified before it is integrated into manufacturing.

2. Assessment

There are two aspects to an assessment here. The first is how the process competencies measure up to the product requirements and market requirements for the new business. The second is how the manufacturing process control system stacks up against industry standards.

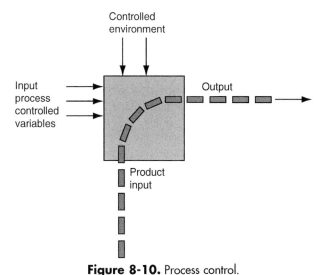

Figure 8-10. Process control.

This is an important question to assess within your operation because the marketplace expects certain levels of quality, and formal process and manufacturing documentation needs to be in place. Companies often rely on "tribal knowledge" to fill the gap. This can work if each and every employee has all of the process knowledge and applies it correctly each and every time. However, when someone is absent one day, the result could be a low quality level in shipments for that day. Back up employees may guess what these steps are, or worse yet, completely ignore the step. The cost associated with poor-quality product reaching the customer is huge compared to the cost to correct the system at the decisive point.

3. Improvement

Accordingly, the competitive issue here is improvement. How fast can we improve the process documentation and improve the level of quality, traceability, and consistency in the manufacturing process? By focusing on improvement and prevention (not additional inspection), you can utilize this to reduce the manufacturing costs. Measurables should be established for this purpose and periodic evaluations should be made.

PROCESS DEVELOPMENT

There are two different ways to develop processes: through internal means and through external means. Either is good as long as the background information accompanies either method. With internal methods, there can be a tendency to have the same problem as with process documentation—only a few select individuals might have the requisite knowledge, and manufacturing's ability to run certain processes without those people may be limited.

1. Internal

The internal method of process development is generally an expensive way to go. This usually involves procurement of tooling, machinery, and fixtures (depending on the process). In addition, training is required to certify the personnel operating the equipment. This takes time and money. However, developing this capability in-house is a good idea when the specific process is critical to manufacturing the product. When the process differentiates your firm from the competition, the process is generally a good idea to have in-house. If it is noncritical and lacks this differentiation capability, there are other options.

2. External

The external method relies on the suppliers' ability to run the processes and document process control. This is a good way to proceed when the company is small, strapped for cash, or is starting out. The vendors or suppliers can perform these operations for a per-unit cost, so the firm does not have significant start-up costs for an individual process. However, the company must ensure that the supplier documents the process parameters for each production run. In addition, the company should communicate this in real time to the customer for incorporation with their other documentation. In addition, using a supplier initially is a fine way for a company to learn about the process and its variations prior to making the procurement commitment for their own system.

OUTSOURCING

This is a topic that many companies deal with in the global economy. Many times, however, the terms *outsourced* and *offshore* are used interchangeably. A distinction must be drawn between the two. Outsourcing is a practice whereby a company has service performed by a firm other than their own. Theses services may be performed by a local supplier or by a value-added vendor. The venue of the services can be local or foreign. When it is foreign, it is generally referred to as offshore. The two do not necessarily go together.

1. Local

Outsourcing allows a firm to leverage its in-house crews by freeing them up for other tasks. The benefit of local outsourcing is immediate access to state-of-the-art capability. By teaming with an industry expert, your firm can save initial start-up costs and keep up with the rest of the industry for per-part pricing only.

2. Global

Offshore production means procuring from a vendor with operations located in a low-labor-cost country instead of a vendor in the United States or some other higher-wage country. Another example would be if you established a manufacturing operation in a low-labor-cost country that replaces a facility you were running in the United States or another high-wage country.

Global outsourcing using low-cost labor can be initially misleading. The fundamental assumption is based on somewhat uniform materials pricing worldwide and that low-cast labor can make the difference. In reality, a firm needs to calculate the *total cost* of offshore outsourcing. Often the cost benefits are calculated solely on the basis of the low labor cost. However, other costs are rarely taken into consideration because they are not allocated to the actual product but are paid for from various other sources and budgets.

One of the biggest costs initially overlooked is shipping to and from the low-cost country (e.g., Asia). A firm can ship either by air or by sea. Most products shipped by sea are packed into cargo containers. These containers are loaded onto ships and spend approximately three weeks on the high seas before arriving at port. The container costs are fixed regardless of how full it is. As much product as possible must be packed into the container to lower the per-unit cost for shipping.

By globally outsourcing you risk product quality issues and, more important, the ability to react to them immediately. A full analysis will measure opportunity costs and side effects associated with an outsourcing initiative. In addition to the total direct and logistical costs, a number of others exist, depending on the product and industry structure. Some of these costs are:

- Additional costs for outsourcing
  - Cultural/communication difficulties: a huge oversight
  - Legal issues and costs to prosecute them
  - Underestimation of the start-up costs
  - Transition costs, such as inventory levels
  - Increasing standard of living and labor costs in the low-cost country

- Cost of additional logistics paperwork
- Cost of employee morale
- Training costs
- Cost of layoffs and severance
- Perspectives on offshore outsourcing
  - Intellectual property protection is much riskier with offshore outsourcing.
  - U.S. products using a highly automated process may not show significant cost savings when produced overseas.
  - New products can have numerous revisions, driving increased quality issues. Transitions made while a shipment is enroute may need rework.
  - Product weight and size can affect logistical costs and infrastructure costs.
  - Products that require scheduling flexibility are best kept in the local venue.

TRAINING OF MANUFACTURING PERSONNEL

1. Certification of Manufacturing Personnel

A. All systems require training of manufacturing personnel

The issue of certification transcends mere training of the team. It requires an interactive exchange of data, information, procedures, and talent development. The manufacturing personnel must be in a position to implement the procedures without direct supervision. This statement might sound nontraditional at first but, in actual practice, a supervisor cannot always prevent mistakes nor can an expensive inspection process catch every mistake. The goal is to have the people doing the work be knowledgeable enough to perform the tasks without mistakes in the first place. This requires training beyond the basics of assembly. Such training must address the issues of yield data, yield pathways to improvement, and data reporting.

B. Certification of the operation stations

The goal of a manufacturing manager is to have a uniform, homogeneous workforce that is trained at the same level in all operations. This scenario allows maximum flexibility when rotating personnel and dealing with interruption of the workforce. To accomplish this goal, the manufacturing manager must train and certify each member of the manufacturing team on all of the operations. In addition, manufacturing must keep the training fresh and updated.

A record of each worker's training and proficiency should be developed and kept up-to-date. Figure 8-11 shows an example of the data that needs to be kept on each employee. The simple record keeps track of the employee, basic training requirements such as safety, and the specific training for manufacturing operations. It confirms that the training was complete and notes the date of the training.

C. Personnel refresher training

The initial training must be followed by periodic refresher training to ensure that the manufacturing process remains consistent over time. As a natural course of events, the importance of certain methods and procedures will degrade. This degradation is due to

Personnel record of certification								
Employee	Date of hire	Safety	OP 1	OP 2	OP 3	OP 4	OP 5	OP 6
1	4/2/1998	Yes	Yes 4/19/2006	Yes 4/19/2003	Yes 4/19/2005	Yes 4/19/2003	Yes 4/19/2005	Yes 4/19/2003
2	5/5/1997	Yes	Yes 4/19/2006	Yes 4/19/2003	Yes 4/19/2005	Yes 4/19/2003	Yes 4/19/2005	Yes 4/19/2003
3	3/2/1989	Yes	Yes 4/19/2006	Yes 4/19/2003	Yes 4/19/2005	Yes 4/19/2003	Yes 4/19/2005	Yes 4/19/2003
4	5/6/1999	Yes	Yes 4/19/2006	Yes 4/19/2003	Yes 4/19/2005	Yes 4/19/2003	Yes 4/19/2005	Yes 4/19/2003
5	6/7/1996	Yes	Yes 4/19/2006	Yes 4/19/2003	Yes 4/19/2005	Yes 4/19/2003	Yes 4/19/2005	Yes 4/19/2003
6	12/3/1995	Yes	Yes 5/20/2006	Yes 5/2002	Yes 5/20/2003	Yes 5/2002	Yes 5/20/2003	Yes 5/2002
7	1/4/1994	Yes	Yes 5/20/2006	Yes 5/2002	Yes 5/20/2003	Yes 5/2002	Yes 5/20/2003	Yes 5/2002
8	2/5/1994	Yes	Yes 5/20/2006	Yes 5/2002	Yes 5/20/2003	Yes 5/2002	Yes 5/20/2003	Yes 5/2002
9	3/6/1999	Yes	Yes 5/20/2006	Yes 5/2002	Yes 5/20/2003	Yes 5/2002	Yes 5/20/2003	Yes 5/2002
10	5/3/1998	Yes	Yes 5/20/2006	Yes 5/2002	Yes 5/20/2003	Yes 5/2002	Yes 5/20/2003	Yes 5/2002

Figure 8-11. Manufacturing employee certification.

the fact that human beings are involved in the process—they possess biases, experience boredom, and have selective memories. Operator refresher training can go a long way to ensure a degree of consistency in the process in spite of degradation.

D. Inspection is not a substitute for certified training

Contrary to an older school of thought, inspection is not a substitute for training and it cannot ensure any level of consistent quality. A quality-oriented attitude must simply be instilled at the beginning of the process. In addition, training teaches and encourages good habits and procedures. At best, inspection can only point out errors. To be effective, the errors must be prevented, even to the extent of taking extra time to train the personnel.

E. Handling of critical components—More than a guideline

One additional element must be considered in the area of training and procedures—the issue of handling critical components. Every product has critical components, from electronic circuits to mechanical assemblies to sensitive materials. Training in handling critical components can reduce warranty expenses and scrap significantly.

Scrap occurs in the factory when mishandling causes the part, assembly, or material to no longer be acceptable in terms of quality or function. Warranty expense occurs when mishandling in the factory causes latent failures in the field, causing customer complaints and lost reputation. Both scrap and warranty expenses cause problems, so the issue of handling critical components must not be taken lightly.

2. Backup Personnel

When staffing manufacturing personnel and training them in the specific operations, it is critical to quality to have them trained in operations other than their primary one(s). This is for two reasons. No manufacturing person should do the same operation indefinitely, and they should change their work content in terms of the operation on a regularly scheduled basis. This helps minimize short-cuts in the process that a person will naturally create, thereby missing steps.

The second reason is logistical. Broadened job responsibilities offer scheduling some flexibility in applying worker talent to a manufacturing line. If one person is not present for a certain operation, another person who is trained can take over seamlessly.

INVENTORY CONTROL

Three different types of inventory are of concern in a manufacturing environment. They are the *existing inventory*, which supports the existing business; the *aftermarket inventory*, which is used for repair parts and spare parts; and the *new inventory*, which is used to build new products. Each class of inventory generally needs to be treated differently because each contributes in a different way to the business and draws certain costs from the business.

1. Existing Inventory

The existing inventory is managed by the day-to-day business. There are industry standards on acceptable turns of inventory by product type, as well as various methods for managing the inventory. The desire is to keep it as small as possible in times when carrying charges are high, and relieving the inventory levels can free up needed cash for an organization. The business needs will generally drive the inventory requirements for the business. Inventory calculations can be made on turns requirements and forecasts. Product configuration, as we will see later, can greatly enhance a firm's inventory position or deteriorate it.

2. Aftermarket Inventory

The aftermarket inventory is treated differently. In this case, inventory items can be already obsolete, difficult to reconstitute, or may have been purchased in a last-time buy whereby the vendor no longer makes the parts. In these cases, the sell price for repair parts and spare parts is generally at a higher profit level and can justify the carrying costs of this specialized inventory. Many times the aftermarket inventory is measured and levels are judged by the same criteria as the existing inventory, but this is incorrect. This inventory philosophy is based on customer satisfaction within the framework of nonavailability. Therefore, companies must factor in the higher costs of inventory levels in their overall business model.

3. New Inventory

New inventory is handled in yet a different way. In this case, inventory can have high costs and few line items. This is because the firm is not purchasing high unit volume and may be amortizing tooling and fixturing costs. In these cases, the inventory must be

Inventory analysis					
	Total ($)	Turns per year	Aging	Financing arrangement	Obsolescence issues
Existing					
Product A					
Product B					
Product C					
Product D					
Product E					
Product F					
Aftermarket					
Part A					
Part B					
Part C					
Part D					
Part E					
New Inventory					
Product A					

Figure 8-12. Inventory evaluation.

procured in volumes sufficient to launch the product. Failure to launch because of inventory costs is a case of saving pennies and wasting dollars. The entire development program was executed to a time line and now is no time to short needed inventory because of costs. Get the inventory in and focus on selling the goods to get paid for it. The new product inventory can be balanced for cost as you ramp up production.

4. Balancing Inventory Needs

As shown in Figure 8-12, a corporation must assess its inventory needs based on its business and operate accordingly. The figure shows a simple way to evaluate the inventory and what happens to it during the year. For an enterprise to be successful in the long run, it must balance the needs of the various business units for an optimum inventory level that is affordable and still serves the needs of the business.

MANUFACTURING RECORDS

1. Types of Manufacturing Information

Given the retention of certain vendor information, what is the appropriate amount of information to keep for general manufacturing data? How much data is enough and how much is too much? These questions must be deliberated to develop a data logging system in manufacturing that can assist in resolving field problems later.

Ideally, data should be captured at the point of labor content. Each operation should log the labor content and the materials involved. The materials and parts should be serialized where appropriate, and the database should capture this serialization. Taking it one step further, if the part is used in an assembly, it should be serialized and fit into a hierarchy of serialization for the end product, as illustrated in Figure 8-13.

As shown in the figure, each critical part of an assembly should be serialized. The database should then post the serial numbers of these parts to the subassemblies. The completed subassemblies should then be assigned a serial number. Each of the subassemblies should have its own serial number, and the list of these serial numbers should be posted to the main product serial number. In this way, a field problem can be traced back to determine critical information required for remediation.

2. Reasons for Retention

Retaining the information just discussed can be very helpful in managing a product liability incident. During one of these incidents, critical parts information is required to determine the company's exposure, both financially and legally. By keeping track of the

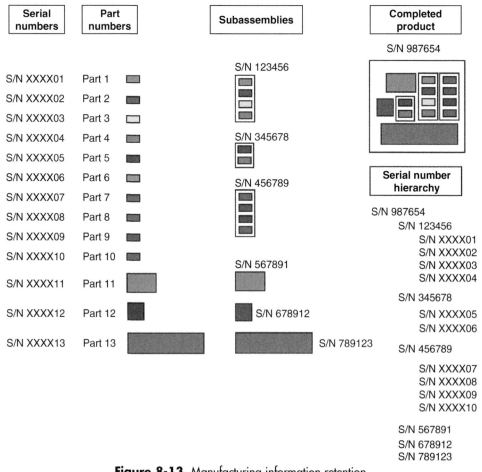

Figure 8-13. Manufacturing information retention.

vendor's serial number(s) and the product hierarchy of serial numbers, the scope of damage can be assessed readily using a database. This information is also useful in determining billing in a warranty situation. The serialized numbering system should be tied to manufacturing or receipt dates. The uncertainty in backtracking depends on inventory withdrawal practices. The parts are to be consumed in the order in which they enter the company.

The result of determining exposure to liability issues is generally a band of noncompliant product serial numbers rather than a specific list of parts. This is simply because of the size of the database that would have to be supported if every part were serialized and logged to the database. These high traceability requirements are part of the reason for the expense of military and nuclear products. Moreover, the better solution is to determine what scenarios might occur and design a practical system to minimize the amount of liability exposure.

3. Data Integrity and Accuracy Requirements

The degree to which the database system can be helpful is the degree to which accurate records are kept and posted. If the database becomes compromised, its usefulness in a product crisis becomes negligible. Be sure to make the database for manufacturing recordkeeping an integral part of the quality requirement for manufacturing. Getting the correct serial numbers posted to the parts, subassemblies, and products is as important as correct assembly. In effect, it is an integral part of the manufacturing process itself.

4. Model for Data, Root Cause Analysis, and Corrective Action

The next section presents a model for the data set, root cause analysis, and corrective action. Although this model will not necessarily fit all circumstances, it does represent and illustrate the philosophy of product control and disposition of field feedback.

The model consists of two basic reports driven by a database. The first report is the record of product return and failure. It represents the source data for the incidents. The second report accumulates information from the incident reports for compilation and analysis. The second report uses the database's capability to sort the various fields to determine trends and the status of specific issues currently affecting the product, as well as trends of historical experiences.

This model is intended to demonstrate the value of a database that can be sorted for information. Depending on the specific information desired, the database could be sorted differently. The real value of this type of database lies in its sorting capability.

Depending on your own information system and data retention methodologies, the implementation might vary somewhat; however, the basic structure remains the same. Figure 8-14(A) shows a sample format for an incident review. As shown, the incident report is divided into five major sections, as follows:

- Incident report information
- Repair analysis
- Repair information
- Failure analysis
- Remedial action

The incident report information cites basic information about the failure or return of the product. It contains logistical information and cites appropriate serial numbers and product descriptions to be used in the database.

As shown in Figure 8-14(B), after the unit is received and logged into the repair department, the repair of the unit commences. The repair analysis record cites detailed information about specifics such as part numbers, quantities, repair orders, personnel involved, production dates, and revision levels. It also records the existing revision level of the product after the repair. The repair information then dovetails into logistical repair information regarding billing and product disposition, as shown in Figure 8-14(C). The disposition might be to repair and return, to scrap the unit, or to repair and place it in rotating repair stock.

Next, as shown in Figure 8-14(D), the failure analysis is performed as part of the repair process. This process translates the repair of the product into specific categories, which might require subsequent action. In a rigorous system, every failure must result in some action to prevent future failures. The failure analysis is designed to determine the root cause of the incident and force it into a specific category so that the failure can have an associated corrective action. In this example, the vendor was determined to be the root cause of the failure. It could have been any of the other causes listed, such as design, customer abuse, or even no fault found. The information cited in these records is used to generate the summary report, which will be reviewed later.

Finally, the analysis must result in some action to remedy the incident or failure. In the sample case shown in Figure 8-14(E), the vendor failure was serious enough to warrant a design change to design around the part or the vendor. A cognizant party reviews the repair and failure analysis and a list of remedial projects is generated for engineering change. The possible action types in this section are:

- Engineering change
- Manufacturing operator refresher
- Alternate vendor selection
- Manufacturing process revision
- No action taken

The chronology of events and the data involved in the incident are included in the incident review record. This information will be used to construct the summary discussed next. The value of the summary lies in the ability to chart the incidents, determine trends, and resolve field problems with the product by sorting the data in various ways to expose the information. Figure 8-14(F) is a general format for the summary. It contains only the information shown in the example record. By itself, the one entry of data does not indicate any trend; however, the configuration of the data and the flexibility to sort it demonstrates its value.

One of the most useful ways to organize data is by sorting all of the incident reports in a multilevel format to expose the information desired. Figure 8-15 illustrates the order of the multilevel sort. For example, one can sort by product description, then by root cause, and then by production date.

This type of multilevel sort arranges the data (and carries all collateral data) by the product description first. Then it goes through each of the product descriptions in the database and groups all similar descriptions together, along with the collateral data associated with

Incident report information					
Report date	XX/XX/XX	Product family	1000		
Reported by *(Internal)*	Employee 2	User company name	BCD co.		
Reported by company	ABC co.	Contact	Mary		
Repair order reference	R 123456	Receipt date	XX/XX/XX		
Reported item description	Product chassis	Reported item number	312-345	Mfr's Serial Number	3025678.01
Description of failure or symptoms	Unit failed catastrophically				

A

Repair analysis					
Analyzed symptom	Unit in operative, fails function list				
Failed item part number	123456	Serial number. ECL[1]	9945.03		
Failed item description	Component	Production Date	XX/XX/XX		
Repair order	R 123456				
Repaired by	1668	Repair date	XX/12/XX		
Technical additional information	Upgraded to Rev 4	Updated serial number. ECL	9945.04		
Part number	Designator	Code	Def. qty	Description	
315-569	R-12	Upgrade	2	Part XX12	

[1]ECL refers to Engineering change level

B

Repair information					
Product upgraded for improved performance at no charge					
Repair disposition			Return	Scrap	Rotate

C

Figure 8-14. (A) Incident report information; (B) Repair analysis; (C) Repair information.

Failure analysis					
Root cause part number	315-569	Description	Part XX12		
Comments					
None					
Defect type	Vendor		Design Process Vendor Life Abuse Setup None Upgrade		

D

Remedial action					
Action by:	1668	Date:	XX/13/XY		
Reviewed by:	1234				
Action type:	Design Change		ECN[1] Op. refr No action Alt. Vendor Mfg rev.		
Action comments:					
Completed design change for production, field returns not yet completed					

[1]ECN refers to Engineering Change Notice

E

Figure 8-14. (Cont'd) (D) Failure analysis; (E) Remedial action.

Detailed analysis				
Production date	Part description	Defect type	Serial number	Repair order
XX/XX/XY	Product chassis	Vendor	3025678.01	R 123456
Receipt date	**Reported symptom**	**Analyzed symptom**	**Repair info**	**Root cause comment**
XX/XX/YX	Unit failed catastrophically	Unit inoperative, fails functional test	Upgraded to Rev. 4	Product upgraded for improved performance at no charge
Root cause P/N	**Action type**	**Action comment**	**Item ECL**	
315-569	Action type:	Completed design change for production, field returns not yet completed	Updated serial number. ECL	

F

Figure 8-14. (Cont'd) (F) Detailed analysis summary.

them. It will further sort within the product description categories according to the root cause of the failure that caused the incident. Next, all of the similar root cause failures within a product description are grouped. A third-level sort sorts this group of root causes by production date. This is shown in Figure 8-16.

A sort like this can link field problem incidents and their resolution. The resolution is shown by a corrective action being put in place and the absence of incidents after the production dates with the fix in place. A simpler sort of product description and root causes yields the raw data for a Pareto analysis. A sort by product description and then defect type can show development performance by charting all of the design-related defects. In short, the multilevel sort can be an invaluable tool in minimizing field problems through early recognition and resolution.

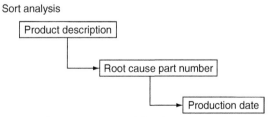

Sort analysis

Figure 8-15. Multilevel sort criteria.

Detailed analysis / example sort				
Production date	Part description	Defect type	Serial number	Repair order
XX/XX/XY	Product chassis	Vendor	3025678.01	R 123456

Receipt date	Reported symptom	Analyzed symptom	Repair info	Root cause comment
XX/XX/YX	Unit failed catastrophically	Unit inoperative, fails functional test	Upgraded to Rev. 4	Product upgraded for improved performance at no charge

Root cause P/N	Action type	Action comment	Item ECL
315-569	Action type:	Completed design change for production, field returns not yet completed	Updated serial number. ECL

Figure 8-16. Multilevel sorts of incident data.

5. Identifying Potential Field Problems

The responsible manager needs to constantly review and monitor field reports. This need is greatest immediately following introduction of a product; however, periodic review is required throughout the life of the product. What are the criteria that constitute an "event"? The answer to this question lies in the acceptable level of field returns and cost of warranty. Ultimately, the answer lies in the customer's aggravation level experienced when using the product.

Product quality events typically take on a characteristic profile. The intensity can vary but the shape of the profile seems to be consistent. Figure 8-17 illustrates the expected profile for a product event.

As shown in the figure, the incident starts with a response from the field through repair or direct returns. As the incidents occur, the data are logged in. In this example, all similar incidents are charted via the sorting mechanism and similar root causes are listed with the production date. Because the organization was responsive, a design change was qualified and implemented on January 10, 1998, to address the root cause of the problem with Part Number 312315.

The database then tracks failures due to root cause part numbers after this date to determine the effectiveness of the fix. The database is then monitored to see whether the field failure occurs after the fix implemented on January 10, 1998. The database shows one entry involving a different root cause part number on December 1, 1998. This would indicate that the product event has been resolved effectively.

The absolute dates are important here, as design changes needed to fix the design are implemented in an absolute manner. If the fix is real, there will be no more repairs or field incidents due to that particular part.

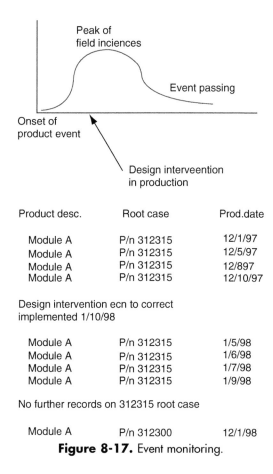

Product desc.	Root case	Prod.date
Module A	P/n 312315	12/1/97
Module A	P/n 312315	12/5/97
Module A	P/n 312315	12/897
Module A	P/n 312315	12/10/97

Design intervention ecn to correct
implemented 1/10/98

Module A	P/n 312315	1/5/98
Module A	P/n 312315	1/6/98
Module A	P/n 312315	1/7/98
Module A	P/n 312315	1/9/98

No further records on 312315 root case

Module A	P/n 312300	12/1/98

Figure 8-17. Event monitoring.

Repairs or incidents with production dates before January 10, 1998, at any point in the future (depending on the type of failure mode) might continue to trickle in. The numbers will rise to a peak level and then die off as time passes, as illustrated in Figure 8-17. To manage the process of a product correction effectively, both types of reports need to be monitored. The profile indicates the financial and company exposure, and the sorted chart of data indicates dates and fix effectiveness.

6. Event Triggering

If the manufacturing process tracks manufacturing performance in terms of PPM, then the field incidents should trigger on an appropriate level as well. Depending on the stability of the design and the stability of the manufacturing process, the trigger points may be adjusted. The operative point is to set the trigger at a point to get an early warning of a product problem so the company can effect a solution and implement it in a timely manner.

PURCHASING AND PROCUREMENT

1. Procurement and Parts Configuration

Having addressed the issues of manufacturing process control and certification of personnel, we now focus on the procurement side of the equation. In order for the process to be in control, the inputs to the process must also be in control. Here, vendors play a key role in contributing to manufacturing quality. In these next sections, we focus on procurement, parts configuration, and vendor certification.

A. Setting up bills of material

The financial success of a product is related to the product configuration. The product configuration is the blueprint for how the product line will be offered to the marketplace, factoring in the different versions, sizes, ratings, and features. How design and manufacturing package the offering can either simplify the inventory management or complicate it to the point of ineffectiveness. The issues extend beyond inventory management and the related expense for finished goods inventory. They include the general confusion factor in manufacturing when there are many versions, offshoots, and forks in the evolution of the product design. Managing these offshoots can soon become unwieldy and begin to consume resources for maintenance and support, as shown in Figure 8-18(A).

The diagram in Figure 8-18(A) depicts a product line that consists of a power electronics section, a logic section, associated software, and some mechanical packaging configuration. In this example, there is one version of a power module, two versions of a logic module, four versions of software, and three versions of packaging. This generates a total of 24 versions of a product offered to the marketplace, as shown in Figure 8-18(B).

If the market demands that a finished goods inventory be in place and on a shelf waiting for disbursement, the product configuration can represent a significant value of finished goods inventory. In addition, there are two versions of logic boards and four versions of software to support. If an improvement or correction is required in the core of a logic board in the hardware, for instance, then two boards will require modification and two locations of inventory must be reconciled. If the issue is serious enough, scrap and rework will apply to the inventory already built.

Consider the product line depicted in Figure 8-18(C), in which the number of versions of logic module is reduced to one and the software is conditionally compiled to allow selection of the four different versions in the factory or the field. In this example [Figure 8-18(D)], the amount of overhead support required is drastically reduced to three versions. The end product can still be available in 24 different and distinct versions; however, inventory and overhead for product support are reduced.

Other points to consider when setting up the bills of material are the design and management of parts cost and vendor negotiations. Figure 8-19 shows an example of a bill of material showing the part, vendor, and other pertinent information that can be used as a management tool. In this example, Assembly 1 has five parts, each having a part number, description, and vendor. Depending on the number of identical parts used per unit, the volume number adjusts accordingly and a total price is placed on the bill of material. Based on the economic order quantity and the product forecast, a negotiated price may be reached. This is the final price that will be used in calculating the manufacturing cost. The difference

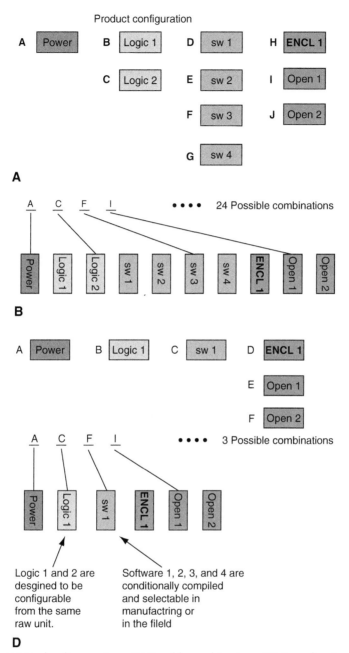

Figure 8-18. (A) Product line versions; (B) Possible combinations; (C) Reconfiguring for simplicity; (D) Resultant configuring.

Bill of material					Initial		Final
Item	Part no.	description	Vendor	Volume Per unit	Bom px-1 Each ($)	Economic order qty	Bom px-2 each ($)
Assy 1							
1	100123	Part A	ABC Co.	5	34	1,000	29
2	100124	Part B	BCD Co.	6	35	200	30.35
3	100125	Part C	CDE Co.	7	36.75	350	43.25
4	100126	Part D	DEF Co.	8	37	750	35.60
5	100127	Part E	EFG Co.	9	38.50	125	32

	Difference each part ($)	Cost impact per unit ($)	Total annual qty	Dollar difference ($)	Labor minutes	Labor cost ($)	
Assy 2						85	
1	5	25	4,500	112,500	10	14.17	
2	4.65	27.90		125,550	11	15.58	
3	−6.50	−45.50		−204,750	12	17.00	
4	1.40	11.20		50,400	13	18.42	
5	6.50	58.50		263,250	14.5	20.54	

	Burden labor ($)	Burden material ($)	Total factory cost ($)				
Assy 123	2.33	1.25					
1	33.01	181.25	373.43				
2	36.31	227.63	461.62				
3	39.61	378.44	737.80				
4	42.91	356	702.13				
5	47.86	360	716.40				
			2,991.37				

Figure 8-19. Bill of material.

between the initial and final cost is calculated and the impact per unit is determined. Factoring in the anticipated volume, a dollar amount is presented for determination of materiality. Noteworthy in this example is the case in which a part (100124) increased in price beyond its initial price; this has a negative effect on the dollars saved.

Next, the labor is factored in terms of time and expense and burdened expense (labor cost shown in white at $85 per hour). Both material and labor are burdened, also shown in white in this example. Depending on the type of accounting system used, material might or might not be burdened. Finally, a total factory cost is calculated. Although this is not a strictly formatted bill of material, it does illustrate the issues that must be addressed in defining one properly and completely.

B. Linking to vendors

With a favorable product configuration and bill of material, the focus now shifts to establishing the part's documentation so it may be procured properly. This can be accomplished by the use of a part card.

The part card identifies all aspects of the part in order to procure it. It must be complete enough so that the wrong part is not procured. It provides the manufacturing linkage from the bill of material and manufacturing process documentation to the vendor and source of manufacture. Figure 8-20 illustrates a typical part card. Each item is outlined with an explanation of its contents.

C. Substitutions and specifications

The issue of substitutions relates directly to the issue of product configuration. The procurement function is responsible, to a large extent, for maintaining the original product configuration. Substitutions made with unauthorized and unsatisfactory parameters (as listed in the part card) can eventually degrade the product or increase its cost.

One basic fact cannot be compromised: The development engineering function is responsible for the selection of a part, not for procurement. Development is the entity responsible for the functionality and performance of the product. Development is also responsible for a product's reliability and quality. This function includes controlling the parts that comprise the end product.

The corporation and the manufacturing system are responsible for creating a system that enforces the selection made by development, but the initial selection is development's and any substitutions should be approved by the development function.

D. Documentation of substitutions

In actual situations, the selection of an alternate component rarely has the exact parameter listing as the original. In many cases, the new parameters are examined and evaluated, and a decision is made to substitute. This is acceptable but the rationale and reasoning behind the substitution should be documented for future reference because there could be interrelationships between substituted parts that affect functionality and performance.

For example, the first part from the bill of material that must be substituted may be substituted with a part that shows minor differences in the parameters. By themselves, this part may not affect the performance or the product configuration. However, when the second part from the bill of material is up for substitution it must be evaluated within the framework of the Part A substitution. In order to keep the record effective, the original Part A substitution should be documented. Such documentation allows the development engineering personnel to evaluate the absolute effect on the product line.

E. Bringing design to cost to the vendor

The philosophy of design to cost should not stop with the company proper; it also needs to extend to a company's suppliers and vendors. The product development team can act as a conduit from the marketplace to the supplier. As the market places pressure on the manufacturer, the manufacturer should allow the pressure to be felt by the vendor. Indeed, this can be a healthy practice, as the vendor can assist with alternatives to meet the factory cost.

Part card		
Identification		
	Part Number and Revision Level	The part number is the company's reference number used to call it out in the bill of material, purchase orders, and material movement within the company.
	Description	This field consists of text to describe the part itself. It can come from the vendor's description, or other reference.
	Reference Product Line	This field references the product or product line the part will be used on. In a data base, this field is helpful in determining where else the part may be used without necessarily going through all the bills of material searching for the part usage.
	Manufacturer's Part Number	This is the manufacturer's model or reference number that is used to identify their product to the marketplace.
Specification		
	Drawings	This is a drawing or a set of drawings and illustrations that describe the part in some detail as required to procure and certify.
	Critical Parameters	These are critical parameters associated with the part. This section of the part card identifies those parameters and tolerances to ensure the procurement process focuses on parameters critical to the form, fit, and function.
	Specifications	These are the hard specifications that define the part's contribution to product functionality. They must include current revision level.
Procurement		
	Sourcing	This section lists the available sources for the product and the vendors associated with those sources. They are prioritized in order of overall preference: function, price, availability.
	Economic Order Quantity	This is the specified quantity that must be ordered from the vendor or supplier in order to obtain the pricing negotiated.
	Longevity of Part	This is the measure of the length of product run. Is the product at the end of its useful product life? Or is recently introduced? How much longer will the product be available?
	Obsolescence Date	This entry forces the development engineering people to ask the question of the vendor and establish correlation between the part obsolescence and the product redesign.

Figure 8-20. A part card.

	Cost	This is the part's contribution to factory cost, considering all of the offsets and above data.
Inventory		
	Revision Level	The revision level calls out the most recent revision of the part allowed in the design. This is under engineering control, and nonqualified substitutions are not allowed.
	Status	This field indicates the current status of the part regarding production and usage. The part is either released for production or is obsolete. Obsolete parts may be used for long-term repair and spares support.
	Inventory Relief Method	This is method used to remove parts from stock. It must be consistent with the rest of the organization's financial treatment of inventory (e.g., first in/first out, last in/first out). In addition, the method selected and implemented must factor part aging.

Figure 8-20. (Cont'd)

The consequences of no action result in the vendor as well as the manufacturer missing out on the market opportunity. Thus, it is in the best interests of both parties to resolve the cost issue in cooperation rather than in a vacuum.

F. Tracking vendor changes

Until now we have centered the discussion of parts procurement and configuration on internally dictated changes, but changes are also forced upon the organization by its suppliers. These changes might in turn be forced on them by their own suppliers or be internally generated. In either case, vendor changes can have a significant impact on product maintenance, as development engineering talent is consumed by qualifying vendor-initiated changes, obsolescence, or pricing actions.

Given their inconvenience and the internal cost, the changes inevitably must be tracked, documented, and referenced. One way to accomplish this is to have the supplier or vendor issue a certificate of compliance on the goods shipped. A certificate of compliance is a formal document that ensures that the goods shipped against a purchase order have been certified by the supplier or vendor to be consistent with the parts specified on the part card and the purchase order. Certification gives the manufacturer some degree of assurance that the parts supplied are in fact equivalent to the specifications on the part card. Furthermore, the certificate of compliance creates a contractual issue with low tolerance for error or oversight.

G. Failed parts disposition

The basic theme of our discussion is to encourage the supplier and vendor to become part of the business process, from specification through development to manufacturing, and including field experiences. When a part fails either in-house or in the field, the vendor will need the failed part as part of its customer feedback mechanism. This requires a process to facilitate the vendor's needs; Figure 8-21 illustrates the general format of that process.

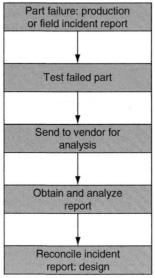

Figure 8-21. Processing failed parts.

H. Traceability systems

The basic concept behind traceability is to generate a data logging system and establish access to information about the component parts and materials comprising the end product shipped to a customer. As the product progresses through the procurement and manufacturing processes and gets shipped to a customer through a sales channel, the ability to backtrack information on component parts and materials might be required as part of a future investigation of a mishap or for other reasons.

Consider Figure 8-22 where, in this simplified case, several parts comprise a product. The product is shipped through a sales channel to a customer. For whatever reason, the product fails and an investigation into the failure ensues. The manufacturer needs to be able to trace the parts backward through the chain, to the source supplier of the part. This information can assist in assessing exposure for other products shipped and in identifying the scope of a recall effort or a field upgrade effort.

Traceability provides answers to questions such as:

- What batch is affected by the root cause of the part failure?
- What date codes are affected, and how do we correlate the product we shipped some time ago to the suspect lot or batch, or to the product received from the supplier?
- Are the parts or subassemblies serialized from the vendor and posted to the product serialization?
- Have there been any other recorded incidents that are similar?
- How do we contain the problem?
- How does the supplier batch process for parts feed our inventory system?
- How were the parts pulled from inventory, and does this correlate to the batch shipped and the volume of product manufactured during the time frame?
- Has this happened before?

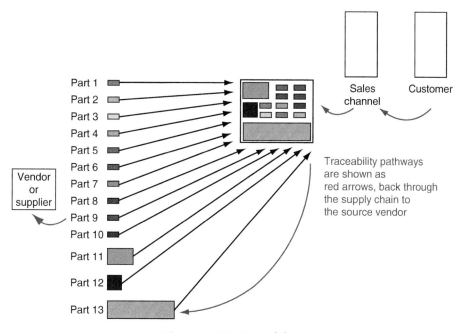

Figure 8-22. Traceability.

The best way to establish the correct amount of data and access required is to project situations and scenarios and determine what questions you will need to answer to determine exposure. Start by gathering this data and fill it in as appropriate. Some will say that there can never be enough data; however, one must design these overhead-oriented manufacturing systems to balance the need for data with the need to run the business. Traceability systems are often used where these subsequent actions cannot be taken lightly. They are used when critical action to contain the problem must be taken and must be effective.

2. Vendor Certification

A. The benefits of certification

The amount of activity required to double-check each and every one of the items we have discussed would absorb human energy to the point of noncompetitiveness. How, then, can a manufacturer ensure that the supplied parts and materials meet the manufacturer's requirements each and every time? The answer lies in vendor certification.

Vendor certification is an activity whereby a supplier is brought into the procurement process with an understanding of its impact on product quality. It entails a series of procedures and internal controls, agreed to by the manufacturer and the vendors, whereby the materials and parts manufactured at the supplier's location have been carefully specified, documented, manufactured, and shipped as part of an overall arrangement with the manufacturer.

B. Model of vendor certification

Figure 8-23 is a conceptual flow chart illustrating a vendor certification transaction. A protocol exists for the vendor/manufacturer interface that results in a higher degree of assurance that materials and parts shipped will conform to development's specifications.

C. Vendor classification program

It simply is not practical for every part to be certified and every shipment qualified. Vendor certification should be used in cases where parts critical to the design affect product performance parameters. When the form, fit, and function of the end product are affected, these parts are absolute candidates.

Parts that affect product liability also require some sort of vendor certification. To make this process work, a vendor classification program may be used. The vendor classification program ties the importance of the part's product configuration to the degree of certification needed to ensure consistent quality levels. Figure 8-24 illustrates the correlation between types of suppliers and certification needs.

At the top of the figure is a gradient of vendor types, ranging from a fully integrated vendor to a commodity-oriented supplier. Along the left is a gradient describing the degree of contractual certification of goods shipped.

The following describes the gradient of four types of suppliers:

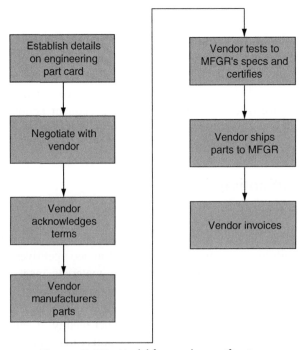

Figure 8-23. Model for vendor certification.

Vendor classification				
	Integrated supplier	Specified supplier	Generic supplier with value added	Commodity supplier
Full vendor certification				
Part number catalog part				
Trade goods multiple sources				

Figure 8-24. Vendor classification.

- *Integrated supplier:* The integrated supplier is one that has an intimate relationship with the manufacturer. What one party does will directly affect the other. The dynamics of the relationship are symbiotic in nature.
- *Specified supplier:* The specified supplier is characterized by a long-standing and stable business relationship between vendor and manufacturer. The supplier is specified along with the part.
- *Generic supplier with value added*: This designation applies when there are many vendors who have a certain capability but only a select few are willing to modify or add value to the parts to suit the manufacturer's needs.
- *Commodity supplier:* The commodity supplier supplies materials and parts that are readily available from many sources. Contracts are awarded based on price and delivery. Performance and quality are assumed and are generally industry standards, either formal or de facto.

Other relevant terms are:

- *Full vendor certification:* This designation applies when there is a full, contractual, interactive exchange of data between the supplier and the manufacturer. There is little left to assume on either side. The transaction is documented completely.
- *Part number, catalog part*: This applies when the supplier has a listing of parts in its catalog and the manufacturer selects a standard part and uses it in its own design. Depending on clarity and configuration controls issues, there is some documentation accompanying the transaction.
- *Trade goods, multiple sources*: This applies to cases in which there are multiple sources of the same part or material. The properties are readily known and individual part certification is not required.

The shading of the boxes indicates a correlation between the criteria: the darker the shading, the stronger the correlation. As can be seen, there is a strong correlation between the integrated, intimate partner relationship between the vendor and the manufacturer and the degree of certification required. The closer to commodity status the vendor's relationship is, the more likely it is that the need for rigorous certification diminishes. However, each vendor should be classified based on your specific requirements, as illustrated in the trade goods/commodity supplier square.

D. Delivery as a prerequisite to qualification

Part of the vendor qualification process involves determining a vendor's track record with respect to delivery. The right product at the right price must be accompanied by on-time delivery performance, or the vendor can cause significant disruption and wasted effort within your organization. Choose the components parts vendors and suppliers carefully. Their performance contributes to your own!

E. Multiple sources

To what extent can the company rely on a single vendor? Companies as well as suppliers desire this type of relationship. With care and some relationship engineering, it is actually possible to achieve it for a time. As time progresses, however, conditions change and reliance on a single source might not be practical in every sense.

As the part is being specified on the part card, alternate vendors are usually called out. A first-tier vendor is listed along with a secondary source. In actuality, it might be better to source from multiple vendors initially, to keep both channels open in procurement. This decision depends on volume, price breaks, and product cost sensitivity, all other things being equal.

Every supplier would prefer to single-source its customers but, from the manufacturer's perspective, this arrangement might not be advisable. A balance must be struck between the needs of procurement and the needs of manufacturing. Set up multiple sources when single-source failure would cause product line failure for your company. In addition, for highly engineered parts, contractually reserve the right of self-fabrication or third-party fabrication in case of primary vendor failure. This should protect the company in steady state so that only the transition must be managed.

F. Source data—Prequalified parts

Traditionally, when a part is specified and qualified for use on a product it generally goes into some medium often referred to as a purchasing book. This is a collection of prequalified and available parts. This medium should be constructed from two perspectives:

- Parts listing
- Vendor profile

The parts listing can be organized by description, part family number with a database link to description, or by some other classification. It should be interactive so that if a development engineer is seeking a part to accomplish a function, it should be possible to scan the available parts currently in use in the company. This would be accomplished by use of the part card discussed earlier in this chapter. The part card is the source data for the database system.

The vendor profile is the reverse view of parts supply, wherein the profile of the vendor is kept on company record. Vendor profiles are helpful for negotiations, future part design, and general business transactional analysis. An example of a vendor profile is shown in Figure 8-25. In this example, the vendor profile can yield a great amount of information

Vendor profile			
Vendor name	ABC company	Information update date	XX/XX/XX
Address	ABC company 1234 N. 56th Street anywhere, USA	Years in business	23 Years
Contact	Mr. Jones, Mr. A, Mr. B, Mr. C	Age of relationship	5 Years
Telephone	XXX-XXX-XXXX	Parts currently supplied	906534-906535-906536
Fax	XXX-XXX-XXXX		906537-906538-906539, 906637-906638-906639
E-mail	ABC@anymail.com		
Web site	WWW.ABC.COM	Annual Dollar Volume	$746,000
General			
Engineering capability			
Manufacturing capability			
Property, plant, and equipment			
Quality index			
Strategic goals			
Attainment index			
Ownership status			

Figure 8-25. Vendor profile.

about the potential supplier. Be sure to also include the sales route to market and the service mechanism, if applicable, to assist in resolving field problems at a later date.

By setting up all of the part cards on an interactive database, this search capability should be feasible. In addition, interactivity could be linked to the vendor for updated information in real time. As pricing and status change announcements come from the vendor, the information could automatically be loaded into the database for retrieval. An example of the workings of an interactive database is represented in Figure 8-26.

In this model, the vendor is linked to the company's management information system. The company has established records for procuring parts for the end product. The procurement record consists of two elements: the part card and the vendor profile. The part card can be the real-time record of updated cost information and lead times. In this way, both technical issues and logistical issues can be updated on a real-time basis.

The vendor profile record can also be linked to the interactive database in real time to update the manufacturer on personnel changes, obsolescence plans, new product lines, and any organizational changes that might affect the vendor/manufacturer relationship. The system should be able to interlink all of this information for all of the parts and the vendors involved with the manufacturer.

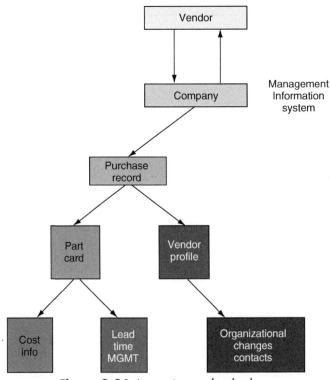

Figure 8-26. Interactive vendor database.

MANUFACTURING THROUGHPUT

1. Forecasting

A. Customer sales, marketing, and manufacturing link disciplines

"Our crew can build the products and get them out on time; just give us a forecast!" So goes the manufacturing lament. The forecasting process is highly important for balancing manufacturing expenses and performance. Incorrectly forecasted demand causes inventory errors and contributes to starts and stops in the manufacturing cycle, which can affect overall product quality. Consequently, forecast data does not begin and end with the marketing commitment to get the product development initiated.

The forecast process is, in effect, the clock rate at which the procurement and manufacturing machines must run. A high forecast drives more materials and demands more labor quickly during the ramp-up process. Specificity in the forecasted product versions brings manufacturing to act on materials. If the materials are not the versions needed in the marketplace, inventories will swell needlessly.

The operative point of discussing forecasts is that neither product development nor manufacturing can be totally effective without an accurate sales forecast. The same care and diligence used in the development of the product should be applied to cementing the real-time link between the sales and marketing forecast and the manufacturing machine.

B. Materials, labor, and throughput

Manufacturing has three basic elements to contribute to the new product's financial success: the procurement of materials, the application of appropriate skilled labor, and the organization to translate the materials and labor into shipments. Throughput allows the organization to reap the financial rewards of the development. The forecast is used to direct these elements so that throughput satisfies the customers.

Consequently, the throughput is only as beneficial to the overall program as is the accuracy of the input. It does not matter what the quality level of a product is, or the specification on performance, or the time required to deliver it if the product is not the one requested by the customer or marketplace—there is no customer satisfaction. Furthermore, the materials, labor, and overhead of the manufacturing machine all have costs associated with them. If the cost is not offset by a proportionate level of revenue resulting in a profit, the manufacturing machine becomes inefficient.

C. The requirement for accuracy

The requirement for forecasting accuracy has already been demonstrated; there must then be some optimal method for generating an accurate forecast. Unfortunately, there really is none. The art of forecasting is a work in progress wherein continuous improvement is the driving force.

From the so-called iron salesmen of the 1970s' automobile companies (who drove finished goods inventory from Detroit into the showrooms of dealers), to the mass-customization of many products (whereby individual orders drive this afternoon's manufacturing), forecasts can contribute to or prevent financial success. The process is twofold. The first step is to feed back to manufacturing what the marketplace needs and wants. It is also a planning means to place goods where the company feels the demand will be. To be successful, the process must include elements for both market intelligence and company planning. Figure 8-27 illustrates the point.

As shown in the figure, the forecast process illustrated by the face "looks" out into the marketplace to determine usage patterns. These "eyes and ears" to the customers must either transcend the sales channel or become an integral part of the sales process. The forecast process also enables introspection into the future desires of the company and makes it possible to work toward a strategic objective. The forecast function then coalesces this information into a usable forecast. The process doesn't stop there, however. See the two-way arrows? The future profile of the company must be driven. This means both determining the marketplace needs and directing those needs toward the company's end goals.

The process is dynamic and changing. Success and accuracy are in the details of continuous improvement.

D. Product configuration: Don't design-in future shortages

As discussed in Chapter 7, there are strategies for configuring a product that can minimize the problems of version forecasting. Not all problems can be eliminated, however. The product design can reduce dependence on forecasts for different versions of a product by incorporating the version selection into the product itself. The result is usually some cost penalty, but if the market can tolerate the cost/price impact, this method can make the issue more manageable if poor forecast information is provided by the sales personnel. Refer to the product configuration discussed earlier to assess the company's critical dependence on forecasting.

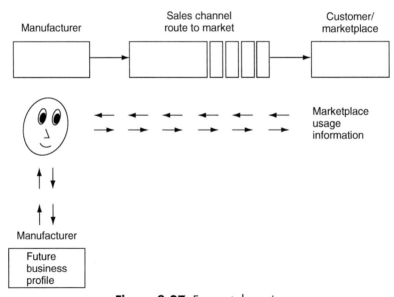

Figure 8-27. Forecast dynamic.

E. Sales program systems

The company should have a sales program and forecast system in place. In fact, these should be components of the new product identification vehicle. If the programs in place are not sufficient to generate accurate forecasts, the company should take time to correct them and make them useful.

To further the point, forecasting for a new product is not just software-based. It is market intelligence. Be sure to have the sales force stay close to the customers to determine usage patterns and forecast levels. Software cannot do this unaided; it is only part of the information system used by management—a medium for communication. The information must be assimilated and analyzed by the sales and marketing personnel.

2. Cycle Time Management

In this section, we review the dynamics of the manufacturing cycle. This cycle deals with the processes of order entry, material procurement lead times, and labor scheduling.

The cycle time of a product involves multiple aspects, many of which are beyond the direct control of the manufacturer. The challenge is to manage the cycle time in such a way as to make it appear (from the perspective of the customer) as though the manufacturer has total control of it. The customers place their orders and trust the manufacturer to complete them in a timely manner.

A. Planning the lead times

"So, what is the lead time on this product, anyway?" This frequently asked question, rooted in the dynamics of intercompany cooperation, has multilevel complexity. The process by which materials are ordered and parts are manufactured, shipped to the primary manufacturer, and assembled into the final product for shipment involves many schedules, priorities, and sequencing. To illustrate this complexity and interdependence, consider Figure 8-28, which shows a feeder supplier to a manufacturer.

As shown in the figure, the process of delivering a manufactured product to a customer is divided into three basic segments. Referring to the customer at the end of the process, the order for the product is placed to the manufacturer. The manufacturer performs the contract review, procures parts, and builds the product.

The parts, however, are probably available from a supplier. The manufacturer then places orders to the supplier. The parts might be standard parts or customized, highly engineered parts. Each follows a slightly different pathway through the procurement process. If parts are standard, pre-engineered parts, the order from the manufacturer placed to the supplier is reviewed in contract review, materials are procured, and the part is manufactured and shipped to the end-product manufacturer for reshipment to the customer. If they are customized parts, the process includes a stage in engineering. The several parts comprising a bill of material are depicted by the arrows shown entering these supply chains to the end-product manufacturer.

Each of these parts then can be sourced by a subcomponent vendor or supplier. This process is shown on the top rung of the illustration. Here, a subvendor goes through the same process to deliver the part used in the bill of material. These parts are depicted as either customized or standard.

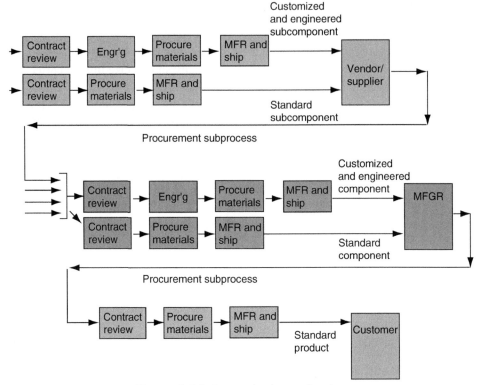

Figure 8-28. Process lead time planning.

Figure 8-29(A) shows a more concrete example of the process that generates lead times in producing a product. In this example, the end product (Module A) consists of 10 parts labeled Part A through Part J. Each of these parts has an associated lead time in procurement. In this example, Part A has the longest lead time (24 weeks). The chart also shows the quantity of parts used in each product and the quantity on hand. The volume of end product that can be produced with quantities on hand can be calculated. This chart also indicates the quantity of product that is currently on order and the date on which it is due. Most manufacturing information systems drive these types of screens. This graphic and subsequent ones in this section are shown for illustration purposes only.

If we investigate Part A further, we see in Figure 8-29(B) that it is composed of five subparts, shown as Parts A1, A2, A3, A4, and A5. Each of these parts has an associated lead time. As indicated, the third subpart, A3, is the part with the longest lead time (24 weeks). This lead time affects the delivery of Part A, which in turn affects the delivery of Module A, the end product. There are additional bills of material for the various Parts B, C, and D, and so on. In this example, Part A was the longest lead part (due to Part A3). By investigating the bill of material further, we see in Figure 8-29(C) that Part A3 is composed of five parts: A3A, A3B, A3C, A3D, and A3E. Each of these parts has an associated lead time. In this example, Part A3C is the part with the longest lead time (24 weeks). This penalizes Part A3, which then penalizes Part A, which in turn penalizes Module A.

Item	Part number	Description	Qty reqd /unit	Qty on hand	Units avail	Qty on order	Due date	Lead time (weeks)
Vendor lead time summary								
Bom		**Module A**						
1	600123	Part A	5	50	10.0	200	XX/YY/ZZ	24
2	600124	Part B	4	73	18.3	220	XX/YY/ZZ	9
3	600125	Part C	3	123	41.0	220	XX/YY/ZZ	21
4	600126	Part D	2	232	116.0	220	XX/YY/ZZ	12
5	600127	Part E	5	99	19.8	220	XX/YY/ZZ	5
6	600128	Part F	4	57	14.3	220	XX/YY/ZZ	12
7	600129	Part G	3	76	25.3	220	XX/YY/ZZ	4
8	600130	Part H	2	28	14.0	220	XX/YY/ZZ	4
9	600131	Part I	5	23	4.6	220	XX/YY/ZZ	2
10	600132	Part J	4	10	2.5	220	XX/YY/ZZ	2
				Max units	2.5		**Max lead time**	24

A

Item	Part number	Description	Vendor	Qty on hand	Qty on order	Due date	Lead time (weeks)
Component summary							
Bill of material		**Part A**					
1		Part A1					10
2		Part A2					9
3		Part A3					24
4		Part A4					9
5		Part A5					3
						Max lead time	24
Bill of material		**Part B**					
1		Part B1					9
2		Part B2					5
3		Part B3					8
4		Part B4					3
5		Part B5					6
						Max lead time	9

Figure 8-29. (A) Vendor lead time summary; (B) Component lead time summary.

Bill of material		Part C		Qty on hand	Qty on order	Due date	Lead time (weeks)
Item	Part number	Description	Vendor	Qty on hand	Qty on order	Due date	Lead time (weeks)
1		Part C1					21
2		Part C2					4
3		Part C3					20
4		Part C4					7
5		Part C5					3
						Max lead time	21
Bill of material		Part D					
Item	Part number	Description	Vendor	Qty on hand	Qty on order	Due date	Lead time (weeks)
1		Part D1					12
2		Part D2					3
3		Part D3					5
4		Part D4					8
5		Part D5					2
						Max lead time	12
Bill of material		Part E					
Item	Part number	Description	Vendor	Qty on hand	Qty on order	Due date	Lead time (weeks)
1		Part E1					2
2		Part E2					4
3		Part E3					3
4		Part E4					3
5		Part E5					5
						Max lead time	5
Bill of material		Part F					
Item	Part number	Description	Vendor	Qty on hand	Qty on order	Due date	Lead time (weeks)
1		Part F1					12
2		Part F2					2
3		Part F3					3
4		Part F4					5
5		Part F5					7
						Max lead time	12

B

Figure 8-29. (Cont'd) (B) Component lead time summary.

Subcomponent level							
Bill of material		PART A1					
Item	Part number	Description	Vendor	Qty on hand	Qty on order	Due date	Lead time (weeks)
1		Part A1A					8
2		Part A1B					3
3		Part A1C					7
4		Part A1D					10
5		Part A1E					5
						Max lead time	10
Bill of material		PART A2					
Item	Part number	Description	Vendor	Qty on hand	Qty on order	Due date	Lead time (weeks)
1		Part A2A					4
2		Part A2B					6
3		Part A2C					9
4		Part A2D					4
5		Part A2E					7
						Max lead time	9
Bill of material		PART A3					
Item	Part number	Description	Vendor	Qty on hand	Qty on order	Due date	Lead time (weeks)
1		Part A3A					4
2		Part A3B					5
3		Part A3C					24
4		Part A3D					8
5		Part A3E					4
						Max lead time	24

C

Figure 8-29. (Cont'd) (C) Subcomponent level lead time summary.

Conversely, by focusing and improving the lead time on Part A3C, the entire lead time of the end product can be improved to the point of the lead part with the next-longest lead time. With some parts, the lead time cannot be compressed easily; therefore, it could become necessary to forecast these parts more accurately or stock them. Often, however, it is difficult to forecast them because they are buried in the product hierarchy and product versions. To forecast these types of parts, you need to forecast the versions of units of the end product, and manufacturing and material procurement should not necessarily be making this forecast—marketing and sales should.

B. Materials management

The basic definition of materials management is the timely presentation of materials and supplies at the point of labor with minimum amounts of inventory, carrying charges, and procured costs, and minimal movement between point of entry and usage. The function of materials management is to confront nonlinear order entry with a laminar flow of materials while achieving the cost requirements set forth in the original design specification.

There are mechanized ways to procure material, but little can make up for informed, interactive communication between the customer interface and the manufacturing element. Frequently, a company might risk the expense of inventory to have parts available for and to anticipate market demand; however, this process is risky if the parts in question are expensive.

C. Workload management

The basic definition of workload management is the timely application of appropriate skilled talent to the labor requirement to transform materials into finished goods. The function of this activity is to match a varying workload (that has varying requirements for talent) with the human resources capable of completing the tasks at hand. Balancing resources against workload requirements can be difficult, especially given the cyclic nature of business. It is difficult to retain personnel in a downturn, and their replacement is not immediate in an upswing, especially considering the demands of training new personnel.

D. Vendor cooperation, component selection

To make the manufacturing cycle somewhat predictable and repeatable, the vendors and suppliers must become integral partners in the process. The development function can assist in the basic selection of parts to a large extent, but the relationship between the manufacturer and the vendor will strongly influence creating the laminar flow of materials and parts needed by manufacturing.

To make the relationship work, there must be value in and benefit from the relationship on both sides. The manufacturer or purchaser cannot always be the benefactor at the expense of the vendor, and the vendor won't continue obtaining business from the manufacturer at the latter's expense. Volume sales and uninterrupted schedules help both sides make money; consequently, both parties need to work toward those ends.

E. Revenues are the goal and require timely shipments

In cycle time management, revenue generation—not necessarily the individual performance, productivity, and efficiency measurements of individual groups within the organization—is the primary goal. Each group needs to work toward that goal by making its specific contributions. It does not serve the cause of revenue generation to expedite only some but not all of the parts with long lead times, or to negotiate price concessions for one part of the product only to be offset by delivery problems and expedite charges on another part.

3. Synchronization

A. The rhythm of manufacturing

The model for the manufacturing process should be rhythmical. Ideally, all of the individual departments and groups should function to create a rhythm in manufacturing. This synchronization is where leverage begins to occur and where the target costs for the product line begin to be achieved.

The product team needs to orchestrate the following issues. In the circumstances specific to your company, there could be additional issues; however, these examples will serve as a starting point:

- Product configuration
- Product design stability
- Quality requirements
- Parts selection
- Field experiences
- Forecast process
- Vendors
- Manufacturing process output
- Personnel consistency

All of these design and manufacturing issues affect the success of the product. Doing a diligent job on each of these will ensure that the product is delivered to the marketplace with the same expectations the company had of the product in the planning stages.

B. The benefits of synchronization

Synchronization does not occur naturally in business; it must be fostered, promoted, and instilled into willing participants in the business venture. Failure to accomplish this results in forces that contribute to disarray, overtime, errors, lower yields, increased costs, reduced quality, increased stress levels in the workforce, and higher rework and scrap costs. The benefits of efforts expended toward synchronization contribute to success in the new product venture.

The same care exercised in the development of the product should be exercised in the development of how the product is manufactured. This ensures that the human and manufacturing energy will be applied to generating revenue for the corporation.

4. Capacity

Manufacturing capacity may evolve in discrete steps rather than linearly. As the need for manufacturing capacity increases, a firm may go to multiple shifts in order to more fully utilize its existing asset base.

A further increase in the demand for manufacturing capacity may necessitate either a new plant must be procured or built, or outsourcing must take place. Alternatively, the product could be built in subassemblies outside the plant with final assembly occurring at the plant. In either case, the addition of manufacturing capacity must be done very carefully so as to protect the finances of the company. Adding capacity too early can incur extra costs, whereas adding it too late causes interruptions in supply. The best way is to focus on an accurate forecast and assess the risk in that forecast, and ramp up capacity accordingly.

QUALITY MANAGEMENT

1. Field Problems and Event Status Monitoring

A. The importance of field monitoring

Field monitoring of product failures or events helps the company from three basic perspectives:

- Financial exposure
- Product reputation
- Company reputation

1. Financial exposure

From the financial perspective, the product problem can be quite a drain on the company. Throughout history, many companies have been able to point to a product problem or field recall that financially stressed the company. Such problems also cause cash redeployment that negatively affects future success. Consider the financial model in Figure 8-30, which shows the initial development investment and the return.

As shown, the original investment is offset by the original planned return. The sales revenue increases early in the life of the product. The field recall issue then causes cash to be spent on remedial activity rather than on funding the investment required in inventory and accounts receivable during the normal growth pattern. This pattern of resource allocation then causes the return on the product to be significantly lower than expected. In addition, the energy and intellect that should be applied to reading the marketplace and generating the next opportunity are spent on the remedial work at hand. This example shows why it is imperative to confront field problems as soon as they appear and act decisively to mitigate their impact on the organization. By acting early, the pool of problematic product is smaller and more manageable.

2. Product reputation

It is also important to preserve the product's reputation as much as possible. This is effectively accomplished by acting to correct any problems as early as possible. A poor reputation attached to a product early in the introduction can decimate the sales effort and the initial launch. In some cases, damage to the product becomes evident through initial feedback from the sales channels rather than from direct customer feedback. The company must act to reduce the impact of bad press through its own sales channels to pave the way for increasing sales after the product problem is resolved.

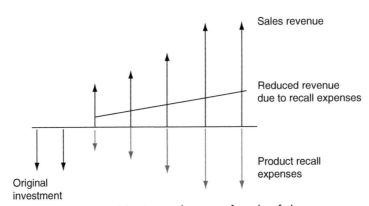

Figure 8-30. Financial impact of product failure.

3. Company reputation

Left unresolved, a product problem can eventually reflect poorly on the entire company's reputation. Given the rhetoric of quality and continuous improvement, failure to act will indict the company for poor performance. The customer base is generally forgiving of mistakes, nonlethal product issues, and issues of inconvenience; however, they have little patience with companies that choose to ignore problems or that fail to effectively address them in a timely manner.

B. The myth of the throwaway product

"We don't have to worry about the product after it is shipped—it's a throwaway product." This familiar remark reflects a manufacturing attitude that provides no feedback loop to assess and improve quality. The useful life of a product might have nothing to do with its required quality level or its acceptance in the marketplace.

For example, a syringe for infusing drugs into a human vein has a limited useful life. It is, in effect, a throwaway product after use; however, the quality requirements for tolerances, hygiene, and protection are at very high standards. Just because the product is disposable is not an excuse for poor quality.

Furthermore, the manufacturer must not ignore field feedback simply because the product is disposable. The initial feedback is invaluable to the manufacturer in terms of improving processes and design criteria; it closes the loop on the entire product offering. For high-volume product lines, failure to heed the warnings in initial field feedback can result in massive recalls, product liability issues, and expensive replacement. The future success of a product lies in the details of how a product failure is handled.

C. Pareto analysis

Pareto analysis was discussed earlier but will be reviewed here for completeness in the context of ongoing field monitoring required to ensure delivery of a high-quality product. The product failure and resolution summary should be consistently mined for trend data and to recognize the onset of product failure problems. The criteria should be set up for a trigger analysis and the reports monitored for the characteristic profile of a problem.

Development personnel should then be deployed to rectify issues immediately as they occur. These practices formulate the product maintenance portion of the development effort.

D. A model for quality management

To tie together the presented concepts into a framework that closes the loop on the product development offering, the product quality system example in Figure 8-31 illustrates the interrelationships and information flow among reports, summaries, and databases.

As shown in the figure, the process starts with the development of the product. (In real-world practice, the complete process is much larger and more involved than this, but for purposes of illustration within the framework of the chapter's subject of manufacturing development, this example should suffice.) Development feeds a fresh design to manufacturing. Along with vendor participation, manufacturing then builds the product and ships it to the customer. The customer then places the product in service. Manufacturing contributes the appropriate records to the database configuration and summary system. Assuming the product does not fail, this example represents a typical transaction.

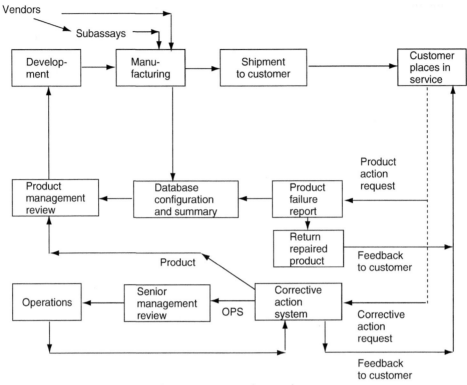

Figure 8-31. Product quality system.

If there is a failure, the product will be returned to the company for analysis via a product failure report. The database is then updated with the product failure information and the root cause is posted for product management review. The output of this information might be a design change, a vendor change, or a part replacement that must be executed by development engineering. Feedback should then be given to the customer about the failure and any product actions initiated.

If the customer issues a formal complaint with the company, the complaint is entered into the corrective action system for senior management review and disposition. Product-oriented requests and complaints are directed to product review. Operational requests are directed to operations. Any changes made are implemented through operations, and the customer receives communication regarding the results of the complaint.

This example shows a general outline of how information should flow for the purpose of acknowledging and resolving a product issue. Depending on the individual circumstances, the flow chart might be more complex or simplified. The operative point in the analysis, however, is to confront product failure issues directly and to resolve them accurately and quickly.

COST CONTAINMENT

1. The Myth of Cost Reduction in Manufacturing

Manufacturing cost reduction is often discussed in industry and at trade shows. In many ways, such discussion is misplaced at manufacturing's doorstep. In general, manufacturing cannot reduce costs on items beyond their control. Given this, not much *is* under manufacturing's control. With a fixed bill of material and hard vendor specifications, purchasing is somewhat limited to suggesting substitution of parts.

Most of the costs are created in development. From purchased parts, to specifications, to processes selected, to platform identification—all are primarily development decisions. The best that manufacturing can do is to reduce labor costs and overhead costs to remain competitive. The point is that if the program is high cost at the launch, do not look to manufacturing to reduce the costs. This is a cooperative effort with a disproportionate part of the costs being contributed by development.

2. Relationship of Labor, Time, and Material Costs

Cost containment is a multidimensional problem. As a natural course of events, production costs will increase with time because everyone likes raises in wages. Personnel will generally be paid more this year than last year, given that all other circumstances are equal. If there are productivity improvements, the methods for assembling and testing a stable product can be improved upon from year to year. Hopefully this can *offset* the increase in wages, and then some.

The material costs may increase or decrease depending on commoditization, proprietary status, and volume. Hopefully, these can be reduced with volume and smarter shopping. Consider the examples shown in Figure 8-32.

Figure 8-32(A) shows three scenarios of a product called A. They are material, labor, and burden components to the total factory cost. The burden is fixed at 2× labor for this example.

As shown in the graph in Figure 8-32(B), there are cases where the material costs may be declining slightly from year to year. This may be due to buying progress, substitution of parts, or favorable currency status. If the productivity offset to increased wages is not enough, there may be a slight increase in labor over time.

In Figure 8-32(C), we show a case where the labor offset is greater from year to year. This worsens the impact of total costs due to the labor component and also the burden. The result is a total cost increase higher than in the first example. In this example, the material cost decreased at a greater rate than in the first example, but it wasn't enough to offset the labor issue.

In Figure 8-32(D), both the material and labor are out of control and this forces the total factory cost out of control.

To achieve true cost containment, one needs to focus on material reductions, improvements in productivity, limits on wage increases, and reduction in overhead expenses. Otherwise, they can all work against you.

Relationship between material, labor, and time							
Item	Description	Scenario	Year 1 ($)	Year 2 ($)	Year 3 ($)	Year 4 ($)	Year 5 ($)
1	**Product A**						
2	Material	Down	20.00	19.5	19.25	19	18.75
3	Labor	Up	9.00	9.1	9.2	9.3	9.4
4	Burden	Up	22.50	22.75	23.00	23.25	23.50
5	**Total**		51.50	51.35	51.45	51.55	51.65
6							
7	**Product A**						
8	Material	Down	20.00	19.40	18.80	18.20	17.60
9	Labor	Up	9.00	9.25	9.50	9.75	10.00
10	Burden	Up	22.50	23.13	23.75	24.38	25.00
11	**Total**		51.50	51.78	52.05	52.33	52.60
12							
13	**Product A**						
14	Material	Up	20.00	20.25	20.50	20.75	21.00
15	Labor	Up	9.00	9.45	9.90	10.35	10.80
16	Burden	Up	22.50	23.63	24.75	25.88	27.00
17	**Total**		51.50	53.33	55.15	56.98	58.80
18							

A

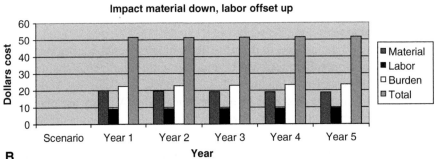

B

Figure 8-32. (A) Cost behaviors; (B) Impact where material is decreasing but labor is slightly increasing.

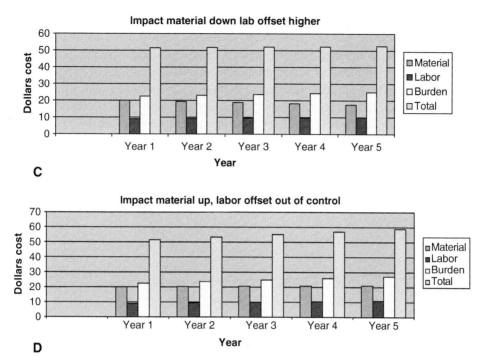

Figure 8-32. (Cont'd) (C) Impact where material is decreasing but labor is increasing; (D) Impact where both material and labor are out of control.

SUMMARY

In this chapter, several topics were reviewed with respect to manufacturing the new product. One of the objectives of this chapter was to establish the link between the development function and the manufacturing development function. Another objective of the chapter was to equip the reader with some rudimentary tools for setting up manufacturing process controls, given the availability of commercially available database and software programs.

The chapter started with manufacturing management and the concurrency of development phases between design and manufacturing. The discussion reemphasized the importance of the interdisciplinary aspect of the new product development process and further emphasized the design for manufacturing philosophy. Next, a discussion of the manufacturing layout was presented within the framework of the product and industry practices.

Next was a discussion of process competencies and how to make the organization competitive in terms of processes. A discussion on outsourcing was presented, followed by a discussion on training and certification of manufacturing personnel. Inventory control was also discussed within the framework of new, existing, and aftermarket requirements.

Manufacturing records were covered next, followed by parts selection, procurement, and constructing the bills of material. Issues such as vendor qualification, documentation, and certification were also covered. The next section dealt with the quality database and managing field product problems. The importance of maintaining product configuration and controlling changes to the product line was presented as a precursor to quality management. Finally, a discussion on cost containment in manufacturing operations was reviewed.

With the basic design, qualification, and manufacturing elements established, the fruits of everyone's labor are in place for launch into the marketplace. This is the focus of Chapter 9, The Prelaunch Checklist.

CHAPTER 9

THE PRELAUNCH CHECKLIST

EXTENSIONS TO ENGINEERING AND MANUFACTURING

1. Product Release Checklist

A. Similarities between launch and flight

There are many similarities between the new product launch and the preflight checklist. The preflight checklist is a methodical review of all the elements associated with ensuring a successful and safe flight. The same holds true for the new product development prelaunch checklist. A protocol must be executed to ensure that all of the proper elements are addressed before launch. This is good practice in the business development arena.

B. The checklist

In Chapter 7, which deals with the actual development engineering, a list of deliverables was presented to ensure that all of the requisite steps are completed during the development process. This could also serve as a checklist at this stage of the project to double-check that all deliverables have been addressed sufficiently and properly.

Often in the course of development, details may be missed, activities may remain incomplete, and specifics may be glossed-over. It is a vocational hazard that can occur quite naturally because of the differences in interpretation of events, procedures, and degree of importance of certain issues.

The checklist ensures that at a single point in time (namely, at this point in the project), all of the specifics of the program get addressed. The list forces the product development team to evaluate each item and assess its degree of completion.

Figure 9-1 presents checklists for business development stages BDP–0 through BDP–5.

C. Risk analysis for missing elements

If the project runs late or increased pressure is placed on the development team, it may be tempting to cut certain activities out of the process to accelerate the program. It is not advisable to do so, and the following recommendation does not endorse this. However, if it does occur, a risk analysis should be performed to determine financial and product liability exposure. The danger lies in rationalizing the selection of many items to be eliminated.

This results in a less-than-complete program. It also exponentially adds to the list of caveats and special handling issues for the program. It complicates the ability to perform these steps after the fact of introduction and to maintain any semblance of a product configuration control system. The best practice is to execute the planned steps before introduction, without compromise.

D. Accounting for deficits

Assuming that certain deficits are identified and the management team decides to proceed anyway, the team must account for the missing items. This means planning and executing in the future, or establishing procedures and documenting instructions to work around the deficit.

BDP – 0 Market and internal assesment			
1	Market opportunity	9	Market dynamics study
2	Competitive analysis	10	Go/no-go
3	Product concept	11	Market-required specification
4	Feasibility study	12	Idea evaluation
5	Strategy for company	13	Value stream map—customer
6	Fitability study	14	Brand label checklist
7	Funding analysis	15	Breakeven analysis
8	Timing		

A

BDP–1 Prototyping			
1	Design specification (preliminary)	13	Preliminary ROI
2	Patent review	14	Design review results
3	Design strategy	15	Packaging optimization
4	Test strategy	16	Capital requirements plan
5	Manufacturing strategy	17	Technology integration study
6	Quality plan	18	Hardware requirements specification, software requirements Specification design
7	Reliability requirements	19	Alpha unit developed
8	Safety requirements	20	Project cost analysis
9	Cost analysis with sensitivity	21	Risk assessment
10	Statutory and regulatory approvals	22	Value pricing
11	Value engineering study		
12	Aftermarket plans		

B

BDP–2 Core development and testing			
1	Criticality analysis	10	Life testing
2	Safety review	11	Design review results
3	Preliminary forecast	12	SPAF
4	Beta unit	13	Pilot run results
5	Institute change control	14	Final production strategy
6	Design verification	15	Final test strategy
7	Design validation	16	Final quality plan
8	Standardization tests	17	Program costs
9	Pilot run plan	18	Cost data

C

Figure 9-1. (A) BDP–0 Market and internal assessment; (B) BDP–1 Prototyping; (C) BDP–2 Core development and testing.

BDP–3 Industrialization			
1	Set up manufacturing	10	Quality database integration
2	Capital procurement and setup	11	Final product cost evaluation
3	Tooling qualification	12	Freeze test, manufacturing, quality plans
4	Final pilot run	13	Field test data and feedback
5	Pilot run certification	14	Freeze marketing strategy
6	Product certification	15	Prepare launch plans
7	Freeze design	16	Initial stock in place
8	Change control active	17	Product design release approvals
9	Procure production materials		

D

BDP–4 Commercialization			
1	Competitive assessment—final	7	Employee certification
2	Final literature	8	Training for personnel
3	Forecasted product	9	Quote database
4	Customer visit schedule	10	Manufacturing yield data analysis
5	New product release	11	Model for corrective action
6	Changes complete & Integrated w/ inventory		

E

BDP–5 Optimization	
1	Field feedback to quality database
2	Manufacturing feedback
3	Expanded production plans
4	Value management
5	Product evolution analysis
6	Cost reduction
7	Product obsolescence

F

Figure 9-1. (Cont'd) (D) BDP–3 industrialization; (E) BDP–4 commercialization; (F) BDP–5 optimization.

Establishing special or conditional procedures complicates the work for the personnel who must embrace the new product and participate in the launch. For example, let's assume the product is incompletely tested in one aspect of application. For the sales personnel and the applications support personnel, special procedures must be invoked when dealing with customers. This is especially difficult in those instances where the customer is specifically inquiring about that particular aspect of the application. The sales personnel must exercise burdensome conservatism and the application personnel must spend additional time qualifying the product for use in this area of the application. With a complete development and testing program, questions could be definitively answered in a timely manner without having to place the burden of uncertainty on the customer.

E. The danger of incompleteness

There is also a more parochial aspect to an incomplete development program—the inherent product liability to which the company is exposed. The nature of product liability exposure is such that one issue can be somewhat controlled or mitigated; two issues could be possibly identified and controlled; but three or more issues begin to statistically approach the point where the risk simply cannot be controlled if there is more than one individual involved with the product. The company then acts to enforce the strictest control over these issues and can run the risk of disenchanting the sales channel with all of the caveats.

PRODUCT CERTIFICATIONS

1. Confirming Agency Certification

A. Agency certifications are an integral part of the development process

Modern product development and marketing demands the use of agency certifications to complete the development process. As a result of government requirements, to determine adherence to specification, and to provide a degree of public safety, agency certifications affecting the industry are becoming more of a factor in the product development process. In fact, certain products cannot be marketed without the relevant agency's approval. In some cases, its absence is also a huge marketing hurdle to overcome.

Securing these certifications is part of the development; it is not added after the development is completed. It is extremely impractical for a product to undergo agency testing and approval after the product is developed and expects things to remain unchanged. The process is a give and take between the agency and the manufacturer, where certain elements are subject to interpretation. Where the requirements are clear, the manufacturer can often anticipate the agency's needs and design them into the product.

When the certification is initiated after the development is completed, the product might require cost adders to ensure compliance. These cost adders can be significant. The agency certification exercise should be a confirmation at this point of the program, not an initiative.

B. Modifications after the beta site test program

Given that the agency approvals are not an initiative at this point, the involvement here is more for the benefit of any changes made as a result of the beta field testing. Certain changes demanded by the customer base may have to be cleared through these agencies. Consequently, the agency approval issues dealt with here should pertain to the specific beta test requirements.

C. Establish relationships with agencies

One of the most effective methods for implementing these last-minute changes is to have an ongoing relationship with the agency involved. In effect, you will be requesting a favor of time to be granted to the company to process these changes. The best way to garner a favor is to have a working relationship established whereby the agency begins to move from a policing orientation to one of consultant orientation and confirmation.

This is where an agency approvals coordinator within the company can provide a valuable service in time management for the project. Their role is that of technical liaison and information expediter between development and agency personnel. They can accelerate the approvals process or delay it, based on their own diligence and drive.

Another method to facilitate integrating the agency into the product development process changes is to immerse agency personnel into the organization itself. Have them communicate directly with development personnel on details and interpretations. This can accelerate results and create a more intimate understanding between the company and the agency.

D. Have certifications in place at launch

It can be quite tempting to launch the product without tying up the loose ends of these agency approvals. However, do not succumb to the temptation, because market pressures and timing could work against your company effort. Once you launch the product, the sense of urgency is diminished greatly. It can be very difficult to keep the pace going once the product is launched.

It is usually best to launch with all approvals in place. It keeps the pressure on both the agency and the company personnel to complete the project correctly and as soon as possible, given all of the requirements.

PILOT RUN OPERATIONS

1. Background

The manufacturing pilot run is a verification of several aspects of the newly developed product. The verification has an orientation that is internal to the organization. Its primary function is to verify manufacturing infrastructure effectiveness and process documentation, and to test manufacturing yield capability.

The pilot run is analogous to a midterm exam in education, a dress rehearsal for a theatrical performance, or an exhibition game in professional sports. It is an opportunity for the company to flesh out the manufacturing processes, to test throughput expectations, to verify the functionality of the quality database system, and to determine the effectiveness of personnel training. It is a smaller-scale test drive of a larger-scale production run for the purpose of verification and correction.

2. Pilot Run Manufacturing

A. Perspectives on the size of the pilot run

Since this is a practice manufacturing run, the size of the pilot run is an important factor in the verification. There is a significant difference between a manufacturing run of 5 pieces and 100 pieces or 1000 pieces. It is easy to hand-build 5 pieces but it is another matter to manufacture 1000 pieces and have them integrated into the testing and quality database.

The problem of lack of manufacturing procedures and systems does not go away if a company chooses to go with a disproportionately small pilot run. In fact, the issue will only get more complex as time goes on. Furthermore, scaling the manufacturing organization up from a small quantity to a large quantity can be disastrous, given the lack of sufficient manufacturing systems and controls. The size of the pilot run should thus reflect normal manufacturing conditions and should be embraced as an opportunity to confirm the manufacturing system's capability.

B. Degree of allowable variability

One of the salient points of the beta test program results is to determine what deviation from the original plan the company and the customer base would be willing to tolerate. There needs to be a balance between the severity of the deviation and the esoteric purity of compliance.

A degree of practicality is the best course of action, in that minor deviations can be resolved quickly during the pilot run. Major deviations should be resolved outside of the pilot run proper. In either case, a large number of deviations (significant or not) indicates that the process setup for manufacturing the product is not in control. A process that is not in control negatively affects the product quality level. The pilot run allows the deviation(s) to be quantified and a correction plan to be developed to bring the processes into compliance.

D. Measures of performance

When the product is first developed, the manufacturing plan establishes certain performance expectations. The pilot run tests the system's data-gathering capability to document performance. The benefit is that it removes the specter of data interpretation by a third party. The data is objective and can be sorted by categories, and corrective actions can be assigned to individuals responsible for their particular areas.

Manufacturing yield data should reflect a rapid approach to the objective with each successive manufacturing run. Given a diligent layout of the entire program, the yield expectations should be achievable.

E. Corrective action

The corrective actions required as a result of the pilot run need to be in place before any subsequent pilot runs. In addition, the corrections should be able to be evaluated individually for effectiveness. The vehicle for correction should be the engineering change as it relates to process changes (assuming that is the only type of change needed). Such engineering changes should be addressed immediately after the pilot run, and another qualifying run should be scheduled as soon as possible.

The degree of risk one may want to take here is one of scale. If the next run is a full-scale production run with large volumes, there is a possibility of significant rework costs to correct manufacturing problems. The temptation is to correct the product as is it being made; however, in its purest form, the pilot run should not require any hand-holding. It should be a process that is deterministic and that can be turned on and off at any time with repeatable results.

F. Timing and staging

The timing of the pilot run is anticipatory to the product introduction. The pilot run must utilize the procured parts and the designed processes. Therefore, it is considered the first production run from a corporate infrastructure perspective.

The disposition of the product coming off the pilot run depends on the level of product conformity to specifications. In many cases, these products are used for the beta test program with certain qualifications, and are possibly even used for initial sales efforts.

There is a tendency to try to race through the pilot run process, its qualification, and the beta test program and feedback integration. The product development team needs to be diligent in its activities so as to complete all of the details. With the sense of urgency increasing, the team needs to bring effort to bear on the details of the tasks and to not treat them perfunctorily.

G. Process improvements

The pilot run results can reveal areas of proposed improvement in the processes used. These improvements should be incorporated into the product line as part of the continuous improvement effort. Barring any show-stoppers in yield, quality, or manufacturability, these improvements can be made as soon as time permits. They should be integrated and measured for overall effectiveness. Because some process improvements require volume product to implement, the team should differentiate between those process improvements that can be implemented in the lower-volume, initial production runs and those improvements that require significantly more volume to implement properly.

FIELD TESTING AND CUSTOMER ACCEPTANCE

1. Beta Testing Program

A. The report card on the entire effort

If the pilot run is the internal measurement of the manufacturing system performance, the beta test program is the external measurement of the product's performance. The beta test program will yield the results of third-party use of the product in a nonlaboratory environment. This process tests the entire new product development and manufacturing process, from one end to the other. It is the report card on the entire product development effort.

The customer base has the opportunity to engage the product, receive it from the manufacturer, place it in service using the manufacturer's documentation, and evaluate its performance against the product claims. It gives the manufacturer feedback on the details of the product function, ease of usage, and performance characteristics.

The beta test program should be an organized evaluation by the selected customer, with a beginning date and an end date established for the test and feedback. Agreement should exist between the manufacturer and the customer as to what test criteria will be used and what type of product feedback the manufacturer will need. After the beta test, the test article should be removed from service and examined by the manufacturer.

These articles are often shipped to customers for testing and may not even be placed in use. Many times a less-than-diligent job done by the factory will result in no feedback at all. Sometimes the manufacturer concludes that no response is good (i.e., no complaint is good news). But in some cases the customer has no response because they have not tested the product yet! In reality, a structured evaluation done in a timely manner that results in specific feedback to the manufacturer makes the beta test program valuable to the manufacturer.

B. Checking all elements of the planned process

There are several aspects to the beta test program, as follows:

- *Customer selection:* The customer selection should be made by factoring in the customer's ability to test the unit and their cooperation in giving feedback, and by the company's liability exposure in case of a product failure.
- *Customer contact:* The customer contact person is the primary point of communication between the manufacturer and the testing customer. Communication should be bidirectional and open.
- *Contractual issues:* The manufacturer should have some form of agreement that the customer will sign and honor for the duration of the testing process. The agreement should cover legal exposure issues and the responsibilities of both parties. It should also cover test duration, timing, and terms and conditions.
- *Waiver of liability:* Due to the nature of the beta test program, there should be some type of waiver of liability established in the agreement to protect the manufacturer.
- *Structure and timing of feedback:* Each type of product being tested has certain requirements for testing. In addition, each project schedule has different demands for timing of feedback. The test program is to confirm previously completed development work and assumptions, not to create a forum for new discovery that will delay introduction. Consequently, exhaustive testing and timely feedback are key to a successful test program.
- *Number of test sites:* Depending on the uncertainty and degree of confirmation required of the testing, the number of test sites may change. The beta test program criteria should be established early in the program, and consideration should be given to the potential sites throughout the development cycle.
- *Conditions to be tested:* As part of the agreement, the specific test criteria and test conditions should be identified. The customer should then agree to perform the tests under these conditions. A customer unwilling to test completely or thoroughly is not a good candidate.
- *Selection demographics:* The customer selection demographics should represent the demographics used in marketing the product. To be thorough, each segment of the customer base should have the opportunity to test the product and give feedback.
- *Recall and exchange terms and conditions:* As part of the agreement, the timing of product recall for the test unit should be identified. If there is a replacement product to be given to the customer after the test period, such terms and conditions should be spelled out.
- *Documentation of feedback:* All customer feedback should be loaded into a database for review and comments. The database should be categorized into critical issues, non-critical needs, and noncritical wants. Critical issues must be effectively addressed before introduction. The database can then be sorted by need or want type, demographics, degree of importance, and so forth.
- *Disposition of feedback information:* The company should have a plan to do something with the field feedback information. There must be a structured plan for incorporating it into the product in the same manner as the initial market needs and wants. Often, if the feedback allows for it, the needs or wants are incorporated in the next revision of the product so as not to hold up introduction.

- *Recommendations:* Similarly, recommendations made by the customer should be factored into the analysis and acted upon.
- *Plan to mitigate any discovered liability:* As a result of the test program, some undiscovered product liability issues may surface. These issues must be addressed in some effective manner before introduction. This does not necessarily mean that the introduction must be delayed; however, the issues must be addressed—either through warnings or instructions.

C. Modifications as a result of beta testing

How should the modifications be selected and made as a result of the beta test program? As stated earlier, the selection must be based on an evaluation that prioritizes issues, from serious, safety-related issues to customer preference. This should be done in decreasing order of importance, as in the original qualification of the specification. Figure 9-2 illustrates the point.

Beta test site feedback							
Customer comments							
			Degree of criticality				
	Feed back item	Action	Preempt introduction	Remedial	Enhancement	Cost impact (%)	Phase-in date
Safety-related issues							
	Item 1		Yes			0.50	Immediate
	Item 2		Yes			0.75	Immediate
	Item 3		Yes			0.30	Immediate
Customer demands							
	Item 1	None					
	Item 2			Yes		2.0	Next revision
	Item 3	None					
Customer wants							
	Item 1				Yes	1.0	Next revision
	Item 2				Yes	1.5	Next revision
	Item 3				No	N/A	
Customer suggestions							
	Item 1	None					
	Item 2	None					
	Item 3	None					
General comments							
	Item 1	None					
	Item 2	None					
	Item 3	None					

Figure 9-2. Beta test feedback.

As shown in Figure 9-2, the beta test feedback shows several responses, some of which should be implemented before introduction because they are safety-related issues. Others are less critical in nature and are desires vocalized by the customer. These can be phased into the product line during subsequent revisions. In this example, Items 1, 2, and 3 related to safety will be implemented immediately and will preempt product introduction. The modifications have minimal cost impact and may be as simple as additional warning labels, warnings, or both in the user's manual.

Still others are suggestions and comments that may or may not be implemented. There is also a cost impact for each of the items to be included. This format serves to document and provide some orderly disposition of the customer feedback and a plan for integration.

D. Series A or B at launch: Weighing the risk

The product team must assess the liability risk of the field feedback. The safety-related comments must be addressed and documented. The risk of accommodating the issue with warning labels to keep the product introduction on schedule must be assessed and decided on. The worst path to take is to ignore the feedback and go ahead with the launch without making any provisions.

Hopefully, the issues returned from the field will be small and not safety-related, allowing the team time to organize the requests and integrate them. If they are serious, then a revision of the initial planned offering or Series A may have to be completed and launched with a Series B device.

E. Units must be installed and exercised

For the beta test program to have any tangible value, the units under testing must be exercised fully. There should be no product hand-holding through this application. The customer who signs on for the beta test is, in effect, signing on to ensure that these test conditions are met. More directly stated, the test is virtually a waste of everyone's time if these specific conditions—designed to stress the product in some specific way—are not met. This should be discussed and agreed to as part of the beta test agreement.

F. Be specific about what you are looking for

In the same spirit, the manufacturer must be specific about prescribing the test conditions to the prospective customer. The customer cannot determine the stress conditions; the manufacturer needs to test, so it is the manufacturer's responsibility to prescribe the test regimen, the timing, the customer's commitment, and the expectations.

G. Selecting the suite of tests

The suite of field tests should exercise the product in normal usage, and somewhat beyond in a few sample cases. The tests should verify functionality, ease of usage, and product performance in reference to the customer's expectations and needs. The test also verifies the engagement of the company and the customer in terms of these expectations, so that the marketing program designed for the product is confirmed.

If a specific instruction set is required to explain the test regimen to the customer, then textbook instructions must be used. If the required procedure is specific to the product and needs clarification, then an application team at the factory should be in place to assist and log in the customer satisfaction or frustration data.

PRODUCT INFORMATION

1. Literature

A. Literature: Term definition

Literature is a term used to describe the management and dissemination of product (line) information. Literature can take several forms and utilize different media. The advent of computer mechanization allows more options for the manufacturer, and will continue to do so. However, the information about the product, published for its use and also for buying decisions, does include several basic elements, as follows:

- *Information to make a purchasing decision:* The information needed to make a purchasing decision can take several forms. Depending on the product, it can be printed matter, electronic media, voice-overs in a video commercial, and so forth. In any case, the objective is to provide the prospective customer with the appropriate information they need to make an informed choice. In some cases, this can extend to a competitive comparison of other products. This can be referred to as a *brochure.*
- *Information to familiarize the user with the product:* The next form of product information is used to familiarize the prospective customer with the product, its uses, and any caveats associated with it. It generally is provided before the sale of the product (or immediately after) and serves as background information. Sometimes this information is merged with the information used to make the purchasing decision. This can be referred to as a *product data sheet.*
- *Information to use the product:* The next form of information is specific information that accompanies the product when shipped to the customer. It is the company's formal document that instructs the customer on how to use the product. It has general information, how to inspect the product upon receipt, how to store the product, the theory or sequence of operation, and other logistical information. It may also include troubleshooting and repair information, but for purposes of this definition sequence we will consider them separately. This is often referred to as a *user's manual.*
- *Information to train about the product:* Some products are rather complex and require training beyond the user's manual. This may include start-up information, procedures, and certification issues for the personnel using the product. These may be entire programs designed to facilitate the use of the product and to mitigate liability to the manufacturer.
- *Information to apply the product:* Depending on the product complexity and configuration, additional information may be required to facilitate its use. For example, some products are components of systems and do not operate without being an integral pert of the system. Consequently, these products require application training to instruct the user about the operative limits of the product. This is to ensure a product is not applied in a system that stresses the unit beyond its intended design parameters.
- *Information to effect repair on the product:* For those products that require repair during their useful life, information is given about repairing the product. As the pathway of information and literature is developed, it can be seen that the information about the various operational aspects of the product follows its lifeline. Sometimes the product is repairable by the customer, and other times it must be repaired by an authorized service center.

- *Information about warranty:* Finally, there is information given about the warranty and how to put it into effect. Included will be the rules of engagement and the terms and conditions.

These are the various types of information generated for the product line. Depending on the specific product line, these types may be combined or issued separately. For some service-oriented products, this information may be negotiated as part of a contract and, as such, there is no standard per se. A summary of this type of information is presented in the form of a matrix in Figure 9-3.

As shown in Figure 9-3, there are five distinct types of product information that can be embodied in six basic types of media. Each product line will demand use of the various types of information and the medium of choice. The product development team should determine the best use of the medium and the arrangement of the types of information to maximize sale of the product.

The following are example outlines of the five different types of information:

TYPE A: Affecting the purchasing decision

A. Introduction/background
B. Features/benefits
C. Application
D. Product configuration
E. Pricing
F. Selection criteria
G. Sizes and ratings
H. Availability

TYPE B: Product usage information

A. Information about your safety
B. About this manual
   - Who should read it
   - Conventions used

Matrix of product information							
Type		**Printed matter**	**Networked communications**	**Fixed DVD**	**Multimedia**	**Video**	**Personal**
A	Purchasing decision						
B	Product usage						
C	Product training						
D	Product application						
E	Product repair						
F	Warranty						

Figure 9-3. Matrix of product information.

C. Getting started
- Overview
- Working life
- Unpacking
- Storage

D. Mechanically-oriented information
- Environmental
- Mounting consideration
- Dimensions and weights

E. Electrically-oriented information
- Wiring standards and codes
- Connection specifications
- Grounding
- Power connections
- Control connections

F. Power-up and commissioning
- Preparation
- Power-up
- Operator information
- Modes of operation
- Diagnostics

G. Specifications

TYPE C: Product training

A. Introduction to product
B. Product application related to operational characteristics
C. Precautions
D. Procedures
E. Certifications
F. Caveats
G. Operator training
H. Continuing training

TYPE D: Application of the product

A. Introduction
B. Application criteria
C. Experience base
D. Limits of the product line
E. Specifications
F. Interfacing of the product line
G. Mechanical
H. Environmental/electrical
I. Human factors
J. Safety considerations
K. System dynamics
L. System performance
M. Significant performance and applications features (SPAF) document

TYPE E: Repairing the product

A. Product configuration
B. Internal repair
C. Field repair
D. Processes
E. Procedures
F. Parts and authorization
G. Certification
H. Turnaround time
I. Flow chart of repair logistics
J. Repair locations
K. Billing coverage
L. Upgrades
M. Referencing the quality database

TYPE F: Prosecuting the warranty

A. Warranty statement of coverage
B. Timing and extent of coverage
C. Product upgrades
D. Explicit versus implied
E. Rules of warranty engagement
F. Third-party arbitration

2. Company Responsibility

The company launching the new product has a responsibility to the end-user to anticipate the user's interpretation of the literature and to eliminate confusion wherever it can. The literature should go through the same design review process as the product because the literature is also the product.

In today's climate where product complexity is great, product documentation is the means for the company to assist the user in applying and using the product. Accordingly, the firm has a responsibility to examine the potential for misunderstanding and ensure clarity in the documentation. This is important from a liability perspective as well as a user acceptance perspective.

COMPANY INFRASTRUCTURE

1. Setting Up the Infrastructure

A. Management support

Managing the development of a new product involves repeated checking and verification of many elements of the program. Now is the time to reassess management's commitment to the program. This is necessary because you are embarking on the initial launch of the product. This launch will consume human energy, time, and funds. Consequently, the effort must have management's complete support.

The support does not end with the investment. Often, management may feel that their part is done when the investment is made; however, their support may be required to adjudicate priorities of collateral departments within the framework of the launch. Management must be made to understand the importance of their continuous effort up to and including the launch. Do not let them abdicate this responsibility.

B. Identify scenarios, develop responses

Marketing a new product requires that a certain infrastructure be in place and functioning at the time of product introduction. This infrastructure exists to accomplish several things, including order entry, clarification of the product's use, application information, availability information, and so on. The specific products will dictate the requirements and needs; however, for purposes of illustration, Figure 9-4 is an example of this.

As shown in Figure 9-4, the interface between the customer and the organization may contain several layers. It can be analogous to the synaptic responses that occur within nerves when a finger touches a heat source. When the finger comes in contact with the heat source, a signal is sent via the nerve bundles to the spinal cord. Depending on the severity of the heat source, a "decision" is made at the spinal column or at the brain to pull the hand away from the heat source. Often the decision is made locally at the spinal column and the hand is withdrawn immediately.

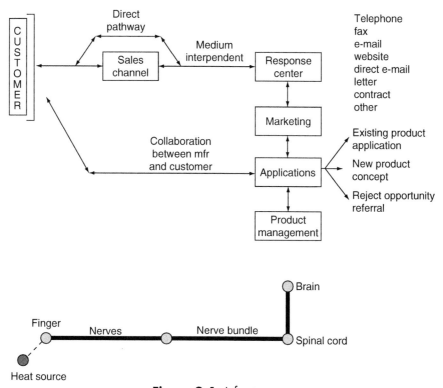

Figure 9-4. Infrastructure.

In a similar manner, the customer sends in a query to the sales organization. This stimulus, which is medium-independent, is recognized by the organization and interpreted to determine the opportunity. The decision to act on the stimulus may be made locally or may be made deeper within the organization. The query traverses the response center through marketing to applications. Applications personnel then confer with the customer to see if a fit exists. Finally, product management oversees the entire transaction.

The response may be to apply the existing product to the opportunity, or generate a new product concept to be evaluated, or summarily reject the opportunity and refer the query to someone better suited to the opportunity. In any case, the operative point is that there is a process, a procedure, and rules of engagement to field inquiries coming into the organization and properly disposition them.

C. Amassing the requisite warmware (The product of people and technology)

Structuring an organization that is responsive to field requests and interrogatories does not stop at the front door of the factory. Nor does the responsibility lie solely with the sales channel. The entire organization needs to be imbued with the sense of urgency and responsiveness that allows the prospective customer to have immediate confidence in the company.

Customers make their decisions based on the product specifications, the company reputation, and their current perception of doing business with the company. This is why it is imperative to make the best possible impression when a customer inquiry is received. The entire company should mirror the responsiveness exemplified at this interface.

D. Automatic feedback and reporting of critical information

Another important point about structuring this interface is to create a systematic method for automatically logging in the customer information. The purpose is to eliminate the company's dependence on the biases and interpretation of the personnel supporting the interface. The idea is to prevent applications and other personnel from summarily dismissing an inquiry lead based on their own opinions. This can cause the company to overlook significant opportunities that are real and achievable but do not necessarily fit the exact form the personnel expected.

In addition, this automated system should log in critical customer comments and feedback so they can be distributed and absorbed by the rest of the organization. Comments regarding safety, ease of usage, misunderstanding, and product failures caused by misapplication rooted in company information or recommendations are all candidates for this critical information requiring documentation and retrieval.

E. Customer/organization interface

The interface should essentially plug the customer into the organization. Customers should become value-added appendages to the organization that provide free-flowing information about the product, company, personnel, and logistics. In its best form, the relationship between the customer and the company should be one of cooperation and open communication. To foster such a relationship, there must first be accessibility; this is what is meant by plugging the customer into the organization.

Young organizations with a smaller universe of customers are very adept at this. The larger an organization gets, and the more diffuse its customer base, the tendency is for the company to essentially isolate itself from the customer. Department titles may not imply this in the company's organization chart and title blocks; however, the reality is often that the operations and development functions, in effect, become isolated. This is what ultimately can contribute to the demise of market share. Stay close to the customer, understand their problems and concerns, and address the latter with your company's products and services.

F. Customer interface is a component of the product

The customer interface is an important component of the product. Without it, few products can flourish in the marketplace. A product that has all of the bases covered in the development process will have a difficult time being successful if the customer interface is flawed. So important is this interface that, many times, mediocre product lines survive and flourish because of the effectiveness of the interface. To ensure that the new product has every opportunity to thrive, pay attention to the interface between the customer and the company.

2. Personnel Training

A. Is the organization set up to handle the new business?

This chapter deals with the late stages of the new product development process. If the organization has not been modified to effectively handle the new product, the team must implement needed changes at this time.

Take a thorough look at the organizational dynamics as they will relate to the new product, not necessarily how they relate to the existing business. The following questions need to be resolved:

- Are there enough people in place to sell, manufacture, support, apply, and service the product?
- Have the objectives for the product line been made clear to the organization?
- Has the organization been oriented to the new product in a balanced fashion, compared to the existing business?
- Are the people in place motivated to make the new product a success?

Figure 9-5 is a training log that will help management determine the skill levels of the personnel associated with the new product. As shown in the figure, the log can keep track of the employees' certifications. It is also useful, from a manpower loading point of view, to see who has what certifications. The manager's flexibility is based on the degree of mobility of the certified personnel. If one area needs strengthening, certified personnel can be moved in from another area.

The chart can also be used to outline the certifications not directly related to an employee's present job description, as well as the employee's desires for the future. Support training and fostering employee growth can then be planned.

B. Sales order entry, service, and parts

Previously, we discussed the company's external and internal interfaces. We also looked at how a customer inquiry is fielded within the organization. In this section, the sales transactional analysis will be reviewed. Not every product will be quoted, nor will every product require service. However, when the company's product is sold to the consumer by a third party, the company must service that third party like a customer.

Training of personnel					
Category	Description	Employee 1	Employee 2	Employee 3	Employee 4
Sales order entry	Quote process	X			
Sales order entry	Pricing	X			
Sales order entry	Sales analysis	X			
Sales order entry	Negotiating skills	X			
Sales order entry	Applications	X			
Sales order entry	Sales order entry	X			
Sales order entry	Order acknowledgment	X			
Sales order entry	Scheduling	X			
Sales order entry	Follow-up communication	X			
Sales order entry	Customer satisfaction training	X			
Service—in-house	Product repair training		X		
Service—in-house	Troubleshooting analysis		X		
Service—field	Customer interface			X	
Service—field	Technical data			X	
Service—field	Troubleshooting analysis			X	
Parts	Spare parts				X
Parts	Replacement				X
Additional skills					
Career path planning					
	Supporting training				
	Supporting training				
	Supporting training				

Figure 9-5. Training log.

For example, the primary customers are issued a quote on the goods, they are issued a delivery date, and their stock order must be entered for processing. Every organization has established practices and procedures for this part of the business. The objective at this point is to ensure that the procedures in place will be sufficient to support the business generated by the new product. The procedures should specify how communication with field organization and the customer should take place. The procedures should also specify how to handle refusals, referrals, and those interrogatories that will not result in an order for the company.

Record retention and retrieval from databases must be tested and verified. This is because critical information coming in at the initial product launch must be analyzed and acted on to ensure a product success.

C. Audit the system

For any system to continue to be effective, it must be audited. The sales order entry system is one example in which a periodic audit can have huge payoffs at the revenue line. Keep in mind that this portion of the company is the primary contact point with the customer. The customer formulates an opinion about the company based on this narrow viewpoint.

To keep this area operating correctly, periodic audits are required. They should be unannounced and structured from the customer's perspective. At the risk of making this sound like a cloak and dagger exercise, the degree to which the organization does not know it is being audited is the degree to which the audit will have value in assessing performance. Take the necessary steps to secure the information without affecting the outcome.

D. Corrective actions

During the course of the investigation and final checkout of the organization, there may be some elements that require correction. Do not rationalize and procrastinate with respect to correcting these elements. The product success is based on certain assumptions, which require that certain systems be in place and functional. To ignore the issue and fail to take corrective action negatively affects the product. This may be serious enough to cause product line failure. If there are corrections needed in the process, make them now. They do not have to be absolutely perfect but they must be effective.

3. Applications Support

A. Response centers and after-hours support

The applications support function should have clear, measurable objectives for supporting the new business development initiative. For example, the response center—a central location for all communication between the field and the company—should have a short throughput time that is measurable and can be evaluated in terms of performance.

The after-hours support function should answer incoming communiqués within some defined period. This is necessary for the inquiring customer to have some idea of when they can expect a response. Establishing a response center that falls short of customers' expectations will cause the company to lose market share rapidly. In addition, many products require a certain amount of support to make them successful in the customer's venue. To this end, the response center must be effective, accurate, and timely in its assistance.

B. Customer support team

Assemble a customer and applications support team to enhance the product's reception in the marketplace. This is for customers who may not be knowledgeable or well-versed in the product. In these cases, an applications support team can enhance sales and establish immediate customer confidence, before and after the sale. The key to applications support is product and applications knowledge. To that end, do not staff this function with personnel who are not fully qualified in the specific area the product serves.

The customer support team should have three basic operational traits: resident expertise, motivation, and thoroughness. The *expertise* will allow immediate feedback and customer assistance, negating the "need to check and get back to you." Such communication delays erode customer confidence. By having the expertise resident in the minds of the support personnel, the information transaction will proceed much more smoothly.

The *motivation* aspect underscores the requirement for the entire customer interface and sales channel. Customers do not wish to interface with people who are bored, lack enthusiasm, or are difficult to get answers from. Consequently, it is critical to the success of the product to have the support personnel motivated to grow the business through increasing sales.

The *thoroughness* aspect protects the company from misapplication and also establishes high confidence levels among the customer base. With the personnel exhibiting diligence and doing a complete job at this level of the transaction, the customer will be more at ease in proceeding to the next step with the company.

C. Technical assistance

In a similar manner, the company's technical support must be accurate and documented. This is required for future possible liability in which the misapplication results in litigation. The technical assistance must support the product applications and warn of misapplication. All this must be effective via whatever medium may be in place between the customer and the company.

D. Simulations

Depending on the type of product and the need for direct communication with the manufacturer, it may be necessary for the customer assistance function and the applications personnel to simulate the product's use in a system. Here again, it is in the best interest of a sale to have these people and their systems in top operating form. The simulation systems should be state-of-the-art and mimic an actual customer service event. In this way, they have value. By establishing the value in the customer's mind, goodwill is created and abuse of the function is kept to a minimum.

As technology evolves and products become more and more complex, the ability to use traditional tools to support the product line might fall short of customer expectations. This is driving the need for expert systems to be implemented. An expert system can mechanize, to a certain extent, the tribal knowledge resident in the applications personnel. In addition, it can analyze a set of conditions and draw conclusions based on the input data and the conditions, according to a prescribed algorithm. This can mechanize a response to the customer and free up applications personnel to develop algorithms for the lesser-known issues. This process continuously improves the development and implementation of the algorithms in the expert system, enabling them to handle more complex situations and thus add value to the service.

F. Model for a customer support database

One of the most effective tools for improving customer assistance and the logistics involved is to create and implement a call-in database. This database will allow the organization to field inquiries from customers in a systematic fashion. The inquiry will be addressed according to a procedure and a protocol. It will also force resolution of the issue if the format is followed diligently by the personnel involved with the process.

Figure 9-6 offers an example model for this database. As stated many times before, the specific business needs will drive the configuration of the database. Figure 9-6 is presented as a model only and will require modification for your specific situation.

Customer support inquiry log/database						
Background information						
Name			**Address**			
Phone			**City/state/country**			
Fax			**E-mail**			
			Customer type			
Reported By			**Call taken by**			
			Action taken by			
Severity	Critical	Noncritical	**Call date**			
			Close date			
Description						
Actions						
Desired resolution						
Service request	Yes / No					
History / recommendation / status	**Most recent issue desc.**	**Recommendation**	**Status**	**Effectivity**	**Next action**	
	Oldest					
Closure						
Call close results						
Follow-up action						
Customer satisfaction	Positive	X				Negative

Figure 9-6. Customer support database.

The database format shows three basic sections in handling an inquiry. The first is the background section that contains identity information and some logistical background information. An important component of this section is the degree of severity of the incoming call. If the call is about a critical situation, it must be acted on immediately with priority over other calls. If it is noncritical, it can be handled in due time.

The second section deals with actions prompted by the call. The customer articulates the desired result of the call. If the product warrants it, a service call may be in order and is noted here. This section is one of the most valuable in the database because it contains the history of the incident(s), presenting the most recent first and listing them in reverse chronological order. This history consists of several elements, including the following:

- *Issue description:* This is the complete description of the problem with the product, as given by the customer. It is the root information that the response center has to go on to assist the customer.
- *Recommendation:* This is the company recommendation given to the customer, based on the response center's description and analysis.
- *Status:* This is the status as a result of the recommendation and some action taken by the customer to correct the problem.
- *Effectiveness:* This is an assessment of how effective the recommendation is in terms of the action taken.
- *Next action step:* This is a best guess as to the next step required in resolving the problem.

The third section deals with closure of the call. It is the resolution in the mind of the customer that the company has corrected the problem and no further communication is required. Its elements include the results at the time the call was closed, any recommended follow-up action, and an assessment of the customer's degree of satisfaction with the incident. This example provides a gradient ranging from a positive reaction to a negative reaction.

4. Field Organization Setup

A. Sales team provides the follow-through on launch

The best marketing program for the best product generally cannot succeed without the field sales organization providing follow-through in the customer's venue. The sales team, in whatever form it may take or whatever channel it uses, is the vehicle to execute the company's strategies. This section deals with getting the sales team positioned and motivated to launch the product and successfully make sales.

There is a certain amount of demand created within the field sales organization concerning the new product while it is in gestation. This generates enthusiasm and healthy apprehension among the sales force. When the product is released for general sale, the product team can leverage this enthusiasm into initial sales volume. In the long run, however, the business team has a challenging job in maintaining this activity level. In addition, depending on the salesforce's compensation structure and the sales channel's means for revenue, both of their outlooks may be shorter than the manufacturer may want to sustain the sales campaign.

The salesforce simply must be directed to carry the torch of new product introduction and ignite the sales channel effectively. There should be no compromises here. The gospel of new business development must be preached and followers must be cultivated.

B. Review the type of field organization in place

What type of sales organization is in place within the company? What are their strengths and weaknesses? These questions (first posed to the business development team in Chapter 1 in the assessment phase of planning) must now be looked at in depth to ensure that the sales group's human energy is focused on the tasks of a launch.

As a general rule, progress is marked by effort directed at the elements of a program that will generate sales. If the sales department is averse to overcoming certain weaknesses, they must be corrected. There can be no finessing of weaknesses in the context of a new product launch. Weaknesses must be transformed into strengths. The introduction and execution of a sales campaign requires a complete, targeted effort. If one element is missing, it cannot be compensated for.

Is this the highly motivated and effective field organization that will carry the launch of the product to secure sales? Is this the organization you had envisioned and made assumptions about at the beginning of the program? If not, then the organization must be modified to meet the requirements outlined in the original assumptions. If conditions have changed that allow for this area to be compromised, then proceed cautiously. However, in reality this rarely happens. The markets and customer needs usually increase to the point where more is demanded of the sales force than was originally envisioned.

C. What are the dynamics required for selling and servicing?

What are the dynamics required in the sales organization to achieve the forecast volume? Do they match up with the customer service team? To answer this question it may be necessary to assess the validity of trying to use the existing sales organization, without improvement, to launch the product. Consider the evaluation tool example in Figure 9-7 and fill it out for your situation.

As shown in Figure 9-7, the example chart categorizes several aspects of the sales organization's skills and evaluates them in terms of priorities, capabilities, cost-effectiveness, technical expertise, compensation, and employment issues. With each aspect, the initial assessment of the organization is compared to the product launch and sales skills required.

The improvements required are noted and an action plan is generated with an assigned completion date. Status indicates the degree of progress toward the goal. For example, reviewing the sales force's prospecting capabilities, the initial assessment of the team indicated a tendency toward route salesmanship rather than a balanced effort that included prospecting new accounts. The initial assumption was that by the time of product introduction (a long way off, at the time) the characteristics would change in the sales force. Now is the time for them to change!

The current state for the product's marketplace demands a fair amount of these skill sets, so improvements are required. Training and practice can help change the attitudes, and this can be accomplished during the introduction for this example product. By using an evaluation and planning tool such as the example in Figure 9-7, the product team can ensure that the sales function will be trained and ready to launch the new product.

D. Ensure the sales team's support at launch

The sales team's frame of mind at the time of product launch is critical to a sustained, successful effort in moving the product to market. There are many barriers that must be

Sales team evaluation tool							
Category	**Initial assessment of existing customer support team**	**Initial assumption**	**Current product require-ments**	**Improve-ment required**	**Action plan**	**Completion date**	**Status**
Interdivisional priorities	Division favoritism evident	Balanced required	10% focus	10% focus	Mgmt decree	At product introduction	Prepare complete
Prospecting capability	Tendency away from prospecting	Needs improvement	Broaden skills in prospecting	Needs major improve-ment	Train and practice	During introduction	Not started
Track record							
Channel management abilities							
Sales per expenses							
Call frequency	Rifle-shot methods currently used	Needs improvement	Needs broad coverage fast	Need incentive for sales	See attached	At launch	40% complete
Mobility	Inside sales only	Need outside sales also	Develop net new accounts	Change personnel	See attached	Just prior to launch	50% complete
Efficiency							
Average territory size							
Technical expertise							
Compensation system							
Motivation							
Driving forces							
Employment factors							

Figure 9-7. Sales team evaluation tool.

overcome by the sales team, and energy needs to be applied to overcoming these external factors. There is no time, energy, or patience to overcome problems in the sales team at launch. All of these problems should have been overcome before launch.

Consider Figure 9-8, which shows some of the barriers to entry the sales team must overcome. The route to market is shown as the manufacturer funding a program specified by product planning. The marketing group, upon completion of the product development, outlines a campaign for the sales personnel. In this example the sales personnel are company employees who work through a network of representatives, indicated by the sales channel. They, in turn, sell the product to the marketplace.

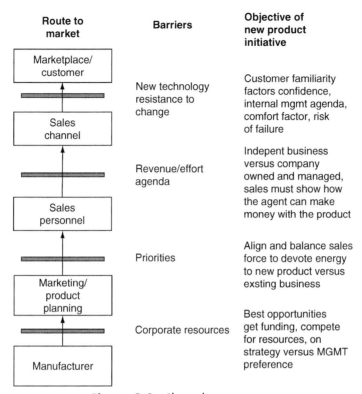

Figure 9-8. Channel management.

Each stage of the process has barriers associated with it. The product planning and marketing group must compete for corporate resources. They must navigate financial and corporate funding politics to get the funding for the best program. They must keep the corporation on strategy in selecting programs. The marketing personnel must reprioritize the sales force's activities to accommodate the new product, and may have to do the so-called missionary work necessary to launch into a nontraditional market. They will have to cultivate new accounts and redirect energy and focus.

The salesforce must be aligned and balanced to support both the new product introduction and the existing business. The sales personnel now have to carry the corporate message about the new product and drive the requirements home through the sales channel. If the sales channel consists of independent businessmen, the corporate salesforce must influence and align their priorities. The new business must receive the appropriate time and energy from the channel. To break down this barrier, the sales personnel must demonstrate how the agents in the sales channel can make money with their new product.

The final barrier in the model is the customers themselves. They have a host of issues, such as unfamiliarity with the new product, technology, and application; their confidence in the product, their fear of failure, and aversion to risk; and any internal existing customer agenda that would preempt their use of the new product. Breaking down these barriers paves the way for the success of the new product.

E. Refine the sales order entry system and reporting

The sales order entry system should mimic the product quality system in that each request for information, each quotation, and each interface with the customer should be cataloged and mined for market data to further the sales of the product and reposition the company for future product offerings.

The system should have sales status (either positive or negative) attached to each query. In this way the database can be updated automatically. The system should feed back data to the individual sales offices to measure their effectiveness against the company's marketing plan. In addition, company management can evaluate the effectiveness of the marketing plan in aggregate.

F. Sales motivation and compensation

Sales compensation can be a complex and difficult issue to administer in a company. The system should be fair and reward those who perform, and management should deal with those who do not perform. It does no good to keep marginal sales personnel on the payroll and attempt to improve their performance and legislate their behavior through compensation. It generally takes too long and has little effect. In addition, the new product launch does not have that luxury of time. The sales force should be operating to an internally generated plan to move volume of product. Price cannot be the issue at this point.

Given these assumptions, individuals should be rewarded for their results. There should be several components in a compensation system, and emphasis on those components can vary based on the type of product. Consider Figure 9-9 as a format for compensation for company-employed sales professionals.

Format for a sales compensation system						
Item	Description	Maintain existing business levels percent %	Expansion of sales volume percent %	Margin preservation	Expansion of available market	Budget maintenance
1	Base salary	60	50			
2	Bonus—volume of new product	10	20			
3	Bonus—volume of existing product	10	10			
4	Bonus—net new accounts	10	10			
5	Bonus—gross margin	5	5			
6	Bonus for corporate goals	2	2			
7	Personal effort	2	2			
8	Net new improvement	1	1			
9	Total	100	100			

Figure 9-9. Format for a sales compensation system.

In this format, the compensation system for the sales force consists of several elements, including a base salary and bonuses tied to performance (both individually and through the achievement of company goals). The key point of the chart lies in the percentages of the sales professional's compensation that results from each of the initiatives. Fundamentally, the sales effort must result in orders, and orders will be generated by sales professionals in response to incentive compensation. If you want to improve their results, structure their compensation system accordingly.

Many companies lose sight of this simple axiom. All of the speeches by management, pleas for balanced selling, and requests for focus on product lines do not achieve the same result as a simple change in the compensation system. The chart shows different focuses placed on the various categories, based on the company's desires. If market expansion is the critical objective, then this category is weighted more heavily toward expansion from a compensation point of view.

5. Tying in Sales Order Entry Systems

A. Remote sales units with bidirectional communications

The remote regional offices should be tied into a network of communications with the factory, whereby information flows freely between the two. This is illustrated in Figure 9-10, which depicts a sales organization that has several regional offices that

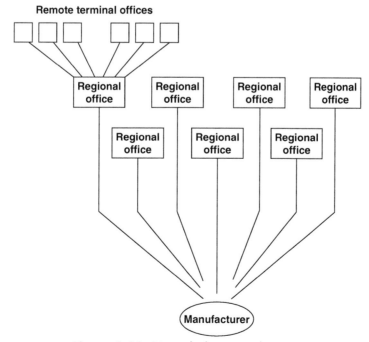

Figure 9-10. Networked remote sales units.

work through remote sales offices. These remote offices can be company-owned or individual businesses. In either case, they get their direction from the regional sales office, which gets its direction from the factory. In this way a uniform new product story can be diffused throughout the sales organization. Policies are communicated in a consistent format and consistent information is fed back to the manufacturer from the regional sales office.

It is very important to have accurate and consistent information transfer between the terminal units and the regional offices and the factory because new product decisions may have to be made early in the introduction to facilitate success. Given this fact, accuracy of communications cannot be an issue.

Because every product is different, and the sales route to market may vary, this model is not intended to be the single solution to all situations. However, it does illustrate the need for accurate and focused communications between the manufacturer and the remote sales offices.

B. Tracking the sales progress

Another basic need during product launch is the ability to evaluate the ramp-up in incoming order rates among the various offices. Figure 9-11 can be used to chart the progress of a new product launch by region. This chart is accompanied by a graph that pictorially shows the sales progress.

As shown, every region shows a nice progression in sales. Some interesting characteristics, however, are shown in Region 2, Region 7, and Region 8. Region 2 showed a rapid increase in orders in Month 2, followed by a decline in Month 3 and slower growth in succeeding months. This engenders questions about what is causing these results. Did one large, unexpected booking cause the rapid growth in Month 2? Similarly, what caused the results of Region 7, which increased in Month 3, followed by another higher increase in succeeding months and another dramatic increase in Month 4? Conversely, what caused the decline in bookings in Region 8 after Month 4 after such a positive progression in the previous months?

This type of sales analysis can give the manager a wealth of information about the market, the progress of sales, the activity level of each region, and comparisons of one region against others. In addition, a look at the specifics of the activities compared to the results will give an indication of a region's performance. This information is extremely useful in assessing the launch progress and each sales office's participation in it.

C. Quotes

Depending on the type of product and marketing arrangement, quotations may be used as a means for selling the products. Although this is not a model for every sales need, it does serve to establish a point about quotations and follow-up of quotes to secure orders. Figure 9-12(A) is a typical model for a quotation database.

As shown in the figure, the sample database allows management to keep track of quotation activity and to follow quotes through to resulting bookings. It contains pertinent information about the activity and walks the sales personnel through the next steps to secure an order. In addition, the percent probability and timing information can be used to generate materials demand in the absence of a formal forecast.

Sales order entry analysis (Units Sold)						
Regional office	Prelaunch	Month 1	Month 2	Month 3	Month 4	Month 5
Region 1	9	11	17	101	81	73
Region 2	11	14	69	48	65	58
Region 3	14	18	26	46	62	56
Region 4	11	14	21	36	49	44
Region 5	13	16	24	43	58	52
Region 6	16	20	30	53	71	64
Region 7	19	24	36	62	84	160
Region 8	24	30	45	79	106	96
Region 9	29	36	54	95	128	116
Region 10	22	28	41	72	97	97

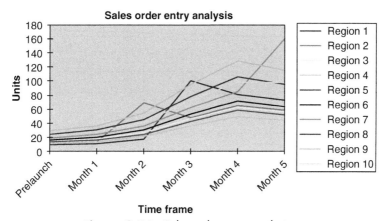

Figure 9-11. Sales order entry analysis.

D. Sales reports

Another required item in the sales management arsenal is the sales report. These reports summarize each project's details and outline the market activity surrounding each project. Consider the sales report format in Figure 9-12(B). It contains several elements but generally follows three main items: background, quotation information, and status.

The first section in Figure 9-13 is the actual sales report. It contains all of the background information about the project being tracked. This section is used by the database for retrieving quotations, general information, and logistics.

The second section is the quotation stage. This section summarizes the listing of firms quoted, the pricing, the calculated margin, and the date quoted. This section also contains the competitive summary of any competitors participating in the contest to secure the business.

Finally, the third section summarizes the project status. It indicates the current project status, the successful bidder, and the particulars about the commercial issues. One of the

Quotation database						
Business segment A						
	Quote no.	Date	Customer	Region	Region manager	Value
	1					
	2					
	3					
	4					
	5					
Business segment B						
	Quote no.	Date	Customer	Region	Region manager	Value
	1					
	2					
	3					
	4					
	5					
Business segment C						
	Quote no.	Date	Customer	Region	Region manager	Value
	1					
	2					
	3					
	4					
	5					

A

Figure 9-12. (A) Quotation database.

Business segment A						
	Project status	Follow-up	% Probability of receiving order	Timing	Sales status	Collateral outlook
Business segment B						
	Project status	Follow-up	% Probability of receiving order	Timing	Sales status	Collateral outlook
Business segment C						
	Project status	Follow-up	% Probability of receiving order	Timing	Sales status	Collateral outlook

B

Figure 9-12. (Cont'd) (B) Forecast by quotations.

Project		Quoted By			
Release date		Employee No.			
Quote no.		Region			
Description					
Quote stage					
				Margin	Date
Quoted to:	1				
	2				
	3				
	4				
	5				
	6				
	7				
	8				
Competition					
	1				
	2				
	3				
	4				
	5				
Status					
	Project let				
	Project pending				
	Project cancelled				
	Successful bidder		Pricing	Margin	Date
	Comments/reasons				

Figure 9-13. Sales report.

most important elements of this section is the Comments area, which requires that reasons be given for securing or losing the order.

E. Demographic marketing data

In order to evaluate sales performance, the management tools discussed in the previous section must be used in concert with the demographic data generated by the manufacturer. There is no point in managing a region that may appear to perform poorly simply because the market may not be available to the extent originally forecast. Assigning forecasts in a fair and equitable manner is the responsibility of the marketing group located at the manufacturer.

F. Lead time management

The product lead times should be communicated to the salesforce in real time. Given the availability of resources to generate the labor portion of the manufacturing equation, the quote database and sales management systems can be used to generate a materials requirements forecast, based on quotations and probability of securing the orders. Although this model may not work for all products, it does illustrate the point that the manufacturing side of the business should be tied very closely to the sales and marketing side of the business to effect on-time production and customer satisfaction.

G. Rapid delivery networking with other agents

One of the ways to alleviate the lead time issue is to stock the product at remote terminal units. By having each of the terminal units containing stock, critical orders requiring shorter lead times (not supported at the manufacturing level) can be fulfilled by shipping product from one remote sales unit to another. In this way the customer satisfaction level can be maintained without undue burden on the manufacturing organization. Local stock at remote terminal units will also encourage market development at the local level to turn those units' inventories.

H. Pricing administration

Administering new product pricing can be accomplished by the sales order entry system. This system can lock-in pricing during the product launch and thereafter, in order to preserve the pricing integrity. This must be accomplished so that the product team can understand the margin dynamics. By holding the pricing fixed via the sales order entry system, the variances in manufacturing and procurement can be resolved more quickly.

If a mechanized network is used to link the offices and manufacturing, the pricing can be communicated to the remote sales units and can be modified uniformly by the manufacturer when the product team initiates a change.

CHANNEL INFRASTRUCTURE

1. Business Case Development

The sales channel is critical to a successful sales program. When evaluating the capabilities of the sales channel, one needs to ensure that a solid business case exists for the company, the members of the sales channel, and the customer. If there is no solid business case for *all* members, the impact of their efforts will not be sustainable. Companies often try to leverage their past reputation or their previous business with a member of the channel—using it to launch a new effort. If the new effort will not generate a benefit, then the business development team needs to rectify the situation. In addition, the business case must be valid from the end customer, back through the channel, to the manufacturer's cost structure.

2. Motivations

As mentioned earlier, the business development team needs to assess the channel members' financial motivation in the sales transactions and how they get compensated for their work. This is especially critical where independent business people are involved in the transactions. They are motivated by their own business and cannot abdicate their own profits for the principal. When a company sales person is the point person in a transaction, the motivation may be different; In such cases, salaries, benefits, mobility, and chance for promotion may drive behavior regardless of financial return. In general, however, compensation drives behavior and the compensation must follow the values added by the members of the channel.

3. Compensation

Another factor in a channel infrastructure analysis is the means of compensation. Whether it is a commission, finder's fee, profit, or a combination, the structure of the compensation should be self-funded by the products themselves. For large companies that have a lot of financial capability, it is less of an issue. For smaller companies, payment method can be more of a problem because it will affect cash flow and profitability. Nevertheless, each member of the channel and each link in the chain from manufacturer to user must add value to the product. If they do not add value, the prospects for long-term success are greatly diminished.

4. Consequences of Nonperformance

When setting up the infrastructure for the sales channel, it is important to have flexibility for dealing with nonperforming channel members. The business development team should measure the effectiveness of the channel members and have contingency plans in the event of nonperformance. If there is a problem, act on it early to save time and energy and, more important, bring sales channel horsepower to bear on the business development enterprise.

ASSESSMENT—PRICE VERSUS VALUE

1. Final Pricing

Setting the final pricing is one of the crucial elements of a product launch. All of the assumptions about the marketplace, tolerance for pricing, and features versus benefits and attaching value to each will be verified or negated by the refined pricing at this time. This section will present several perspectives on this important step.

A. Field feedback helps set pricing

Just as the product cannot be developed without knowledge of the marketplace, neither can the pricing be developed without marketplace information. The information available from the field personnel, both in sales and service, is used to refine the product pricing. To that extent, do not ignore the feedback received from the field. Their perspective can yield clues to product reception from the pricing standpoint.

When the sales force is company-employed and giving honest feedback, they are providing valuable information within the context of the company's position in the marketplace and their take on how successful it may be. If, for example, the product is targeted at a premium price strategy in a commodity marketplace and it has no collateral value added, the sales force could see potential problems in their area that may not be evident to the corporate marketing staff.

Factor in their input on pricing, make a decision on the pricing, and implement consistent pricing at product launch. This underscores the primary philosophy about pricing: consistency. Do not change the price during the launch. This confuses the field people and undermines the launch effort. Wait to determine the full effects of the price selected and then the corrective action after launch, if required.

B. Verify your position in the marketplace

Throughout the development program, time in the marketplace is passing. Competitors are participating with their own programs and product introductions. Now is the time to evaluate the product and company position in the marketplace and ensure that the pricing that was originally set can be held.

In all of the perspectives offered in this section, the objective is not to rationalize a price cut just before introduction. Remember, this is what will generate profits to pay for the initial product development investment and to fund the next program. Instead, the idea is to ensure reasonableness in the pricing to maximize the impact of the product launch. Reverify your position in the marketplace to make sure the product offering can support the pricing.

C. Reverify your product strategy

In the same manner, reverify the original product strategy. Do the features and benefits generated in the embodiment of the product still support the price? Are the target markets and the target volumes still reasonable? Has anything in the marketplace caused a change in strategy? In short, now is the time to confirm the pricing from the product strategy perspective. Ensure that the product can still command the price, given all of the other changes that occurred in the market during development.

D. Rerun the return on investment calculation

The investment in the new product is essentially sunk and the costs are unrecoverable. It is an interesting exercise to rerun the return on investment (ROI) calculation, using the actual figures instead of the estimated figures. How does the program appear now? Has it changed for the better because of lowered development and product costs, or is it worse, to the point of not being sellable to management if it were a fresh program?

Use this data as an operative lesson in future programs to see which assumptions work and which do not. These are the lessons in new product development that are learned over several programs, and pertain to how a product development team is grown and nurtured over several projects.

E. A low-volume specialized product or a high-volume standard product?

At this point, it is wise to check out the product offering and how the customer base, the market, and the sales force perceive it. Does the product still have the initial high-volume appeal—serving a wide range of demographics with the same embodiment—or has it degenerated into a universal platform, attempting to be all embodiments to all customers with little or no standardization?

These are the perspectives that one must understand just before launch. Did the team lose sight of the new product goal and deliver a product that will be an uphill battle to launch and sustain in the marketplace? Can changes in pricing offset these issues? Consider these issues carefully and set the pricing accordingly.

At this point in the program, the price is one of the last elements of the product offering that can be modified. When setting the pricing, consider all of the aforementioned factors and lock onto pricing for the launch. Given all of the diligence exercised throughout the program and the final checklist evaluation, the pricing set should be workable in the marketplace. The customer base will tell you if the pricing is out of line. If you have done your homework, it won't be.

3. Market Impact

Do not underestimate the impact the product introduction will have on the marketplace. The marketplace is a dynamic arena, and a business introduction or product introduction will not go unnoticed—there will most likely be a reaction. This could be competitors changing pricing or terms to offset your firm's introduction. Accordingly, your business development team needs to think through all of the possible scenarios for the market reaction and make sure you act accordingly. Do not box yourself into a corner that you cannot get out of in pricing or terms. In addition, because the marketplace is supplier-led in most instances, the pricing is relatively inflexible upward.

4. Limits

Establish and understand the limits of pricing action throughout the life of the product. When setting final pricing, understand that there will be a reaction that you must have anticipated already. The product has certain limits in pricing adjustments, and certain assumptions about cost reductions that haven't been met yet. Accordingly, price movement has limitations.

PRODUCT RUN PREPARATION

1. Materials Procurement: Preparing for the Ramp-up

A. Ordering materials in advance of the pilot run

In any new product environment, there is always risk management. In the later stages of new product development, the risk in procuring longer-lead-time material in advance of the pilot run is financial. This financial risk is associated with unusable inventory if the results of the pilot run and testing program necessitate changing one of these long-lead-time or specialized parts. Given the assumption that the product team is diligent in its selection of components during each stage of the program, this risk can be mitigated. In any event, the fundamental requirement for the pilot run is to have the processes and material in place. The risk of a material change should be low in comparison to the risk of not being able to initiate the pilot run and subsequent beta testing.

B. Risk analysis of changing parts after the pilot run

The issue of inventory and disposition of obsolete parts, based on a change after the pilot run, is one of percentages. Given a complete bill of material, if a small percentage of the parts change (necessitating scrapping the parts), their value pales in comparison to the value of lost revenue in arriving late to the marketplace. Even if they are expensive parts, the issue is one of loss mitigation at this stage of the project; time is more valuable than material, for the most part.

C. Allocation of common parts within standard inventory

When introducing a new product to the manufacturing arena, another concern for the product team is common parts usage within the existing business. The existing business, which consists of products that can use some of the same parts used in the new products, can be a problem when increased volumes in existing products may rob parts from the pilot run or initial production runs of the new product. In these cases, it is important to reserve parts with some type of new product restriction code implemented in the company's management information system Or even physically segregate them within the manufacturing facility.

D. Blanket purchase orders to vendors with ramp-up

Another element of the materials equation is to establish agreements with the vendors to utilize blanket purchase orders for annual quantities of material. This material could then be delivered in stages, as required by a monthly forecast. This enables the procurement function to take advantage of quantity pricing breaks and allows the vendors to plan for their production requirements.

For a blanket purchase order to be effective for a vendor, the same requirement for the manufacturer applies. The product on the blanket purchase order is committed to and must be shipped. Failure to do this will result in cancellation charges that will absorb material, negate any advanced production done for efficiency reasons, and disrupt schedules. The same holds true for the vendor supplying parts to you. The commitment is made at the time of the purchase order. If you change parts, you own the original blanket order quantity that is unused.

E. Purchasing manufacturing cues versus standard parts

If the parts being procured are not catalog parts, the blanket order secures a manufacturing cue at the vendor. There is also vendor material risk in a blanket order that may change. In this case it may be a good idea to secure, through purchase order, a manufacturing cue at the vendor. In this way the parts you may need are secured by virtue of a reserved manufacturing time and duration. However, the downside of this arrangement may be the material procurement at the vendor level. If it has a long lead time, little may be gained because the vendor's material will not arrive in time for the manufacturing cue to be used up.

FEEDBACK SYSTEMS

In preparing for a product or business launch, it is important to have feedback systems in place to evaluate each aspect of the new business. In this way the business development team can monitor early results and signs and take corrective actions. Systems should be set up for the following operative areas (these are lists of the items you will want to keep track of and should have a system in place to report and correlate the data):

1. *Sales*
   - Total sales
   - Breakout of sales by territory and group
   - Characterize sales transactions, how long to make sales, and how it is done
   - Comparison of volume ramp-up versus original plan
   - Obstacles in sales efforts
   - Successes in sales efforts

2. *Channel absorption*
   - What channel member is performing and why?
   - What are the barriers for channel members?
   - What is the key measure of training initiatives?
   - What is the key measure of launch initiatives?
   - How has the individual channel member management embraced the product or business?

3. *Quality*
   - First-pass yield data
   - Total number of returns
   - Field returns
   - Percentage of no-fault-founds to total returns
   - Customer comments

4. *Functionality*
   - Customer comments
   - Feature usage patterns
   - Time to use

5. *Acceptance*
   - Customer satisfaction metrics
   - Repetitive sales (if applicable)
   - Customer endorsements

6. *Competitive pricing action*
   - Each competitor response to launch
   - Time to response
   - Amount of response if it is a price action
   - Rebundling of values

SUMMARY

This chapter can best be characterized as a checklist for new business and product development. The development phase is complete and manufacturing arrangements have been made, and now the focus has shifted to the preparation work for product introduction.

The chapter started out with the checklist for the new product line. It then moved into the agency approvals and pilot run manufacturing. The product receives external validation in the customer beta test program. The information system associated with the commercial aspects and usage of the product was described.

Next was a discussion of the product information and the various forms is may take, followed by the examination and confirmation of internal support systems to garner business generated by the product. These systems involve training of the personnel, the application support personnel, and the field sales organization. In all of these functional areas there is a change in the operations that must be introduced and cemented into the organization.

Next was a discussion of the channel infrastructure requirements and how to integrate them into the organization. Final pricing is established at this point because the program is very near launch. To effect a successful launch, the sales order entry system must be functional and effective.

Next was a discussion of the material requirements in the stages of transition, from development prototypes through initial production. Finally, the use of feedback systems for the business development team covering the various aspects of the business was reviewed.

Each stage of the development process brings closure and anticipation, simultaneously. With the completion of the elements discussed here, the product is poised for the transition the product team has planned for—The Launch.

THE PRODUCT LAUNCH

PRODUCT LAUNCH

1. Objectives

The product launch is the formal release of the product to the marketplace. This can occur on a worldwide basis or it can be targeted to specific niches, or something in between. The objective of the launch is to leverage the preparatory work into a mass announcement of the new product or business. The most positive notoriety is available to the business development team at this point. Also, the most heightened sense of awareness on the part of the competitors occurs at this point. Aside from all that, the product launch is your show—make the most of it.

2. The Rollout

During the development phase of the project, a constant balance is struck between maintaining the secrecy of pertinent development areas and the need to promote the program to various parties. The rollout is the time for all efforts to be directed to introducing and promoting the product. The next section will review the various aspects of the rollout.

A. The rollout formally launches the product

The product rollout or introduction is the formal means to launch the new product. It is specifically designed to get the word out to the customers and the marketplace that the new product is available for sale, and that it consists of the certain features and benefits. The rollout should be timely and pervasive. Unless there is a compelling reason to have a phased introduction, the rollout should simultaneously announce the product in all sectors of the marketplace.

The rollout should also be a time to generate enthusiasm about the product. This enthusiasm needs to pervade the sales group, the sales channel, and the customer base. In some industries, the rollout is a media event anticipated by the marketplace because of pre-introduction activities. This anticipation helps cultivate the enthusiasm needed to carry the product.

B. Media usage is critical to launch plans

Media usage is critical to the launch of a new product because the media, in whatever form, create leverage. This leverage enhances the effect of the initial launch. It brings the company and product name to the forefront of the marketplace. Without this, the manufacturer could not devote enough energy, time, and funds to create the same effect. There are several different types of media to use in the initial product launch; the following list represents a few of these types, along with their explanations. (Each product and business will have its own needs and requirements for media usage. These examples are just for reference.)

- *Trade press:* The trade press involves those publications and other related media that participate in the industry the new product will serve.
- *Newspapers:* Newspapers can assist locally by introducing the new product to the general population that reads the papers. They provide completeness in the introduction and often spark the interest of other publications that comb regional news sources for new information.
- *Direct mail campaigns:* Direct mail campaigns target the customers directly in the hopes of soliciting interest in the new product. If the nature of the product is such that the details of the product are already known, direct mail campaigns can focus on its availability.
- *Seminars:* Seminars are an effective way to introduce a new product to a specific market segment. They are based on the assumption that a somewhat generic product presentation and problem/solution development can be imparted to a group of people participating in an industry. Seminars can also be specific to only the product introduction.
- *Professional awards:* The ability to win a professional or industry award can be an effective tool in product promotion. The award legitimizes the product as well as the launch.
- *Trade shows:* Trade shows effectively reach industry and market segment individuals. While these people are at the show and in the booth looking at the product, you secure their entire attention outside of their traditional decision-making venue. This gives the manufacturer an advantage in introducing and promoting the new product.
- *Internet, web site:* Explosive growth in the use of these media and offshoots can get information to the customer directly and immediately. This assumes the customer requests the information and knows where to secure it. The use of a web site still requires the customer to search for the information or log onto a known location. Certain links can be used to also direct a generic request for information to the company's site.
- *Magazine articles:* Magazine articles about general issues that point out the problems and generic product solutions can assist in promoting the product. When a reputable magazine publishes such an article, it adds legitimacy to the product, the personnel, and the company.
- *Vocational white papers:* White papers or generic subject primers are useful in training an audience in a new product area. Since such papers are generic, they cannot be overt advertisements and still be called white papers. However, they do position the authoring company and the personnel involved as experts in the field and a source for solutions.

C. Telling the new product story

The new product story must be effectively told to the marketplace. At first blush this may seem like a daunting task; however, the strategy, tactics, and execution requirements are all in the original business plan. Refer to this plan for the various aspects of the marketing program that will be needed at this time.

Traditional product marketing generally follows a certain pattern (i.e., problem solution development). This means that the marketing program outlines the problem to the prospective client and then presents its products as the best alternative to solve the problem or meet the need. This has been and will continue to be an effective method to promote new products. There are, however, different motivations for the use of products, and thus the marketing of these products should be altered accordingly.

D. Establishing momentum

The issue of momentum is as important in the product launch as it is in the development program. The product launch is analogous to lighting a candle. Energy is applied to the candle to light it. When the energy or match is removed, the candle should stay lit. The same holds true for a product launch. The company invests energy in the form of funds and human energy to initially launch the product through a campaign to generate enthusiasm. The sales channel must keep it lit. An even better analogy for a large launch effort might be a fire, which is ignited by a match. The small amount of energy is leveraged into a larger and growing fire.

E. Incentives

For a sales organization to be effective, a clearly spelled out and equitable incentive program should be in place and functioning. This incentive program should also account for a new product introduction. If it does not, there may be a need to create a separate incentive program to launch the new product. There is a danger with this, however, in that certain windfalls may occur or, conversely, there may be a total lack of incentive participation in certain circumstances. The incentive program ideally should factor in the new product promotion so as to drive a successful introduction.

3. Promotional Activities

Promotional activities are tied to the promotional budget. If there is a large budget there can be more activities, such as travel, on-site training, and customer visits. If the promotional budget is small, then the business development team must make do with what they have available. This means leveraging resources to gain the maximum amount of promotional benefit.

The promotional budget is part of the sales budget for the product. Some products may need a lot of promotion to get placed in the mind of the customer. For newer technology requiring considerable user or consumer education, the promotion cost is comparatively very large. If the product is already known and some new version is being introduced, then the budget can be devoted to pure awareness rather than awareness and training.

4. Mass Introduction

In a global marketplace, what is the best way to launch the product? Should it be rolled out on a regional basis, a national basis, or an international basis? The rollout needs to be coordinated with the product's manufacturing capability. A wide, international mass rollout being supported by a single-point manufacturing situation could be disastrous for a product. Contingencies need to be worked out to accommodate this type of introduction.

In addition, the feedback systems for this type of mass rollout need to be in place all over the world and fed back to the business development team for analysis and action. Would you use an open-loop feedback system for a worldwide launch, whereby the team only monitors but is not interactive with local feedback? Or would it be better to operate a closed-loop feedback system whereby the business development team can monitor and have influence at the local level? These decisions must be made well in advance of a launch in order to set up the system properly.

PRODUCT PROMOTION

1. Product Promotion and Customer Visits

A. The need for and value of customer visits

Very few products are able to sell themselves. Unfortunately, products do not engage customers, establish business relationships, resolve problems, and apply solutions to meet customer needs. Sales people and application people need to perform this function. Consequently, placing company personnel in front of customers is a healthy activity. It gives the product team primary feedback on the product offering, remedies incorrect assumptions in development by observing the customer reactions directly, and removes development interpretation of customer issues in the absence of customer interface.

From the sales promotion perspective, it allows the sales personnel and the channel to reinforce the promotional objectives success factors. If third-party channel members sell the product line, it affords the company-employed sales personnel to observe the promotion in action and effect improvements or changes in their approach.

B. Participating in initial sales transactions

The customer visit creates bidirectional awareness between the company and the customer. The company learns about the issues facing the customer within the framework of the new product offering, and the customer learns about the benefits afforded by the company and its products.

There is another, more selfish reason why the customer visit is necessary. The initial marketing story is developed in somewhat of a vacuum of the widespread customer base. It is developed with a select few customers and then a comprehensive promotional campaign is prepared. The customer visit helps validate this marketing story and refine it for optimum performance.

By directly interfacing with the customer, the product group can learn which issues are (and are not) important to the customer, regardless of the original assumption. In this way the promotion can be made more effective.

2. Local Promotion

There is a difference between a promotional package that is designed for a local or domestic introduction versus an international one. The differences can be in the entire communication package and in the value proposition. It is critical to have the business development team understand the venue they are playing in and how information will be received and acted upon.

A. Marketing considerations

A domestic introduction can leverage certain collateral events and activities to capture market interest. For example, local- or domestic-oriented news or events can be used in a promotional campaign to create immediate identity with the intended audience. This identity can be leveraged to tie into an ad campaign.

B. Pricing considerations

The issue of pricing on a domestic level is generally straightforward because there is absolute homogeneity in the marketplace, currency, and transportation logistics, and no customs requirements are established. Internationally, it is a different story.

3. Global Promotion

When a product is marketed globally, the attendant effort is more complex. Assumptions made about pricing levels and the value proposition may be quite different from location to location.

A. Marketing considerations

The value proposition and product story may be completely different in an international location versus a domestic one. Internationally, there are different currencies, business practices, accounting practices, and consumer mentalities. What may be an absolute feature and benefit domestically may not be even desirable in a foreign location. Domestically, we tell a product story using colloquialisms and phrases that can easily resonate with our marketplace. Overseas, there is no recognition of these terms and a product introduction may fall flat.

B. Pricing

Pricing is another important issue. Certain markets may require a different pricing/value package for the product. It is a fact that all markets are not homogeneous in terms of value proposition. A product constituted for the U.S. market will not necessarily sell in a foreign market unchanged and with the same marketing story. In a foreign market the value system, expectations, warranties, and temperament are clearly different. The business development team needs to understand the prevailing market conditions and trends when setting pricing, terms, and the offering.

CUSTOMER VISITS

1. Role of Corporate Personnel

The corporate customer visit is a very important element of the promotion process, from three perspectives:

- Product focus and promotion
- Quality assurance in channel operations
- Primary check of business development assumptions

A visit to a field office by a member of the corporate business development team can have a great impact on product focus. All channel members have more things to do than time to do them. This means there is competition for the channel member's time. Even within the same company this can happen. A joint sales call by the channel member and the corporate representative impacts the local market and is a positive factor in product promotion.

Second, such a visit allows the corporate representative to observe first-hand the channel in operation. No reports, analyses, or presorted information—just direct contact in assessing the quality of the channel activities and the degree to which they are integrating the product initiative into their plans.

Third, a direct customer contact visit allows the corporate representative to directly assess how the marketplace is absorbing the new product story. The representative can determine how credible the customer believes it is and how well the story is working. A corporate customer visit is one of the healthier activities a member of the business development team can do.

2. Feedback

Feedback from a customer visit must be handled very carefully because several things are in play during and after a visit. First, the channel member is going to try to make a good impression during the visit. This will include visiting favorable customers and accounts, and representing channel activities focused on the product in question. In reality, the focus may not be all that great.

Second, there is a danger of overinterpreting the comments and feedback from the customer and acting on them immediately. This can introduce instability into the promotion equation and cause product introduction problems. It is best to take a measured approach to analyzing the feedback and act slowly in modifying the marketing plan.

3. Modifying the Marketing Plan

In keeping with the philosophy of the previous discussion on feedback, do not overreact to a specific visit and modify the marketing plan right away. If all of the work you have done up to this point has been accurate, then reacting to a single visit may be a mistake. Visit a cross section of customers to validate a need for a change, and test a few cases first.

INITIAL SALES EVALUATION

1. Initial Monitoring of Results

A. Models for new product rollout

There are several models for a new product rollout. There can be a *regional* rollout in which the introduction is limited to a certain geographic region. There can be a *national* rollout in which the introduction is within the nation's boundaries. Or there can be an *international* or global rollout in which the product is introduced across the entire global marketplace.

Each has a cost associated with it and results that can be expected. There is also the company side of the rollout, which is separate from the geographic side. This focuses on the product introduction from the depth of the sales organization and route to market (Figure 10-1).

As shown in Figure 10-1, the matrix of rolling out a new product is two-dimensional, with one aspect being geographic and the other focusing on the sales channel's route to the customer. Additional effort is required to establish a more pervasive introduction geographically as well as organizationally. For example, pushing the envelope in the channel requires training on the product and how to sell it. Pushing the envelope geographically will require promotional funds and time.

B. Monthly monitoring

After the initial rollout, there is a need to monitor results on a periodic basis. Depending on the specific product and business, this may vary. However, for illustrative purposes (and, more often than not), the monthly monitoring model will most likely be adequate. This is represented in Figure 10-2.

As shown in Figure 10-2, the promotional efforts are under closed-loop control. This means the results of the effort are being monitored via the market reaction and compared with the initial requirements. The assessment is made and the promotional effort is adjusted accordingly. If the results show lack of sales at a given effort level, then more effort may be required to equal the launch initial requirements.

To place this in more quantifiable terms, consider Figure 10-3 as an example of a launch rollout evaluation in which there are five basic company regions that need monitoring. Each region is evaluated in terms of investment and results. They can also be evaluated in terms of how they performed in comparison with each other.

Matrix of rollout					
	Regional rollout	National rollout	International rollout	Rollout by application	Rollout by allocation
Company sales organization					
Sales channel depth					
Marketplace					
Customer base					

Figure 10-1. Rollout matrix.

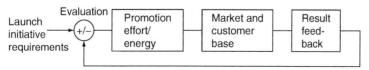

Figure 10-2. Monitoring results.

Initial feedback monitoring						
	Actual expense	**Budgeted expense**	**Actual time (hours)**	**Total avail (hours)**	**Actual sales ($) (units)**	**Budgeted sales ($) or (units)**
Region 1	$20,000	$35,000	600	800	400	500
% Budget	57.14		75.00		80.00	
Region 2	$22,000	$36,000	450	800	600	550
% Budget	61.11		56.25		109.09	
Region 3	$33,000	$40,000	590	800	850	750
% Budget	82.50		73.75		113.33	
Region 4	$44,000	$35,000	230	800	190	550
% Budget	125.71		28.75		34.55	
Region 5	$66,000	$50,000	120	800	250	1,250
% Budget	132.00		15.00		20.00	
Total	$185,000	$196,000	1990	4,000	2,290	3,600
% Budget	94.39		49.75		63.61	

Figure 10-3. Initial feedback monitoring.

Figure 10-3 shows the performance of five regions participating in the new product launch. The numbers represent the raw data and the percentages. There is diversity among the regions. Region 1 achieved 80% of the sales requirement but spent only 57% of the allocated funds. They expended 75% of the available time during the period.

Regions 2 and 3 performed the best of all five, since they delivered more than the budgeted sales. Region 2 even outperformed Region 3, expending only 56% of the allocated time and only spending 61% of the funds.

Conversely, Region 5 spent 132% of the funds and delivered only 20% of the sales required. In addition, they showed little participation in terms of expended time. This would need immediate correction if Region 5 is to contribute effectively to the launch.

The data also show that with 94% of the funds expended and 50% of the time devoted to the launch, 63% of the result was obtained. This indicates that action and intervention are required from headquarters to get the program on track. A section on measurement later in this chapter outlines additional criteria for evaluation.

C. Making the strategy work

There is a difference between formulating a strategy on paper and executing tactics to actually make it work. During the launch phase of the project, it is important to focus on those tactics that will implement the overall product strategy. Refer to the business plan for the strategy, and formulate the tactics to execute it at regional and local levels.

Each region or office knows its customer base and marketplace well enough to formulate and execute successful tactics. If the tactics employed do not affect the results, one of two things may be happening: The local people have not bought into the program, or the strategy is unachievable given all of the prevailing conditions.

The problem of buy-in can be resolved from a personnel standpoint. The other issues are much more far-reaching in scope. The strategy under which the program was launched is suspect. This should be a very rare occurrence and certainly not pervasive, since a diligent job of assessment throughout the program prevents misunderstanding in the global sense.

D. Rechecking the price/value matrix

The launch is a good time to recheck the price/value matrix that was constructed earlier, when the product was being specified. The initial launch will indicate how the market responds to the product, its features, and the value placed on it. The idea here is not to alter the promotional campaign or to redirect it, but, rather, to take the necessary steps to refocus on certain features that the market may find more appealing than originally understood. This is designed to only enhance the launch effectiveness, not to correct it.

E. Is the story getting out?

One of the most nagging issues in managing a new product introduction is determining whether the new product story is being diffused accurately into the field. Is the story, cultivated internally within the new product planning group, being used in the marketplace to secure sales, or are the individual sales people generating their own market presentation? Consider Figure 10-4, which offers a feedback system to confirm this.

In Figure 10-4, the product pricing and the value associated with its features are presented to the marketplace. The customers establish purchasing patterns based on these factors. Through after-sales assessment and lost sales assessment, it can be determined whether the presentation is being used in the regions and whether it is working correctly.

F. Is the sales force preaching to all?

The other factor to consider in the sales assessment is whether there is adequate homogeneous coverage of the marketplace, or if there are voids (structural flaws) in the coverage. This factor is illustrated in Figure 10-5.

G. Overcoming obstacles

New product development is a process of overcoming obstacles. There is uncertainty in marketplace dynamics, uncertainty in development engineering, uncertainty in manufacturing, and uncertainty of product acceptance during the promotion. These obstacles must be effectively overcome to launch the product properly. In addition, the bulk of the effort in producing a new product is behind you; do not let problems in promotion hold back the business.

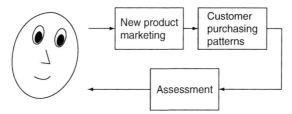

Figure 10-4. Price/value matrix.

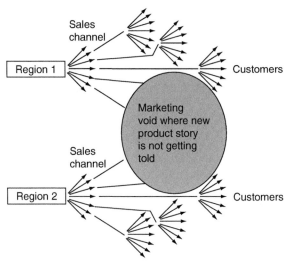

Figure 10-5. Market coverage.

If the obstacles are physical, work around them. If the problems are people, replace them. Now is not the time for a conservative approach. The product launch is intense and short-lived. Use the energy to create product momentum.

H. Use the energy at launch

As you progress through the product development process and execute the various phases, the scope and the commercial terms of the product offering become relatively defined. As you near the launch, there are very few changes that can be made—perhaps only a very slight reorientation of direction and goals. The only real weapon you have at launch is human energy to bring to bear on the launch itself. Focus that human energy potential directly on the tasks at hand. Establish and instill the enthusiasm in the sales channel to leverage the introduction. Create and apply the energy to show quick, decisive results. This alone will help promote the enthusiasm needed for a successful launch.

2. Early Modifications for Success

A. The least desirable thing to do at this point is change the product

So the greatest product in the history of the world has been launched, and early results indicate that the product is foundering in the marketplace. What do you do? An excuse often used by customers to not accept a new product is to cite a certain lack of features or package of values. The temptation is to immediately change the product for the exact approval by the customer. Unfortunately, there may be other reasons for the lack of interest, and even these added features may not sway the customer into a purchasing decision.

If the product group was diligent in their activities, there may be no compelling reason to even consider changing the product. Changing the offering significantly is the least desirable action to be taken during the product launch. It sends a very mixed message to the sales channel, the marketplace, and the customers. It adds confusion to the product line. It virtually destroys the legitimacy of the marketing story that is used to launch the product.

Changing the product contributes to a sense of disorientation in the company and its relationship to the prospective customer base. The operative point here is: Do not be too quick to change the product offering.

B. Give the launch a chance

It is important to take care when considering changes to the product in the wake of a launch that fell short of expectations. There are several factors that may have contributed to the lackluster results, some of them obvious and some latent. Be sure to clearly understand the issues before making changes, and also determine whether the issues affect the launch in a homogeneous manner. This means you need to determine whether the same issues produce the same results in various demographic situations. What you might feel is harming you in a certain demographic may actually be assisting you in another.

C. Repackaging values

If a decision is eventually made to alter the product, one of the easiest ways to create the alteration is to redefine the package of values offered. This adds value to the product offering without necessarily adding a lot of cost. For example, offering a package of hardware and different software sets can enhance the salability of the product. If a product change is anticipated, reconsider it as a first step. If you missed the mark on the first try, with all the planning and execution, what makes you think you will hit the target in the middle of an introduction? Stay the course and see the production through.

Another means to induce purchase is the use of rebates. These are simply cost reductions that are employed temporarily and allow the initial pricing to be somewhat preserved. Other promotions can be used as well. The basic idea is to leverage sales without incurring too much additional cost.

D. Combination of different products leverage the existing base

Another way to leverage the existing customer base is to offer the new product along with the existing, repetitively purchased product. This capitalizes on the familiarity the customers have with the existing product and allows the new product to tag along, in the hope that the customer will see the value and start using the new product. Each case depends on the individual circumstances.

E. Product recall: If necessary, do it!

It is no secret that one of the greatest fears in new product development is a product recall. A product recall is one of the worst things to administer during a product launch. Whether the launch needs to be completely halted depends on the level of confidence in the product correction and the stage of correction implementation. If the correction is made and is implemented in production, the launch can continue. If not, the launch most likely must be terminated and rescheduled for a later time. In either case, the issues are very difficult to navigate around.

If the problem warrants a product recall, plan the steps and execute the recall decisively. Many products are successful after relaunch. The marketplace can forgive a mistake and an inconvenient correction. It is less tolerant of negligence.

A product recall has a specific protocol that should be followed. The issue of product recalls will be discussed in more detail in Chapter 11.

F. Mitigating the damage

From a marketing perspective, there is a critical need to contain or mitigate the damage in a recall situation. Whatever steps can be taken to preserve the company and product reputation must be taken. The company reputation is the first priority and the product line reputation is second priority. In either case, the idea is to pave the way for tomorrow's business from the same customers who are having problems with your product today.

There is one absolute element: The marketplace and customer expectations of newly introduced products are ever-increasing. Customers have very little tolerance for product problems, so the more successfully you can reduce the impact on them, the better.

3. Measurements and Benchmarking

Upon market introduction and feedback of initial results, the business development team should evaluate the planned results versus the actual. This is the essence of benchmarking the business development enterprise.

When a company plans a program, all of the factors for success are to be considered (i.e., finance, timing, costs, market opportunity, development effort, and results). Therefore, if we look at this best-effort plan and compare it to what the company actually did, we can get a measure of the company's performance. In subsequent development programs, your planning will have the benefit of certain experience and improvements. By comparing subsequent programs, improvement is factored in automatically.

As part of the comparison of planned versus actual, trigger points need to be established in order to know when action is necessary. If the initial sales are less than expected, a trigger point for program modification needs to be established. If the sales input order rate is in excess of planned quantities, then a different action plan needs to be triggered.

PRODUCTION VOLUME FORECASTING

1. The Myth of the Hockey Stick Forecast

A. The aggressive sales forecast

Remember when the new product was being justified for investment? At that time it was easy to be very aggressive in the forecast. This aggressiveness and optimism helped justify the program. Now, however, it is time to deliver the sales, and the tendency toward an aggressive forecast will give way to something that is more achievable. The traditional hockey stick forecast (discussed in Chapter 5) is something that is generally very difficult to achieve. This is due to the dynamics of a sales campaign and the mathematics of a geometric progression. Also, all of the original assumptions to drive the growth must be in place and effective. This is not always the case in the time frame between initial forecast and product introduction.

B. Why very few products have characteristic growth

Figure 10-6 presents a comparison of planned and typical actual sales progressions. The format is based on several assumptions. The first is that the planned forecast has the company sales personnel leverage their activities through the sales channel by successfully engaging 10 salespeople from the channel. This corresponds to a 100% acceptance rate in the sales channel. Each of these sales people has 10 customers available to them to sell the product to (i.e., the available market). The other assumption is that it takes five periods to achieve a success rate, indicated in the success rate column. In this ideal, planned example, the expectation is for a 100% success rate.

Growth projections									
Planned sales									
	X	Channel sales	Customers	Success rate			Customer		
					Period 1	Period 2	Period 3	Period 4	Period 5
Sales 1	10	1	10	100%	2	4	6	8	10
		1			2	4	6	8	10
		1			2	4	6	8	10
		1			2	4	6	8	10
		1			2	4	6	8	10
		1			2	4	6	8	10
		1			2	4	6	8	10
		1			2	4	6	8	10
		1			2	4	6	8	10
		1			2	4	6	8	10
				Total	20	40	60	80	100
Actual sales									
					Period 1	Period 2	Period 3	Period 4	Period 5
Sales 1	6	1	10	50%	1	2	3	4	5
		1			1	2	3	4	5
		1				1	2	3	4
		1				1	2	3	4
		1					1	2	3
		1					1	2	3
		1			0	0	0	0	0
		1			0	0	0	0	0
		1			0	0	0	0	0
		1			0	0	0	0	0
				Total	2	6	12	18	24
				% Actual/ planned	10	15	20	22.5	24

Figure 10-6. Comparison of planned and typical actual sales progressions.

Figure 10-6 shows that the first channel sales person achieves a sale to 2 out of their 10 customers in the first period. Therefore, the first sales person's volume is two. In subsequent periods, more customers are added at a rate of two per period. Additional sales persons generate sales in the same manner. In addition, all sales personnel in the channel start selling to their 10 customers, each at the same time. The progression shows a volume building from 20 to the total available of 100 within five periods.

The actual projection is somewhat different and probably represents a projection that is more typical. In this part of the example, the hit rate of success with the customer is 50%. In addition, there are only 60% of the 10 salespeople in the channel that will support the program. This may happen indefinitely or just at the beginning. In either case, the effect is quite damaging to the program.

Another real-life effect is that not all channel sales personnel will immediately embrace the program. More often that not, they will have a tendency to delay their entry in the promotion as they observe others' success or failure. They often do this to protect their customers from newly launched products that could have reliability or other product problems.

The 40% not participating are represented by the italicized 0 entries. The delayed participation is represented by the shift in time. The 50% hit rate is represented by the value 1 versus the value 2 in the planned portion of the example (e.g., for Sales person 1 in Period 1), 2 versus 4 in period 2, 3 versus 6 in period 3, and so on. The impact of these adjustments is significant, reviewing the bottom line where the percent actual to plan shows that results are poor.

C. Planning for realistic results

The operative lesson is to forecast and justify the program with achievable forecast numbers. Eventually the product team will have to deliver these numbers, so be realistic in the forecast and allow no complacency in the execution to obtain the numbers. At the outset, ensure that the forecast can be met. At the launch phase of the new product program, reexamine the initial assumptions made and operate the new business to a reasonable forecast that can be achieved in the time frame outlined.

INVENTORY CONTROL

1. Forecasting and Building Inventory

Inventory is a great cost to the corporation. Whether it be finished goods inventory or raw inventory, its cost to the working capital of an organization is high. Accordingly, the business development team must be very careful in matching the inventory levels to the customer delivery requirements and raw inventory lead time.

In general, there are three types of forecasting:

- Field-generated forecast
- Internally generated forecast
- Historically generated forecast

The *field-generated forecast* can be the most accurate in that it can reflect the actual requirements coming in from the field. The forecast should be done in units rather than dollars in every case. Without the right checks and balances, the field forecast could be abnormally high.

The *internally generated forecast* is one where an individual or group analyzes field data, input, and outlook, and assesses the number and type of units that will be sold in these applications. This type of forecasting is somewhat dangerous and can lead to high inventory levels and obsolescence issues.

The *historically generated forecast* is one where next year's forecast for materials is assessed based on this year's usage. This is even more dangerous because it does not factor in the field requirements that are evolving and changing now.

2. Meeting Customer Demand

The key is to create a match between the market requirements and the company's ability to deliver. The reason why this is so important is that during product introduction the firm is trying to build momentum. The momentum is broken with an interruption in shipments due to inventory shortage. The products are not getting out and applied, and repeat customers (if applicable) are not being acquired. In addition, the leverage from satisfied customers (as discussed earlier) is restricted.

3. Controlling Inventory Levels

The level of inventory is driven by several factors:

A. Product configuration

The product configuration can have drastic effects on the inventory levels. If a modular approach with a different mix of modules comprising the end product is used, there can be a reduction in the amount of inventory. If each product is completed with no leveraging, the finished goods inventory can be huge.

B. Sales initiatives

If the sales group starts promoting changes or modifications to the product, this can have an impact on inventory levels. The special product with modifications now becomes a standard using different inventory, or more "specials" are generated.

C. Rollout

How the product is rolled out to the marketplace can inflate remote warehouse inventory levels.

D. Market demand

If the market is demanding a version different from the one in stock (both in raw inventory and finished goods), the inventory can be impacted. Fundamentally, the key at product introduction is to get the proper level of inventory to track the ramp-up and allow the organization to secure and execute the orders in a timely manner and capture market share.

SALES CHANNEL INITIATIVES

1. Communication, Agreement, Commitment to Objectives

A. The essence of any new product development

Coordinating a new product launch involves many facets of corporate and human behavior. The same characteristics needed for the development are needed for the launch. Each part of the organization must contribute to the launch and, as such, requires communication, agreement, commitment, and understanding. This is something that cannot be finessed, sidestepped, worked around, or ignored. The participants, if they want to be players, must effectively and repeatedly demonstrate these basic tenets of group activity.

B. Communication to all parties

There must be swift and accurate information disseminated throughout the sales channel as it pertains to the new product launch. This is needed because the organization is evaluating the effectiveness of the product offering in the marketplace, the pricing, the marketing tactics, the strategy, and the market's consumer as well as competitive response to the product introduction.

C. Agreement on policy, procedures, plans, and goals

At the time of product launch, there should be no disagreements on product policy, implementation procedures, the company marketing plan, or overall corporate goals. Keep in mind that all human and corporate energy must be devoted to the success of the product, not to infighting or territorial power plays. The corporate objectives govern the individual actions, and the new product launch is an integral part of the objectives that cannot be ignored.

D. Understanding and tolerance of problems and temporary obstacles

The product group must establish a spirit of cooperation between the various departments involved with the new product launch. They need to instill a pattern of perseverance in the sales personnel as well as plug the entire product development group into field organization feedback. The product group must work with the sales channel to overcome any obstacles that may arise.

It is management's responsibility to create and foster an environment of tolerance and understanding among all the players in the launch. This is needed because any uncertainty left in the product development program must now be absorbed by the sales function and the product group. Both must work together to secure incremental market share.

E. Commitment to objectives

Similarly, all elements of the organization must support the product group during this launch phase. The commitment to the corporate objectives must be real and lasting. The organization and its people must lock onto them and execute them. With the entire organization committed and in lock-step, the new product will easily attain orbit!

SUMMARY

This chapter transitioned the business development from planning to transaction in the marketplace. It is the phase of the process where initial results are due and senior management watches for early returns. Consequently, a basic tenet of this chapter was to guide and ensure a successful new product launch.

The chapter started with the rollout of the product line and the various means for a successful rollout and building momentum. Next was a discussion of product promotion. A distinction was drawn between local and global rollouts and their respective differences in marketing. Following that was a discussion of customer visits and the importance of such visits and use of feedback information.

Then, initial monitoring of results as used to evaluate the execution of the plan, the performance of the channel, and critical success factors were discussed. This was followed by discussion of product modifications for success, corrective actions, and product recall sequence and loss mitigation. Initial sales of the product within the framework of realistic expectations of unit volume were also discussed.

Transitioning from a newly launched product to a supportable business in terms of inventory and materials was reviewed as well. As an integral part of the launch process, the all-important customer visit was covered in terms of needs and tools available.

The chapter ended with a discussion of the basic elements required in any endeavor involving a group of people: communication, agreement, understanding, and commitment to the corporate objectives.

Given a successful launch, management of the new product as a business becomes the focus of Chapter 11, The Pursuit and Product Management.

THE PURSUIT AND PRODUCT MANAGEMENT

PRODUCT PORTFOLIO

1. Caring for the New Product in the Marketplace

The business development manager's role is to nurture the business. The product manager is the focal point of responsibility for the corporation, advocate for the customer, and architect of the product and business model. Accordingly, the responsibility resides with the team and its leader for a successful product run that repays its development costs and generates profitable revenue for the organization.

2. Price Pressures and Market Share

A. Markets are dynamic

New business development occurs in dynamic and changing market conditions, which include performance and pricing. In the field of product development and marketing, for every product action—an introduction, a pricing action, or a feature enhancement—there is a market reaction, but it may not be equal and opposite. In certain circumstances there may be a larger response or a backlash. The best that could be hoped for is a competitive response clouded in obscurity or confusion.

Practically speaking, however, the market does respond to product actions. The business development team must learn to anticipate the reaction and plan for it as part of their initial action.

B. Responding to the response

This being the case, the manager needs to formulate a response to the market reaction; otherwise, the company may be left in a vulnerable position. The battle for market share is almost analogous to a chess game, with each move counteracting another and an overall strategy that is implemented with tactics. The manager should be aware that once the game commences there is no time for complacency. You are in the game and the game must be played or forfeited. The same holds true with product management.

C. Review the product marketing tactics

With each advance and retreat there is generally a needed response, which may not always be price; it can be some package of values along with price. The expected responses are somewhat peculiar to the individual market players. If we reexamine the model presented in Chapter 3, there are four major classifications of market players:

- Defensive orientation
- Direct pursuit orientation
- Oblique pursuit orientation
- Opportunistic orientation

There is an expected response from each of them, based on their position and outcome of an interchange. Figure 11-1 summarizes the expected responses when one player takes on another. This is not to say that under certain conditions and circumstances the response will always follow this form; however, it does represent a typical expected response, given the nature of the players.

As shown in Figure 11-1, the reactions to the market actions vary. Below the chart is the legend of the various actions the players can take. PA, for example, represents a pricing action. One can either raise the price or lower the price to effect certain responses. The company can introduce a product enhancement, indicated by PE. A new rendition of the product can be a leapfrog of the existing product, indicated by LP.

Market dynamics				
	Defensive orientation	Direct pursuit orientation	Oblique pursuit orientation	Opportunistic orientation
Primary Market Player	This is the accepted leader in the marketplace. They have the largest market share.	This is usually No. 2 trying to become No. 1. The contest is the battle of the Titans.	This player generates product opportunities that attack market opportunities obliquely and which are difficult for the entrenched competitor to counter.	This player discovers opportunities and acts to secure them. Will leave market at a moment's notice to preserve capital in a fight for market share.
Defensive Orientation	AC, AP	AC, AP, PE, PA	AC, AP, PA, HR	AC, AP, PA
Direct Pursuit Orientation	LP, PE	AP, PE	PE, PA	PA
Oblique Pursuit Orientation	PF	PF	PF	PA
Opportunistic Orientation	LP, LM	LP, LM	LP, LM	LM, HR

PA, price action +/– ; PE, product enhancement; LP, leapfrog product; LM, leave market; HR, harvest residuals; PF, change playing field; AP, acquire product line; AC, acquire company.

Figure 11-1. Market dynamics.

One can choose to abandon the market altogether, indicated by LM. If you are an entrenched player and are clearly outmaneuvered in the marketplace, you may choose to only harvest the residuals (indicated by HR). If your company is clever enough and has the influence, it can change the playing field to its advantage; this is indicated by PF. If your company has the commitment and wherewithal, it may choose to acquire the product, indicated by AP. Finally, if your company wants immediate results, it may simply acquire the company and integrate it into its own operation, indicated by AC.

The traditional company roles would indicate that an opportunistic player normally would not take on a defensive player. With all of the ownership changes and acquisitions in the market, an opportunistic player could be acquired by an offensive player and funded to take on a leader with the backing of the offensive player.

The chart is organized such that the left side of the chart indicates the action against any of the players along the top of the chart. Their tactics are listed at the intersection of the horizontal and vertical rows and columns. For example, an opportunistic pursuing a defensive player has the options of leapfrogging the product introduction to attempt to acquire customers dissatisfied with that product, or leaving the market altogether.

D. Is the price right?

"How's my price?" is often asked in the selling process. Is it worth cutting price to obtain business? How low should the company set prices to introduce the new product? Not all of these questions will have simple answers, depending on the circumstances, but the company must take a position on the questions and act accordingly.

There are extenuating circumstances in which a company may want to lower prices to get business, even at a slight loss for a short period. However, one thing is certain: Once the company lowers the price in the marketplace, there is little chance of it ever being raised because the expectation has been set in the marketplace and competitors will react to the price change in some manner. If they reduce their price, you are now in the position of responding to the reduction. Good luck trying to increase the price under these circumstances.

Aside from the strategic and market-driven reasons for preserving pricing, there is a pure financial perspective. Consider Figure 11-2(A), which shows what happens when prices are reduced to drive product volume and increase profit.

As shown in the figure, the product generates revenue and profit. This profit offsets the fixed cost of production. A reduction in the price lowers the profit, forcing additional unit volume to generate the same profit dollars. Conversely, an increase in price increases the profit margin dollars per unit and requires less unit volume to generate the same profit. Breakeven is where the profit generated is equal to the fixed costs.

In this example, the product normally sells for $500. The unit cost is $275. Under normal conditions, the profit for a single unit is $225. The revenue and profit are shown for unit volumes, from 0 through 1100 units. The fixed costs are $120,000. The figure shows what happens if the price is lowered or raised by 10%. As can be seen, the breakeven point shifts from the normalized value of 533 units to a low of 436 units, given a 10% price increase, and a high of 686 units, given a 10% reduction in price. A price reduction requires a 29% increase in unit volume to generate the fixed cost offset. Lowering prices to generate volume can be an entry into a race that cannot be won.

Figure 11-2(B) displays these points on a graph. It shows the change in breakeven points based on the pricing action. It indicates wide fluctuations in volume to get to breakeven.

Financial impact of price reductions with elastic demand

Unit Volume	0	100	200	300	400	500	600	700	800	900	1000	1,100
Base Revenue ($)	0	50,000	100,000	150,000	200,000	250,000	300,000	350,000	40,0000	450,000	500,000	550,000
Profit Normal ($)	−120,000	−97,500	−75,000	−52,500	−30,000	−7,500	15,000	37,500	60,000	82,500	105,000	127,500
Profit +10% ($)	−120,000	−92,500	−65,000	−37,500	−10,000	17,500	45,000	72,500	100,000	127,500	155,000	182,500
Profit − 10% ($)	−120,000	−102,500	−85,000	−67,500	−50,000	− 32,500	−15,000	2,500	20,000	37,500	55,000	72,500
Price normal ($)		500										
Price Increase 10% ($)	0.10	550										
Price decrease −10% ($)	−0.10	450										
Variable.cost /unit ($)		275										
Fixed cost ($)		120,000										

	Unit profit	Break /even	Volume effect	
Gross Profit Normal ($)	225	533		
Gross Profit +10 % ($)	275	436	18%	Less
Gross Profit −10 % ($)	175	686	29%	More

A

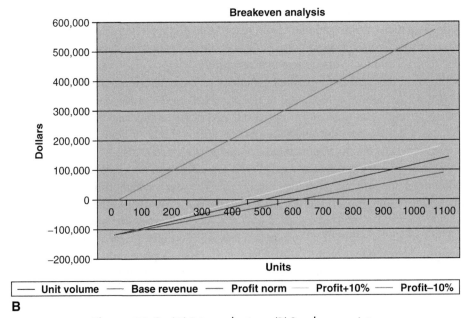

B

Figure 11-2. (A) Price reductions; (B) Breakeven points.

E. Holding prices and adding value to preserve margin

If a pricing action carries this liability, how can the business development manager compete in the face of changing market conditions? The response to this question lies in bundling the product values that support the existing price. In other words, give more for the same price. The challenge is to do this without incurring too much additional cost, which will deteriorate the gross margin and effectively cause the same problem as a price reduction. This is because it is, in fact, a price reduction.

F. Changing the playing field to your advantage

If the circumstances allow it and if the idea is clever enough, change the playing field of competition to your advantage somehow. Create some offering that the competition cannot respond to easily or with equivalency. As stated before, this is a chess game that is won or lost based on these moves. Make better, more informed moves than the other players and you will be successful.

GROWTH STRATEGY

1. Business Growth Strategy

A. How will the business grow?

The responsibility of product management includes maintaining a certain degree of continuity of the business planning cycle. In the same way that the product features and technology have an evolution, the business model must have an evolution, too. At the onset of the program, a comprehensive business plan was developed to guide the development and the marketing of the product. The product, however, represents only a small part of the process of growing the business. The follow-on activity must now be directed toward growing the business and satisfying the original business plans with market modification. Go back to the overall plan that was developed and start to work on the follow-on programs and products. This follow through is what will create the anticipated success. Take the market feedback from the first product introduction and develop the next generation of the business plan.

B. Remediation

If the original assumptions of the business plan are incorrect or need updating, fix them and begin to execute them. The saying "Do not argue with success" is true enough, but it is absolutely imperative to understand the reasons for success. They are the bases for future actions. Conversely, if a midcourse correction is needed, do not hesitate to make it—decisively and completely. In summary, learn from the successes as well as the failures. Build on the successes and remedy the contributions to failures.

2. Market Development

A. Expanding the available customer base

This process encompasses identifying the market opportunities that are on target with the product launch plan, as well as identifying new opportunities for small changes in the product configuration. The market development aspect gives true definition to the term *customer engagement*. It is most effective when the customer is an active participant in the communication. It takes detective work to uncover new customers and market segments for the product.

The next sections will discuss the various aspects of the market development process and how they apply to growing the business.

B. Market development flow chart

There is a protocol that can be used in the development of the market. The basic philosophy is to sell the product, learn from the interaction, and reapply this new learned knowledge to guide in selling in the future. Figure 11-3(A) represents a simplified version of this protocol, in flow chart form, as it relates to market development of new products.

As can be seen in the figure, there are two basic paths. The first is where the product fits the marketplace without any modification. The second is where the product may need minor modification to be marketed in an application. This example assumes the product is somewhat fixed but modifiable in design and construction to accommodate the new application. It requires intimate, but not extensive, development work to accomplish. The term *intimate development* refers to the change in configuration, bills of material, parts, components, and/or assembly methods.

Figure 11-3(B) illustrates a case in which the product is initially designed to be modifiable either in the field or by factory personnel. The product is designed on a basic generic platform, which is set up to serve the target applications. It is also flexible enough to serve other applications that expand the business by product modifications. It can be modified by altering some element inherent in the product, such as software. The best test to determine which category the product belongs to is to determine whether it can be modified back to its original state. In the first example [Figure 11-3(A)], it cannot. It is engineered and manufactured to be in this state.

As shown in Figure 11-3(C), this product has two basic components: a hardware function block and a software function block. The hardware block has generic interfaces from three basic perspectives. They are the human interface (how the user interfaces with the product), the control system interface (how a hierarchical control system would tie into the product), and a vocational or functional interface (how the product relates to the environment it acts on). The software block serves as a repository for the "personality" of the product and functional modification. The product can be made to serve a wide variety of customers and needs by modifying the basic function, the human interface, the control system interface, or the vocational interface. Input and output can be also reassigned to facilitate this transition.

In this way a generic design can serve a growing list of applications with a small, incremental effort. For example, the hardware of an engine or transmission could be modified for performance, mileage, or comfort by using the same basic platform and altering software in the engine control computer. In a broader product application, standardized generic car platforms can be used to generate a variety of products serving different market segments for automobiles by varying the configuration and accessories.

C. Use creativity to develop new customers for the product

The marketing professional should use creativity to enhance their product's market share and expand the available market. This is done by engaging the customer again with the

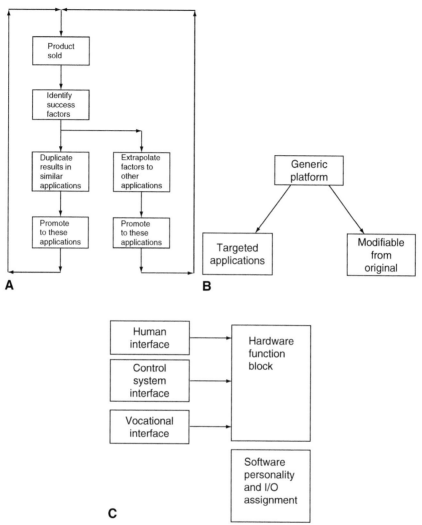

Figure 11-3. (A) Market development by application; (B) Market development by platform; (C) Market development by configuration.

standard product offering and broadening the scope of the product to serve the extended needs of the customer base and to uncover new applications for the modified product.

Figure 11-4 indicates that the product's available market is of a certain size, illustrated by the width of the channel on the drawing. Implementing the flow chart and exploring additional applications increase the available market dramatically. The activity driving this desired result is shown in the flow chart on the left of the drawing. It starts with engaging the customer and creating newly served applications generated by the interaction and dialogue. To accomplish this, redefine and manage the scope. Then the concept is implemented and refined. Finally, it is locked onto and promoted to a new collection of customers now available for using the product. The result is a wider pathway to more available markets.

The term *pathway* can best describe the illustration because nothing can last forever, and the product's sale and growth of market share is thus a pathway. The width cannot remain

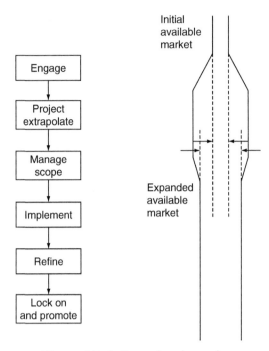

Figure 11-4. Expanding the market.

fixed forever. The idea is to manage the process faster than the competition and carve out a business segment that is defendable.

D. Tiered markets and classes of customers

By implementing these concepts while widening the available market, it is possible to create classes of products that serve different classes of customers. The terms *tiered market* and *tiered product line*, which serves that market, refer to a fit between the product and the customer. In many cases, simple changes can create better product identity in the mind of the customer.

Behind this approach is also a selfish motive for the corporation. By creating tiers, additional margin can be realized in the product line by commanding increased pricing while adding little incremental cost.

E. Segmenting the market

In summary, the basic process is to segment the market during the promotion phase and productize additional offerings to widen the available market and expand the business. As a starting point for this exercise, go back to Chapter 3, seek out all of the ideas for products and versions, and test them in the marketplace. Implement the ones that require little change.

3. Application Development

A. Apply technology building blocks

The issue of application development requires technological engagement with the customer, in addition to commercial engagement. This is where the basic building blocks of technology embodied in the product are available for redistribution and reconfiguration into different products altogether, from the factory perspective. However, they are an on-target solution to the customer's problem. An example of this reconfiguration of features is to think of the original part as a technology toolbox. Both development and manufacturing can be used to solve different customer problems in different applications.

Another example of this is a pizza parlor where pizza is made and sold to consumers. The infrastructure for storing raw materials (foodstuffs), assembling them into desirable portions of food, baking them in an oven, and delivering them uses the manufacturing technology and the vendor base to serve new products to new customers. What other products could be made using the same materials and marketed to the same class of customers?

B. Degree of modification constraints

The manager needs to be wary of going too far with modifications and segmenting to serve too many customers. This will dilute any economies of scale and destroy any leverage that was created. In addition, too many product versions are difficult to track, implement, support, and procure components for, unless each generates significant volume and profit to absorb the overhead needed to implement them.

C. Search for latent issues solved by a rendition of the product

As the product is being accepted into the marketplace and is being used, there may be latent needs not met by the product. They may be functional or related to human factors ease of usage. The commercial and promotional effort should also focus on seeking out these issues and producing a version of the product that does meet these needs. The timing of this depends, of course, on the seriousness and degree of change. This is nothing more than attempting to meet all of the customer's needs with the product, which should have been done throughout the product program.

Another way to promote the product (if it can be easily modified) is to engage the customer to determine what other collateral equipment is required with the product to make the application work. Then embody the collateral equipment or its function into the product at a lower total cost, making it easier to use. This will simplify the customer's use of both products and give your product the advantage.

4. Coalescing Market Feedback and Structuring Modifications

A. Watch the feedback

Now that the product is out and has an established base, the product group must watch the market feedback carefully. Earlier in the business development process, we discussed the need to refrain from trying to be all things to all people. It simply cannot be done well and still support any initiative for product leverage. The retained market feedback must be organized to evaluate the potential return for each added or exchanged feature. These requests can be evaluated like the original opportunity, with the exception that they represent incremental business. It is necessary to ensure that they do not drain resources to the point of disadvantaging the base product.

The product group may even want to take an active role in soliciting feedback from the customers and users to make certain determinations themselves for planning and product maintenance purposes.

B. Select categories for specific items and group changes

In the same manner that the original product features were qualified and selected for incorporation, the specific market feedback can be organized into the following three basic categories:

- *Minor modifications:* These are simple changes. Sometimes even cosmetic or user-support changes can be made to improve the product's acceptability in the target application. In some instances, these minor modifications can be effective enough to open new opportunities for the products.
- *Product feature enhancements:* These are more involved changes from a development and manufacturing point of view. They involve product changes and requalification, and are scheduled into the development hierarchy of activities.
- *Major platform changes:* These are based on the market feedback, which defines the next generation of product. They are significant enough to warrant a complete redesign and possible selection of a new platform.

All of the market feedback must be evaluated and categorized for decision and implementation purposes. The benefits of each of these implementations can then be evaluated in a financial manner, along with the strategic implications, before action is taken.

C. The next generation of product

This feedback related to major platform changes and, to some extent, the enhancement requests define the next generation of product. The time to start formulating this next generation is now. Keep in mind that these requests represent the customer's input to the product development process. It is only one component of the process. The strategic objective and the technological and cost initiatives also must drive the plans for this next generation of product. In the same manner as the product evolution flow chart was completed initially, the chart must now be updated with the customer feedback.

In addition, the two elements should be compared to each other. The first element is the market feedback; the second is the original plan for what the follow-on generation would look like. In this way, marketplace trends can be catalogued and the rate of change can be tracked to leapfrog the next versions of the product accurately.

D. Market response is only one component of the business direction

The market will give the overall direction but the business development team must select and qualify the details. It cannot be overstated that the market feedback indicates what the customers desire now. However, it cannot forecast where the competition may be, what the future cost structures will be, and what competitive moves should be made by the company to be a player in the future.

E. How can you improve your position?

The gathering of market feedback and the analysis of data are directed toward only one goal: to improve the company's competitive position with the products available. The staging of the product introductions, the improvements, and the new products are designed to improve competitive position and gain advantage. They are not for esoteric purity, line completion, or any other singular reason; rather, they are part of an overall plan to secure market share and build a profitable business.

PRODUCT MAINTENANCE

1. Mechanics

We learned in Chapter 6 that as more and more products are added to the portfolio, the need for product maintenance in engineering increases dramatically. If left unmanaged, the effort devoted to product maintenance will eventually outweigh the effort devoted to product development. Accordingly, one of the issues facing business development mangers is that of critical usage of development time and how it is applied. A simple method is to asses the needs of supporting the products already existing and apply the remaining hours to the new initiatives. Going forward, however, it is critical to develop product programs that will take only a minimum of effort to support.

2. Managing Product Change Requests

One of the product manager's key roles in effecting success with the new product is to maintain product configuration control and a certain amount of stability with the product's design, manufacturing, and marketing. Not all change is good. Each proposed change must be considered and acted on within the framework of the overall program and its goals. Repetitive changes that are not well-orchestrated and have questionable impact tend to destabilize the product within the corporation and eventually in the marketplace. The manager must rein in these changes and implement only those that will offer a concrete contribution.

This is a difficult task, since people associated with the product (from vendors to manufacturing personnel) will have a tendency to change the product for their convenience. For example, the vendor will have a tendency to raise prices and the purchasing people may have a tendency to respond to this pricing by changing parts. Supplier obsolescence generally yields to availability, given added component cost. Manufacturing may add steps to a process, thereby increasing the labor. Corporate initiatives that are implemented in isolation of product objectives can also contribute to added costs.

3. Managing Engineering Change Notices

Several forces tend to drag a product line off its base configuration, as discussed earlier. The product manager must evaluate the requests for change and determine what changes will be integrated, and what the potential benefit will be. In some cases there will be no choice in the matter. In other cases the choices are discretionary. The product changes fall into the following three broad categories:

- *Safety related:* Safety-related issues are further subdivided into three categories, as follows:
  - **Product:** Product-related issues can be serious enough to cause production to stop, or less serious such that the changes could be phased-in at a future date.

- **Application:** The application issue concerns those cases in which the product is applied in certain circumstances in which there may be a safety concern. In these cases, the product should not be applied and the product stakeholders (ranging from the sales people to the end-users) should be effectively notified.
  - **Documentation:** This issue pertains to the product documentation containing a recommendation, practice, or procedure that may render the product unsafe or may incur a safety concern.
- *Performance-related:* Performance-related issues concern the operative parameters of the product. Information may come to light in which the product performance falls short of the expected or advertised performance. In these cases, a certain retraction must be communicated to the people using the product or potentially procuring the product. This is to remain in effect until the correction is implemented. These actions fall into the following three categories:
  - **Stop production:** The stop production order is the most aggressive position in that production of the product is halted until the correction is in effect. A recall of the product may or may not be required.
  - **Immediate change:** An immediate change is just as aggressive, with the exception that the product is not halted in production. However, a change is designed and implemented in the same expedient manner.
  - **Scheduled change:** The scheduled change is less serious, in which a performance-related issue is identified and then scheduled for implementation without suspending normal operations. The issue may or may not be communicated to all parties, depending on the circumstances and seriousness.
- *Enhancement-related:* Enhancement-related issues are similar to the schedule changes in that they are implemented in a scheduled time frame within the scope of normal operations. They do not necessarily fall under the category of design flaws or corrections. Some manufacturers have used this terminology in the marketplace to describe some of the performance-related issues!

4. Learning Curve Cost Reduction

A. Progressive cost reduction

Progression down the learning curve pathway toward cost reduction is necessary and is accomplished through several mechanisms, which will be discussed in this section. The issue is multidimensional and can be accomplished in a variety of ways. The product manager must keep a watchful eye on product costs and manufacturing throughput. Left unattended, these factors will deteriorate to the unfavorable side.

Consider Figure 11-5, which illustrates the learning curve cost reduction. It consists of the three basic components of a product's cost and illustrates the deterioration that can occur when volumes are not met. In this case, the burden is calculated based on the labor content. Figure 11-5(A) shows the actual versus planned volumes experienced in the marketplace. Materials increase because of the shortfall in volume, and labor increases (either as actual hours or in the cost per hour). In this example, increased non-value-added labor crept into the manufacturing equation.

Since the burden is levied on the labor, the creep in labor has a multiple effect when considering the burden and the labor together. This can make cost containment a losing

Learning curve cost reduction								
	Material		Labor				Burden	
	Planned ($)	Actual ($)	Planned Hours	Actual Hours	Cost/ Hour	Labor Cost ($)	Planned ($)	Actual ($)
P/N	1,000 Qty	750 Qty	1,000 Qty	750 Qty	1,000 Qty	750 Qty	1,000 Qty	750 Qty
							1.2	1.3
A1	10	11.5	0.1	0.15	50	7.5	6	9.75
A2	15	16.5	0.1	0.15	50	7.5	6	9.75
A3	16	16.5	0.1	0.15	50	7.5	6	9.75
A4	17	17	0.1	0.15	50	7.5	6	9.75
A5	19	19	0.1	0.15	50	7.5	6	9.75
B1	22	22.1	0.2	0.25	50	12.5	12	16.25
B2	33	33.2	0.2	0.25	50	12.5	12	16.25
B3	44	45	0.2	0.25	50	12.5	12	16.25
B4	55	55.3	0.2	0.25	50	12.5	12	16.25
C1	9	9.5	0.15	0.2	50	10	9	13
C2	10	10.1	0.15	0.2	50	10	9	13
C3	12	12.3	0.15	0.2	50	10	9	13
C4	14	14.2	0.15	0.2	50	10	9	13
Total	276	282.2	1.9	2.55		127.5	114	165.75
Variance		6.2		0.65		32.5		51.75

A

	Year 1	Year 2	Year 3	Year 4	Year 5	Year 6
	Planned	Actual	Actual	Actual	Actual	Actual
Material ($)	276	282.2	292.1	302.3	312.9	323.8
Std Cost ($)	276	276	276	276	276	276
Labor ($)	95	127.5	132.0	136.6	141.4	146.3
Std Cost ($)	95	95	95	95	95	95
Burden ($)	114	165.75	171.6	177.6	183.8	190.2
Std Cost ($)	114	114	114	114	114	114
Total ($)	485	575.5	595.6	616.4	638.0	660.3
Std Cost ($)	485	485	485	485	485	485
Total Variance ($)		90.5	110.6	131.4	153.0	175.3

B

Figure 11-5. (A) Learning curve cost reduction (B) Cost variances.

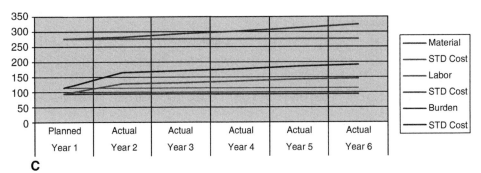

Figure 11-5. (Cont'd) (C) Cost comparisons and trends.

proposition. The data shows the variance numbers. The burden numbers applied to labor change from 1.2 to 1.3 times labor because of unabsorbed labor to cover manufacturing expenses. Given a fixed set of manufacturing expenses based on 1000 units, a reduction to 750 pieces means that the manufacturing expenses must be applied over the 750 units' labor content (i.e., burdening the reduced volume piece part even further). This describes a cost scenario of the planned and the actual as the product exists in development.

Figure 11-5(B) presents the effects after production, where the product is now in the marketplace and has no effective cost containment measures. In Figure 11-5(B) the costs of material, labor, and resulting burden increase with each successive year. The table shows the change in total cost as $90.45 from planned Year 1 to actual Year 2. This is the difference of $575.50, shown as the total cost in Year 2, minus the planned standard cost of $485 (shown below it and also as the total in Year 1).

By looking at the previous chart comparing planned and actual, one can see that the $90.45 consists of $6.20 in material variance, $32.50 in labor (due to non-value-added time in manufacturing), and $51.75 added because of unabsorbed burden. Figure 11-5(C) graphically illustrates the deterioration in successive years. The data in Figure 11-5(C) show that in Year 1 through Year 6 the material labor and burden increase dramatically, even though each year represents only a 3.5% increase by itself. This lack of cost containment can cripple a product line quickly and limit the number of available options for the product manager.

B. Methods of cost containment

The following is a list of methods, along with their summaries, used to achieve cost containment:

- Procurement-related cost containment
- Manufacturing methods and labor
- Manufacturing venue
- Currency fluctuations (foreign content-related)
- Other methods

1. Procurement-related cost containment

The basic focus of this first method of cost containment is to simply buy more effectively and to leverage other buying power. A natural tendency for a product left unmanaged is for the manufactured cost to increase. This occurs for several reasons, primarily that the procurement of materials becomes rote and vendor uncertainty begins to be absorbed by the manufacturer rather than the vendor. For example, a vendor-obsolete part is generally available for some period at a premium price. Left unmanaged, this part remains on the bill of material for the product. The design becomes financially compromised because the changeover to the newer, more available part is not implemented. To effectively manage this, the pathway toward materials cost containment requires active participation.

2. Manufacturing methods and labor

The focus on labor reduction is to determine easier ways to produce the product, with less labor content. This can be accomplished by implementing additional tooling and capital equipment. The capital equipment will require an investment to secure. One of the more proactive ways to manage the labor issue is to design the product in such a way as to be forward-compatible with more fixtures and automated manufacturing methods. The less-mechanized means can be employed initially and, as the market volumes dictate, the more-mechanized equipment can be phased in.

The caveat here is that the volume should be well-understood with few surprises occurring in the marketplace. Given this, the capital equipment required for mass manufacture should be in place at the initial launch. Manufacturing capacity issues should not be a major constraint at this juncture. Mechanizing the manufacturing content will decrease the labor and, if it is in balance, can generate additional throughput (given no other constraints). Labor costs tend to increase due to the increasing cost of wages (production personnel desire wage adjustments, just like the rest of the corporation). This must be offset with a decrease in labor time applied to the product. In addition, there may be a tendency to add quality control checks at various trouble spots in the process, further aggravating the problem by adding labor time. The material and labor initiatives are designed to widen the cost/revenue gap to maximize profit.

3. Manufacturing venue

One way to alleviate the pressure of increasing costs is to relocate manufacturing to a foreign venue where labor costs may be significantly less. Although this may well be the case, relocating the manufacturing venue can also create manufacturing problems. Process training, process control, and process improvement are not assumed to be foreign manufacturing capabilities. Foreign workers must be trained and evaluated just as are the domestic personnel.

It may be more difficult to instill values and process control at these remote locations. If the materials are procured at these locations without the corporation's direct control, there could be additional performance and manufacturing problems introduced by substituting components. Local purchasing without the benefit of corporate resources and technical guidance can lead to a degradation of features and performance. Considering that the corporation is subject to the vagaries of a domestic supplier's offering changes, imagine the loss of configuration control at the remote level.

Local foreign practices, laws, and customs may also interfere with the manufacturing of a product with known configuration control. Other considerations are the export compliance laws and transporting technology to foreign nations. The decision to use foreign manufacturers should be based on an overall strategic goal and long-term plan, not simply the difference in a labor rate. In many industries, recent trends in design and manufacturing have significantly reduced the labor content to a small percentage of the product's cost, thereby further weakening the argument for a foreign venue simply for the sake of labor rate reduction.

4. Currency fluctuations (foreign content-related)

Currency exchange rates can significantly affect the product's profitability. Global products are affected by global conditions and currency fluctuations that can work to the company's advantage or disadvantage. If all transactions surrounding a product—from the procurement of materials to the collection of payment—occur instantaneously, then currency fluctuations have small impact because prices and terms would automatically be adjusted to accommodate the fluctuation. However, when time delay is introduced into the equation, either side can get caught short in an unfavorable position, or can benefit from a favorable swing in currency.

C. Risks and rewards in currency fluctuations

In a typical manufacturing scenario, the company deals with this issue on both the buying side and the selling side. It occurs on the buying side if the component parts used in the product are of foreign manufacture. It occurs on the selling side when the product is being sold to a foreign company. This is illustrated by the example shown in Figure 11-6(A).

Here, a company purchases a component from a Korean firm. The exchange rate from Korean won (KRW) to U.S. dollars (USD) is given at 929 baseline. At the time the procurement was negotiated, 1 USD bought 929 KRW worth of this Korean company's goods.

	Seller's price (Sold in this currency)	Exchange rate[1]	Buyer's cost	Change in exchange rate	Adjusted buyer's cost	Effect (%)	Actions for the buyer	Actions for the seller
Currency risk in US buying								
Buyer's Upside	18,998 KRW	929 KRW/USD	$20.45	985 KRW/USD	$19.29	94.31 Favorable	Buy KRW at 929/USD	Raise prices in KRW
Buyer's Downside	18,998 KRW		$20.45	800 KRW/USD	$23.75	116.13 Unfavorable		Lower prices in KRW
Currency risk in US selling								
Buyer's Upside	$20.45	929 KRW/USD	18,998 KRW	800 KRW/USD	16,360 KRW	86.11 Favorable	Buy USD at 929 KRW/USD	Raise prices in USD
Buyer's Downside				985 KRW/USD	20,143 KRW	106.03 Unfavorable		Lower prices in USD

[1]KRW, Korean won; USD, U.S. dollar

A

Figure 11-6. (A) Impact of currency fluctuation.

	Actions for the buyer	Actions for the seller
Currency risk in US buying		
Buyer's upside	Buy KRW at 929/USD or anything over	Raise prices in KRW
Buyer's downside		Lower prices in KRW
	Actions for the buyer	Actions for the seller
Currency risk in Korean buying		
Buyer's upside	Buy USD at 929 KRW/USD or anything under	Raise prices in USD
Buyer's downside		Lower prices in USD

B

Figure 11-6. (Cont'd) (B) Transactional analysis.

1. The U.S. buyer's perspective

The currency risk associated with buying materials from the Korean firm is shown first. The upside for the buyer occurs when the strength of the dollar increases with respect to the KRW. If the transaction was negotiated at 929 KRW to 1 USD, the component cost the company $20.45 USD. If the exchange rate goes favorable to 985 KRW to 1 USD, the component only costs the company $19.29 USD, or only 94.31% of the original.

If the exchange rate flips the other way, it can be a disadvantage to the U.S. company. In this case, when the dollar weakens with respect to the Korean won, and at 800 KRW to 1 USD, more dollars are required to purchase the 18,998 KRW ($20.45 USD) part. This represents a price that is 116.13% of the original price.

2. The U.S. seller's perspective

From the seller's perspective, if a part now sourced by an American company costs $20.45 USD, the Korean company can procure that same part for 18,998 KRW. This is at the baseline rate of 929 KRW to 1 USD. If the exchange rate goes to 800 KRW to 1 USD, the buyer's (Korean firm's) upside is that, after the negotiation, they only need to pay 16,360 KRW to procure $20.45 USD worth of goods. If the exchange rate deteriorates to 985 KRW to 1 USD, the $20.45 USD part will cost the Korean firm 20,143 KRW instead of 18,998 KRW, which represents a price that is 106.03% of the original price.

One of the ways to alleviate the problem of sensitivity to currency is to hedge against these fluctuations to your advantage, depending on the circumstances and your position in the transaction. This means buying foreign currency when it is favorable in the hopes of using it when rates go unfavorable.

Figure 11-6(B) summarizes actions the buyer, the seller, or both can take in the buying case, and actions the buyer, the seller, or both can take in the selling case. This chart summarizes the actions that either party can take to minimize cost or preserve profit in response to the fluctuation.

In considering the currency risk in buying, the buyer can hedge currency fluctuation by purchasing Korean won with dollars when the exchange rate is 929 or anything over 929 (e.g., 985 as listed in the table). The reason to purchase Korean won at 929 is to protect against a weakening of the dollar. In this manner, the buyer can buy the part at 18,998 KRW and it will cost $20.45 USD, nothing more.

Hedging is designed to protect against the risk of an unfavorable change. In the case where the exchange rate goes to 985 KRW, the seller can raise prices. This is because the actual dollars needed to purchase the part at 985 KRW would still be $20.45 USD (outlay of hedged cash in Korean won, however). In the downside case, the seller may have to lower prices to induce the company to purchase the part from the Korean firm.

Switching to the other side of the transaction, if the Korean firm was to purchase the part (priced in dollars) with Korean won, a decrease in the Korean won-to-dollar ratio will allow the Korean company to procure the $20.45 USD part for fewer Korean won. They may hedge their currency by purchasing dollars at 929 or anything less (e.g., 800 KRW). This means less Korean won outlay for them. The option for the U.S. company is to raise its price in dollars to increase profit, because it will mean the same outlay in Korean won to the Korean company. If the exchange rate goes the other way and the Korean won-to-dollar ratio is 985 or anything over 929, the U.S. company may have to lower its prices to continue to secure the business from the Korean company.

In addition to changes in currency that affect cost and pricing, there is also a domestic factor of inflation that must be accommodated. In periods of low inflation, wages, costs, and prices are relatively stable. In periods of high inflation, each of these excursions can severely affect the profitability of a product. Wage changes affect not only the costs but also the split between material and labor, thus affecting burden as well.

Inflation not only affects the cost and pricing of a product, but can also affect the company's ability to invest in new development. If the company's products and transaction speed do not keep pace with inflation, the resulting shortfall may prevent investment. A very high rate of inflation may also make other non-company, non-development-related investments appear more lucrative. These situations take a company off strategy and are very harmful in the long run.

D. Other methods

There are a host of other alternatives for lowering costs, including partnerships and sub-contracting arrangements. The fundamental issue with these arrangements is to look for the situations in which the arrangement is beneficial for both sides. A one-sided arrangement will not last very long. Relationships that do not have longevity can cause uncertainty and instability in the product line.

Another method for cost reduction is to reevaluate the feature set of the product using a value management approach. This approach is fundamentally an exercise in examining the elements of value transferred with the product to the customer, and attaching value to each of the features by examining the benefits to the customer associated with them. It is possible in some cases to deplete the product of certain nonvalued features, assuming they result in a cost reduction. If so, these depletions will result in certain cost reductions that can result in effective margin enhancement without jeopardizing customer acceptance. It is important to note that if these are significant, the product team has essentially missed its mark in defining the product.

5. History of Product Problems

A product history book (the history of the product, field comments, changes, and reasons for changes within a single, accessible three-ring binder) is another product maintenance tool that is quite helpful in maintaining continuity during the product life. It is especially important when a field complaint occurs and the organization responds to it with a product change. The product is changed but the reason may not be uniformly disseminated within the organization. Without the history and summary of events leading up to the change, it is likely to generate another change if another complaint comes in. This lack of background makes assessment of the field feedback difficult to prioritize. When multiple changes or even reversals of engineering changes occur, the product configuration is out of control and uncertainty in performance and safety are being injected into the product line.

QUALITY MANAGEMENT (PRODUCTION)

1. Quality System—Field Data Analysis

Previous discussions in this chapter have centered on the various performance issues associated with the new product. We have discussed the technological performance issues, the commercial performance issues, and the financial performance issues. In this section we will discuss quality performance issues. These performance issues are measurements of the overall product line's ability to meet customer expectations.

A. Quality system objectives

The objectives of a quality system are as follows:

- *Data collection:* Data collection is the gathering of intermediate and summary data throughout the product's life cycle. It includes design, manufacturing, and field data.
- *Manufacturing monitoring:* This is a measurement of the manufacturing performance in terms of efficiencies, speed, and ability to absorb uncertainty and produce conforming product.
- *Field feedback:* This is the measure of how the product satisfied the customer's expectations in terms of performance, reliability, and use issues.
- *Corrective action:* This is the active participation by members of the product group to improve overall performance and correct mistakes. It is also a measure of the performance of the development/manufacturing value chain in producing the product.
- *Measurement of effectiveness of corrective action:* This is the measure of the effectiveness of the product group's change decision making. Are they actually correcting

issues or are they introducing added, intertwined issues that will be more difficult to untangle later?

- *Degree of normalcy and stability:* The quality system can indicate the product line's level of stability. A product line with many changes (one following another and even reversing previous changes) is an indication that the product configuration is out of control.

B. Manufacturing as performance art

The historical and prevailing attitude toward manufacturing is that it is a machine of human operators that can be turned on, turned off, sped up, slowed down, and can change course and produce perfect, zero-defect results consistently. Companies delude themselves that, with enough procedures, documentation, drawings, and training, they can magically expect this performance. In reality, it is more accurate to think of manufacturing as an art form. A typical pattern in a manufacturing environment is that some form of training temporarily increases performance in speed and accuracy, but this is followed by degradation in performance resulting from boredom and other fatigue factors.

This being the case, there is a sweet spot of elapsed time in manufacturing where the tasks are new enough to be interesting and challenging enough for the operators. A constant introduction of design- and manufacturing-related changes resets the manufacturing cycle to the earlier, uncertain time where training is a major factor and mistakes are more common. If there is a steady diet of this, the manufacturing group never achieves the second stage where speed and accuracy significantly benefit the company.

C. Mechanics of a quality system

The basic mechanics of a quality system are illustrated in Figure 11-7. The specifics of how to obtain the data are system- and company-specific. The overall mechanism is more universal.

As shown in the figure, the process begins with the design and development of the product. Upon completion of the project, the product moves into the manufacturing environment. The quality system is designed to monitor and record data needed to support the product. The product is shipped to the customer and the customer places the product in use. In the unfortunate event of product failure, the customer returns the product to the factory for repair. This can be a warranty or nonwarranty repair.

Alternatively, there may be local repair shops authorized to do the repair. In either case, the repair data should be fed back to the factory for analysis. This is very important for the initial return so that the factory may confirm failure modes and determine whether any corrective action needs to be taken with the design.

The repair analysis is used to make the repair and also to feed specific information about the failure modes back to the various functional groups involved in creating the product. For example, the corrective action recommended by the repair department may be safety-related, may deal with fine points of the design, or may highlight fundamental flaws. The feedback for manufacturing may be process-related, venue-specific, or procedural. Feedback for shipping may include packaging-related issues and shipping guidelines. Finally, in-use issues might include clarification of the user's manual and other application information.

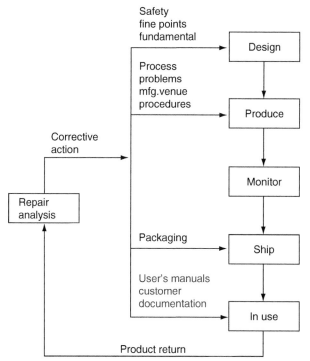

Figure 11-7. Product quality feedback system.

D. Loss mitigation

The overall objective (besides financial) behind reducing warranties is to improve customer satisfaction. The financial cost of a warranty, however, can be high. At each stage the product goes through, from component parts through manufacturing, to shipment and in-use, the cost of remediation multiplies. Therefore, the earlier the problems can be captured, traced, and resolved in a deterministic manner, the lower the exposure to the company. The quality system therefore should be designed to capture and report these problems at the earliest possible point of occurrence.

E. Warranty exposure market and dollars spent

In every company, from large to small, a warranty is thought of in either the cost of quality perspective or the customer satisfaction perspective. The cost of quality perspective is oriented toward the financial exposure involved with product failures, returns, repairs, scrap, and labor costs. The customer satisfaction perspective is a measure of the marketplace's acceptance of these failures, and a response to the company's ability to effect speedy reparations.

F. "Show me the data!"

"Show me the data!" is the battle cry of quality control. All too often, opinions creep into the discussions pertaining to how a product is doing. This happens especially when there may be a recent run of failures. Human memory patterns seem to focus on the bad and do not correlate very well with the level of shipments, historical rates, and other data. This is why it is important to chart field data on returns, product failures, and shipments. Serialization also will help in pinpointing exposure on certain problems. As will be shown later in the section dealing with recalls, it is important to be able to identify the end of a field problem. Consequently, when it comes to quality levels of a product, the team needs to converse in terms of hard data.

G. Setting the warranty limits: external versus internal perspectives

Intentionally or not, every organization essentially sets limits on what they will tolerate in warranty returns. They may be driven by the cost of quality or the cost of lost market share. The important point to remember is that there are two tolerance limits on product warranty: internal limits and external limits. The internal limit is the amount of warranty the corporation will accept before taking aggressive corrective action. This aggressive action may be a selective or total recall of the product. When the level of return is less than this number, little is done in the way of corrective action. When it exceeds this number, the entire organization is aware of the problem and swings into action.

The other limit is the patience the market has for the product problem. Below that limit, the customers' experiences will be scattered and easy to maneuver around in a sales situation. This can happen where the customer base is diverse and unrelated, vocationally or geographically. Above this limit, however, the product begins to get a bad reputation. This is common where members of the customer base are in the same vocational group or the same industry, where the bad reputation gets passed from one customer to another even if they personally have not had a bad experience with the product.

H. The exposure to the organization

A significant amount of direct and indirect labor is associated with supporting a product that has problems. With each successive problem, more and more effort will be required to conduct normal operations because each product problem has its own caveats, rules, and issues. As the problems are piled on, more rules (now affecting the other rules in addition to the problems) are added. Figure 11-8 illustrates the point.

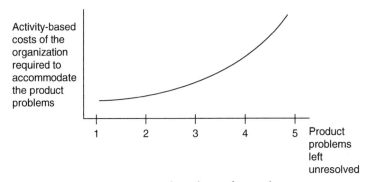

Figure 11-8. Activity-based costs for product issues.

As can be seen in the figure, the effect of supporting additional problems becomes exponential. Left uncorrected, this requirement for manual effort outside of the normal operations and system begins to dominate the quoting, selling, development, and manufacturing operations. This adds non-value-added labor to the product and the business. The non-value-added labor slows down an efficient process, while adding uncertainty and the potential for error.

2. Business Quality System—Corrective Actions

Not all issues that come in from the field are product-related. Many times they are business system-related, and a change in the business system may be needed to improve operations and keep customers happy. Certain corrective action systems allow this to occur and provide a means for logging the issue and making changes. The key to the corrective action system is that, in addition to correcting the specific customer issue, it can contain the problem and prevent it from expanding, and also examine the process and prevent future occurrences. These containment and prevention steps are key to attaining world-competitive performance in the business enterprise.

PRODUCT RECALLS

1. Recalls: A Business Nightmare

A. Seven steps of product recall

One of the more distasteful tasks (and the true measure of a product manager) is deciding on and executing a product recall. Although it is one of the most disheartening actions to take, a product recall may be the only way to save a product in the long run.

A product recall forces a direct about-face in the present pathway of a product. It allows the company to reset its reputation and generates a fresh start for the product. The product recall engagement must be well-considered, decisive, and speedy. Lack of any of these attributes can cause the recall to fail and prompt the marketplace to abandon the new product altogether. A product recall is a public admission that the company failed to produce a good product. It is a step that can draw criticism and admiration at the same time.

The key to any successful recall is to face up to the product problem, decide on the recall promptly, and execute it with diligence and speed. The market has tolerance for some inconvenience but has no patience with ineptitude on top of product problems. There are seven elements in a recall; the protocol, along with the explanation of each step, is included here:

- Contact
- Explanation
- Actions required
- Procedures
- Acknowledgment
- Compensation
- Closure

1. Contact

The contact portion of the protocol involves the company contacting the customers in some organized fashion to advise them of the recall. This can be a registered letter with return receipt indicating delivery and receipt, telephone, fax, e-mail, or some other verifiable means.

2. Explanation

The explanation outlines the specific defect, the potential effects, the impact of the defect on the user, and the degree of seriousness of the defect. It must identify the scope of impact that will require appropriate and timely action by the user.

3. Actions required

This is a detailed outline of the actions required to resolve the recall issues for the customer. The actions define the scope of the customer and company requirements, the specific actions to take, the timing of these actions, and the responsibility of each. It should also prescribe the terms of the recall if expedient actions are not taken. A flow chart of events might need to be included as part of the actions.

4. Procedures

The procedures spell out the disposition of the product in the customer's possession and the arrangements for any planned upgrades. Specifically, the procedure section involves one of the following elements: exchange of product, replacement of product, repair of product, substitution of product, or upgrade of product. One of these should be selected and the procedure for the selection should be spelled out completely.

5. Acknowledgment

This is a very important part of the recall process. The acknowledgment provides evidence of the company contacting the customer, the customer responding to the contact, and the product disposition being executed. Any arrangement of bounce-back cards, telephone, or some other means is demonstrable proof of the recall engagement showing each party's responsibility and fulfillment.

6. Compensation

Depending on circumstances, there may be compensation involved. This compensation may also be included in the explanation portion of the recall. It outlines the compensatory steps the company will take in securing customer satisfaction.

7. Closure

This is a formal record to close out the recall. It provides demonstrable proof of the company's diligence, recall engagement, resolution, and proof of mitigation. Depending on product circumstances, it may be included in other sections also.

B. Determining whether a recall is necessary

Determining whether a product recall is necessary is an inexact science. Aside from the protocol necessary in a safety-related recall or a blatant misrepresentation of the product because of performance, the recall must navigate the perceptions of the customer and the marketplace. The customer perceptions are unpredictable in many cases. The result of this navigation is a corrected product with its reputation for quality intact. In fact, the company issuing the recall exemplifies high regard for quality during the recall. At this point the customer lacks confidence in the product—all they can rely on is their confidence in the company's quality reputation. Only after the product is corrected can the product's reputation be restored. This is why it is critical to be decisive and correct in determining the recall. If the company is not, its reputation will also be damaged.

Several factors are involved when determining the necessity of a recall, including:

- *Known bad component(s):* This is where key components can cause product failure or safety compromises. Once the manufacturer knows of these issues, a recall is most likely in the product's future.
- *Known bad process:* This is where manufacturing process used in construction can cause product failure or safety compromises. Once the manufacturer knows of these issues, a recall is most likely in the product's future.
- *Latent product risks:* This is where components will fail at a future point. They are presently functioning but, as time and use accumulate, the latent flaw will surface and cause premature product failure.
- *Safety-related issues:* This broad category applies to any safety-related issue of the product that is addressed by design or manufacturing issues, as opposed to application issues. These issues generally require immediate action, with specific execution plans and time lines.
- *Blatant misrepresentation due to a product performance shortfall:* This is where an extreme oversight, inadvertent product changes, or a complete specification violation places the company's reputation in extremis.
- *Failure to deliver based on contractual performance:* This is the situation where the product or group of products was sold to the customer under some contractual performance guaranty. Failure to achieve the performance will necessitate some means of correction or recall.

If your specific situation meets the basic tenets of these criteria, there is a good chance that a product recall is needed.

C. Internal actions: Hold production until resolved

In a recall situation, several internal moves are required to execute the recall properly. Several external moves must also be made. These moves are designed to facilitate the recall and also to ensure some measure of protection in case of product liability.

In addition to the logistical implementation, production must halt immediately until the problem is defined, analyzed, and resolved. Production can only resume after the fix is in place. Personnel must be selected to implement the recall, including a designated recall officer. They must be trained and audited. An infrastructure specific to the recall should be set up; it must be easily accessible as well as responsive.

D. Designating the recall officer

This position is a product management position specific to the recall. The person in this position designs all aspects of the implementation and activates these systems. He or she makes critical decisions about the product line to minimize the reputation damage and also to mitigate the financial exposure.

The recall officer should have a strong product manager profile and be able to act decisively and swiftly. He or she must go through barriers and have a strong marketing and customer orientation. The original product manager for the product line's development is usually a good choice unless there is some vocational weakness that prevents him or her from completing the task.

E. External actions

The external moves to be made depend on the organization's size, route to market, and the problem's degree of severity. Field service, customer support, and local sales offices may be involved in the process. Ideally, however, the company should be the focal point of the recall, with primary customer contact and the shortest pathway to and from the customer to provide expedient service to the customer, given the inconvenience of the recall situation.

F. Communication

One's first instinct in implementing a recall may be to keep it as low-profile as possible. This is true from the reputation perspective, but it is difficult to keep any product issue hidden for any length of time. Instead, the company should be careful in deciding on the recall but, once the decision is made, they must communicate it effectively to all parties concerned.

The issue of communication can even be the subject of a potential litigation. There is a responsibility associated with getting the information to the right people in the shortest period of time. The communication should permeate the company, satellite offices, customer support function, and manufacturing.

2. Mitigation of Liability

A. Protection from liability during a crisis

A fiduciary responsibility of the product manager and the business development team is to mitigate the financial impact of a recall on the company. Diligently assessing the issues and prosecuting them with speed and accuracy generally accomplish this. The best interests of the customer and the company must be kept in mind. With safety as the primary motivator and customer satisfaction as the closing signature, a crisis can have a positive outcome and the long-term financial impact to the company will be minimized. Open, well-considered communication is desired; cover-up tactics cannot be allowed.

B. Involving the organization early and completely is key

A recall is a time when the organization can assist the product manager in correcting the product. The manager must rely on the organization's infrastructure to support the recall. Although the manager is the architect of the recall, it must be a company-wide effort, not an individual one. To that end, there must be company-wide communication. Action plans must be made known and supporting people must be trained, motivated, and in place. Once the decision is made to recall the product, involve the company immediately. It will serve the recall effort, the product, and the customer base.

C. Define an end to the crisis and move on quickly

The operational concept behind a recall is to design what constitutes an end to the product crisis and drive the events toward this end. If the recall means to get every unit back to the factory and repair it, then the end of the crisis is when this is accomplished. If the recall means to inform every product owner and schedule a refit at some service location, the end of the crisis is when this is complete. In either case, the recall team must exercise speed and diligence to this end.

The product team and the company need to monitor for an end to the field problem. This means to look at the quality database and the returns coming back. The quality database will tell you if the fix is effective by indicating no returns past the production date when the fix was implemented. The warranty returns will indicate diminishing labor and materials associated with the recall. This indicates that whatever product is coming back is now returned and the financial exposure is minimal, given no latent issues.

D. Activity-based costs associated with recalls

In a similar manner, product problems incur significant costs for the added activity to support a recall effort, as discussed earlier. To that end, it is desirable to end the recall as quickly as possible and get on with the product and the business it generates.

PRODUCT EVOLUTION

This concludes the basic product management section of this chapter where general product management issues, cost reductions, quality initiatives, and general product maintenance issues were reviewed. The focus now shifts toward the pursuit of a successful product that is oriented toward growing the new business; follow-on planning; market development and application development; pricing issues; and growth strategy. The next section will explore these basic tenets of new product marketing within the framework of product management, to grow the business and pave the way for follow-on new product initiatives.

1. Line Completers

An initial product introduction may only be a small apart of a business development initiative. Launching the first product may set the stage for future line completers and expansion of the product offering. Accordingly, the launch of line completers should be scheduled to take the best advantage of the newness and market response to the original introduction. The idea is to create a business situation that creates momentum and allows the company to achieve the desired forecast as soon as possible, and keep the competition on the run. In this way you capture the revenue needed and keep the competitors from establishing their own momentum.

2. Product Evolution Flow Chart Development
A. Present market feedback versus the original assumption

Most product evolution flow charts are developed as part of the process of launching into a new business development. Unfortunately, if they are not integral to the business development process through to product introduction, they soon fall by the wayside and lose their impact and guiding direction. Without such flow charts there can be a tendency to react to competitive tactics and market dynamics, running the risk of a program that lacks direction.

Comparing the new product to the original product and market projection is very important. It serves as a basal temperature reading on the speed at which the market is reacting, the competitive resolve, and the level of company performance. If the projected product demand falls short of the actual market feedback, the market is moving faster than anticipated. If the projected product demand is far ahead of the actual requests, the market is more lethargic than anticipated and leapfrog introductions may be too advanced for the customer base to accept.

Consider Figure 11-9, which compares the original plan and the current market feedback. As shown here, the product evolution flow chart indicates an initial product offering and a second one is planned for after the end of Period 3. A third offering is planned for after the end of Period 4. The products represented by First Offering, Second Offering, and Third Offering are the original plan. The height of the product boxes indicates their feature/benefit, their performance, or both. The taller the product diagrammed, the more features and performance it has.

The actual product introduction for the first product version was a little late and is shown accordingly. The accelerated market response is shown where the market expects more features. The lethargic response is shown underneath it, where the market expects fewer features and less functionality. The dotted line at the point of measurement factors in the late product introduction. If initial market feedback outlines a product like the product represented by Market Response Exceeds Plan (with significant features and performance requirements exceeding your planned offering early), you are in a fast-paced market and must act accordingly to keep up.

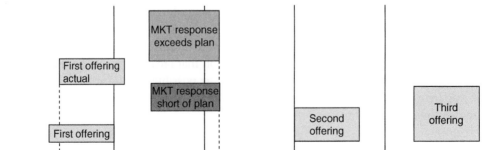

Figure 11-9. Product evolution comparison.

B. Adjust the leap

The comparison will detail the response of the next product action. The leapfrog must be enough of an improvement to position the company effectively. Even at this point, be sure to factor in ample improvement or enhancement so as to be in a favorable competitive position, even after the product enhancement is developed. With knowledge doubling faster and faster, it is especially important to target properly. This is especially important in a fast-paced arena because the competition is not sitting by while you launch the product, gather market share, and develop follow-on enhancements. In fact, the introduction itself ignites a product capabilities race to secure market share.

C. Attacking yourself as the next step

Let an introspective philosophy permeate the organization to stay current with the product. Constantly attack yourself before the competition does. If your product planning personnel begin to think like the competition and work in your best interests, they will guide your moves to always be ahead of the competition.

3. Projecting Technology Integration Needs

Once the product evolution analysis is completed, the organization will have a good idea of the product requirements needed to support the business development initiative. With the product requirements defined, you can identify the technology needed to support the initiative. Technology in its raw form cannot be easily deployed in an organization. It must be understood and become second nature to the personnel that employ it. Therefore, procuring technology to support the business development initiative is much more than buying the technology. It will take time for the organization to absorb it and be able to platform it on a product. Accordingly, it cannot be bought like a computer or some other piece of equipment.

The technology must be used to secure the correct product platform on which to build the various versions of product. The best way is to ensure the correct platform is to stage the technology integration along the same axis as the product evolution and time it so that the technology is procured and integrated well in advance of its being used. In our earlier model of R&D, this is generally the function of research. By looking at the product platform breaks along the product evolution flow chart and backing up in time, you can set up the technology, procurement, and integration schedule for the company.

In the same way, installing personnel and training them in the new technology is another component of the product evolution. In order to implement the technology, it is important to have personnel trained in the technology. Accordingly, the product and new technology implementation flow chart must also have personnel training integrated.

4. Navigating the Pathway

It is easy to articulate a plan for the training and expense in procuring technology. It is another thing to carry it off amidst the trials and tribulations of daily business. For example, we discussed the issue of product maintenance and how it can displace the effort in new product development. Time and funding must be allowed for the training and use of technology so personnel can become proficient in applying it. It is an investment just like anything else. Often, we expect performance from our personnel without giving them the time to absorb technology properly. A rushed pathway here will result in problems later. Ensure that the pathway is funded in terms of time and money.

PRODUCT LIFE CYCLE MANAGEMENT

1. Life Cycle of a Product

A. Evaluating new product programs

How has the new product development program performed against plan? Is it a success? Is it marginally successful? Did the business development team deliver the results promised? There are several ways to answer these questions; indeed, several factors can be used to evaluate the degree of success or failure. Fundamentally, beyond the strategic initiatives, a new product development is supposed to generate funds to the corporation at a rate that far exceeds the rate of funds consumption. This profit is used to fund new programs. The dynamics represent the corporate covenant between the product group and corporate management.

A simplified method of evaluating the degree of success of a new product program from a purely financial perspective is to compare the original expectations and the actual results. Figure 11-10 outlines an example.

As shown, Figure 11-10 summarizes the various scenarios in the program planning stage and compares them to the actual results. The various case scenarios allow the actual results to be placed in perspective with the original business plan. The chart shows an investment of $1.5 million, a product price of $500, and a cost of $250 in the planning stage. A total of five cases with different volumes were generated to assess the internal rates of return (IRRs) and the net present values (NPVs). The actual results indicate an increase in costs as the product life cycle continued with the best-case volume. Actual NPV and IRR are much less than planned.

This method can be used to evaluate your own programs in a simple, objective manner.

B. Opportunities and threats during the cycle

Previously, we discussed several threats that can endanger a product program. However, several opportunities also become available during the business cycle. If the company has difficulty in focusing on goals and demonstrating the resolve to attain them, the opportunities that befall the company can also become threats. These are the worst kind of threats because they foster complacency and demand the most costly extravagance, namely, time and energy.

As important as the vocational aspects of new product development are, the corporate characteristic traits are equally as important. If the program is on target and business development is diligent, the real competitive threats will be handled effectively in most cases. If, however, the corporation has a weakness for the prettier opportunity, it is virtually impossible to carry a program through to completion, let alone achieve success. Focus and execute throughout the business and product cycle. Do not be driven off-course by seemingly better opportunities. They detract from your primary objective.

Evaluation								
New product plan		Year 1	Year 2	Year 3	Year 4	Year 5		
Price each		500	500	500	500	500		
Cost each		250	250	250	250	250		
Case 1 volume		1,000	2,000	3,000	4,000	5,000		
Case 2 volume		800	1,600	2,400	3,200	4,000		
Case 3 volume		600	1,200	1,800	2,400	3,000		
Case 4 volume		400	800	1,200	1,600	2,000		
Case 5 volume		200	400	600	800	1,000		
	Investment ($)						NPV based on 5% inflation ($)	IRR
Gross profit case 1 ($)	−1,500,000	250,000	500,000	750,000	1,000,000	1,250,000	1,563,427	30%
Gross profit case 2 ($)	−1,500,000	200,000	400,000	600,000	800,000	1,000,000	965,027	22%
Gross profit case 3 ($)	−1,500,000	150,000	300,000	450,000	600,000	750,000	366,628	12%
Gross profit case 4 ($)	−1,500,000	100,000	200,000	300,000	400,000	500,000	(231,772)	0%
Gross profit case 5 ($)	−1,500,000	50,000	100,000	150,000	200,000	250,000	830,172	−17%
Actual results	Investment ($)	Year 1	Year 2	Year 3	Year 4	Year 5	NPV Based on 5% inflation ($)	IRR
Price each ($)		500	500	500	500	500		
Cost each ($)		250	290	330	370	410		
Case 1 volume		1,000	2,000	3,000	4,000	5,000		
Gross profit case 1 ($)	−1,500,000	250,000	420,000	510,000	520,000	450,000	323,807	12%

Figure 11-10. Evaluation of a new product plan.

C. Continuous evolution is integral to the product life cycle

The product life cycle demonstrates the need for vigilance and continuous product improvement to nourish the business. When the product volume is building up, the costs are going down, and the profits are beginning to be realized, it is easy to get lulled into a sense of complacency. However, the time to embark on the new version or the next product is when the business is riding high. The cash position is healthy, the profitability is peak, the market presence is good, and there is only one way to go if nothing is done: down! Therefore, the best time to question your market strength and do some business introspection is when the natural tendency is to be complacent. The fact of the matter is that this is the very time the competition is making their next move. This is shown in the chart in Figure 11-11(A).

Life cycle management									
New product plan		Year 1	Year 2	Year 3	Year 4	Year 5	Year 6	Year 7	Year 8
Price each ($)		500	500	500	500	500	500	500	500
Cost each ($)		250	250	250	250	250	250	250	250
Case 1 volume		1,000	2,000	3,000	4,000	5,000	3,500	2,000	500
Case 2 volume		800	1,600	2,400	3,200	4,000	2,200	1,000	200
Case 3 volume		600	1,200	1,800	2,400	3,000	1,500	750	175
Case 4 volume		400	800	1,200	1,600	2,000	900	400	125
Case 5 volume		200	400	600	800	1,000	400	200	75
	Investment								
Gross profit case 1 ($)	−1,500,000	250,000	500,000	750,000	1,000,000	1,250,000	875,000	500,000	125,000
Gross profit case 2 ($)	−1,500,000	200,000	400,000	600,000	800,000	1,000,000	550,000	250,000	50,000
Gross profit case 3 ($)	−1,500,000	150,000	300,000	450,000	600,000	750,000	375,000	187,500	43,750
Gross profit case 4 ($)	−1,500,000	100,000	200,000	300,000	400,000	500,000	225,000	100,000	31,250
Gross profit case 5 ($)	−1,500,000	50,000	100,000	150,000	200,000	250,000	100,000	50,000	18,750
Actual results	Investment	Year 1	Year 2	Year 3	Year 4	Year 5	Year 6	Year 7	Year 8
Price each ($)		500	500	500	500	500	500	500	500
Cost each ($)		250	290	330	370	410	410	410	410
Case 1 volume		1,000	2,000	3,000	4,000	5,000	3,500	2,000	500
Actual gross profit case 1 ($)	−1,500,000	250,000	420,000	510,000	520,000	450,000	315,000	180,000	45,000

A

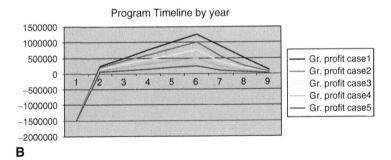

B

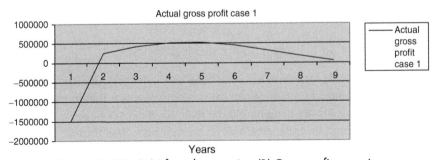

Years

Figure 11-11. (A) Life cycle scenarios; (B) Gross profit scenarios.

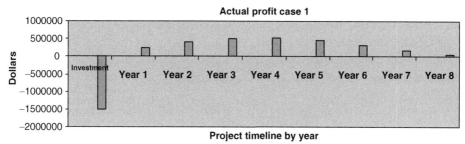

Figure 11-11. (Cont'd) (C) Actual gross profit scenario.

The graph in Figure 11-11(A) shows the buildup of volume under the five cases outlined in the business plan. Since Year 5 is the peak of the product line in volume, product management should launch new product efforts sometime before the end of Year 5.

Figure 11-11(B) shows the gross profit, under the plan scenarios, which will be used for the investment. Figure 11-11(C) shows the actual gross profit of the product run, based on volumes and updated cost data. It indicates that the peak of the product's contribution to corporate profits occurs earlier in the cycle, and reinvestment should probably occur nearer to that time rather than in Year 5.

As shown in these various ways, there is a point on the product life cycle's characteristic curve where reinvestment must be made. The shape of the curve may change from product to product, but the operative lesson is to recognize the peak of the product cycle and take effective action to preserve it in the long run.

D. Maximizing profit

The business development manager's responsibility is to maximize profit for the corporation, if possible. This directive consists of three essential elements, all things being equal:

- The volume of product must be pushed. Meet or exceed the target volumes but do not fall short!
- Maintain the product pricing in the marketplace. Take every possible step, including value enhancement to reduce the need for rolling back pricing.
- Make a conscious attempt to contain cost increases and work to reduce costs. By reducing costs, the company is in a better position to offer enhanced value or meet competitive pricing, should it be required.

2. Product Obsolescence

What is the company's product obsolescence policy? Is it different for legacy (older) products than for new products that change faster with changing market conditions? How much inventory should be kept to support these products? These are questions that require answers as the business grows. Because of the product maintenance requirements, companies cannot afford to devote an ever-increasing amount of time to older products. Development personnel must be freed up to pursue new technologies and platforms. The new products should be able to significantly outperform the older product's functionality in order to allow sales and marketing to convert customers to the newer product line.

Many companies have standardized policies for obsolescence and ramp-down of production, and phase-in of new product(s) to convert customers. Sometimes all these plans go awry when the customer (Who is large enough?) demands that the product continue. It is important to know your customer base and prepare them for phase-out in a timely manner and understand the impact it may have on them. With product obsolescence, communication is key.

Sometimes the company is the mercy of a specific vendor who wants to obsolete their own component that is being used in the product. In these cases, the company may attempt to institute a design change to utilize a different component, or redesign the product to eliminate the component, or upgrade to the vendor's new component. Either way, a seemingly small decision regarding obsolescence can have huge market impact.

3. Inventory Reserve for Repairs of Obsolete Products

In general, the company needs to establish a policy based on market needs and newer version product convertability, and to carry inventory to support that policy. If a product is to be supported for 10 years after the date of obsolescence, then the product universe needs to be identified and an assessment of conversion rates needs to be done; the proper amount of inventory for the long-term users of the product can be factored in.

If parts are readily available throughout the projected 10 years, then inventory can be bought as needed. If last-time buys are needed, the 10-year support can have huge inventory implications. An alternative can be to redesign around the component for repair purposes, but this takes engineering time and funding.

The following describes a decision tree for the inventory management of an obsolete product. The tree starts with the decision to obsolete the product. Then the company must decide how long to support the product. Next, ask if there are parts available. If the answer is yes, you buy them as needed and support the product in repair operations. Then you check inventory and ask whether additional parts are needed. If they are, go back to the beginning. If parts are not available, then you determine if there is a last-time buy for the parts. If there is, you need to assess the requirements to support the product in repair operations for the period you have decided to support the product to the installed customer base. If you can execute a last-time buy, do so at that number and continue to support the product. If you cannot execute a last-time buy, you may need to redesign in a substitute part. If you choose this path, you will have to requalify the original design with the new part.

This represents an example of the philosophy for long-term support. Your operation may need to enhance this decision tree to meet your specific requirements.

4. Decision Management

Managing business development and product management is like any other position of responsibility in life. It requires diligence, perseverance, decision making, and decision management. It is easy to render a decision in many cases, but quite another to manage the impact of the decision toward a specific goal.

This is the essence of business management. Gather the facts as best as you can and make an informed decision within the framework of the business's mission and goals, and see it through to completion.

SUMMARY

This chapter represented the vocational capstone of the business development process. The chapter was presented in two sections. The first was oriented internal to the organization and was focused on a discussion of the mechanics of product management. The second section centered on the external focus of promoting the new product in the marketplace.

The chapter started out with a discussion of the role of the business development manager as champion of the product. Next was a discussion of the growth strategy for the business. This strategy needs to be interactive with real-time market feedback. Product maintenance was considered next, with a discussion of the various aspects of product maintenance needed to execute the product life cycle.

Then followed a review of how to define and implement the quality system for the organization. Next was a discussion of product recalls and how to mitigate the corporation's liability and still maintain customer satisfaction.

The evolution of the product and the behind-the-scenes requirements were reviewed next. This important activity charts the future and paves the way for core technology development. Finally, the product life cycle and how to manage the product within the framework of corporate goals, obsolescence, and long-term inventory requirements were reviewed.

With the business development basics in place, you are now ready to embark on the new product journey of your own! To assist you in this endeavor, Chapter 12 (Business Development Records Format) presents perspectives on the product development process and outlines a framework for you to organize your own project. Good luck in developing your new business!

BUSINESS DEVELOPMENT RECORDS FORMAT

This chapter consists of two parts: a strictly mechanics-oriented section and a section outlining final perspectives on new product development. The mechanics section is designed to show the reader how to structure their files so that information and data are easily stored, presentable, and retrievable in an organized manner. It can also serve as a checklist for the various aspects of business development process. The second section reviews operational perspectives on new product development as a vocation and a career pathway.

ORGANIZATIONAL FORMAT

Background

1. General Usage Patterns of Product Types
2. Trends in Industry and the Marketplace
3. Driving Forces in the Functionality
4. Driving Forces in the Technology Employed
5. Product Opportunity Summary
6. The Niche/Advantage/Uniqueness of Position
7. Statement of Fit with the Enterprise
   A. Idea evaluation
   B. Feasibility study
   C. Product fit study resolution
   D. Strategy for company resolution

8. Brand Label, Joint Ventures, Acquisitions
   A. Miscellaneous topics to agree on
   B. Brand-label checklist
      1. Brand-Label Narrative
   C. Diligence checklist

Market Data and Dynamics

1. Primary Market Data
   A. Market research and profile
   B. Market opportunity assessment
2. Secondary Market Data
   A. Value stream mapping—customer values
3. Market Segmentation Profile
   A. Market dynamics characterization
4. Product Mix—Analysis
   A. Product/services mix
      1. Products and Services Narrative
   B. Weighted evaluation of opportunities
   C. Value pricing
      1. Narrative on Background of Pricing
5. Pricing Levels and Trends
   A. Creep impact on cost
6. Estimated Costs
7. Volume
8. Demographics
   A. Value engineering
9. Classification of Customer Base
   A. Model for a sales transaction

Communications

1. Internal Correspondence
   A. Sales department
   B. Development
   C. Finance
   D. Manufacturing
   E. Quality
2. External Correspondence
   A. Customers
      1. Account Development Model
      2. Action Plans
      3. New Items/Issues
      4. Problems/Solutions
   B. Agencies
      1. Qualifications
      2. Testing
      3. Approvals
   C. Reps/Distributors/Agents
      1. Contract
      2. Assigned promotional Territory
      3. Sales Goals
      4. Forecasts
      5. Issues

Pathway to Market

1. Flow Chart of Route to Market
2. Funds Flow for Transaction
3. Percent Cost Added through the Channel
4. Sensitivity Analysis through the Channel
5. Analysis of Transactional Pressure Point through the Channel
6. Alternate Routes to Market
7. Impact of Alternative Routes

Customer Interface

1. Outline of Marketing Plans
2. Customer, Agent, Reseller Matrix Profile
   A. Agent reseller management
3. Time Line for Customer Visits
4. Status Report for Each
5. Link to Communications Section
6. Site Visit Report
7. Customer Profile

Competition

1. Competitive Analysis Comparison Form (Internal and Customer), Alternative Point of View
   A. Competitive comparison chart
   B. Competitive comparison with customer alternatives
2. Spider Web Chart for Each Competitor
3. Pricing Analysis
4. Route to Market
5. Sensitivity Analysis
6. Technology
7. Trends
8. Strategy Against Each Competitor

Cost Data

1. Bill of Material
2. Cost Roll-Up
   A. Design to cost versus cost-plus
   B. Parts cost roll-up
3. Cost Basis and Assumptions
   A. Cost systems
   B. Product cost analysis
   C. Profit sensitivity with cost
4. Vendor Links
5. Agreements with Vendors
6. Labor Analysis
7. Vendor Plan
8. Labor Plan
9. Methods Plan

Financials

1. Product Life Cycle Forecasts
   A. Pressure cycle forecast
2. Revenue with Mix
   A. Profit in backlog
3. Factory Cost with Learning Curve
   A. Learning curve cost reduction
4. Development Cost
   A. Project cost analysis
   B. Project evaluation worksheet
   C. Expense and cost profiles
5. Tooling Costs
6. Capital Equipment
7. Sales General and Administrative Costs
8. Return on Investment Calculation
   A. ROI calculation worksheet
   B. Internal rate of return
   C. Comparative ROI worksheet
9. Return on Net Assets Calculation
10. Income Statement
    A. Breakeven analysis
11. Balance Sheet
12. Funds Flow

Product Configuration

1. Product Versions: Scope of Product Range
   A. Sample catalog number breakdown
2. Options and Product Configuration
   A. Configuration worksheet
3. Product Evolution Flow Chart
4. Catalog Number Breakdown

Product Evolution Flow Chart

1. Conservative Development of Product Line
2. Aggressive Development
3. Most Likely Scenario
4. Detail of Present Product Plans and Disposition
5. Specs of Future Units
6. Product Life Cycle Control Sheets
7. Time Line with Contingency

Marketing Requirements Specification

1. Format and Outline
2. Preliminary Forecast
3. Design Validation
4. Product Concept Plan
   A. Background
   B. Industry trends
   C. Market opportunity
   D. Tie to strategic plan
   E. Scope of product line
   F. Component parts and product configuration
   G. Functional sequence of operation
   H. Performance requirements
   I. The operating envelope
   J. Standards
   K. Cost target
   L. Timing to introduction
   M. Human factors engineering
   N. Safety
   O. Longevity, product life
   P. Service plan
   Q. Field replacement parts
   R. Corporate standards

Design Approach and Approach and Specification

1. Format
2. Design Strategy
3. Technology Integration Study
4. Test Strategy
5. Manufacturing Strategy
6. Quality Plan
7. Reliability Plan
8. Safety Plan
9. Outline
   A. Scope of product line
   B. Component parts and product configuration
   C. Functional sequence of operation
   D. Performance requirements
   E. The operating envelope
   F. Standards
      1. UL Submittal Package Format
      2. Statutory and Regulatory Approvals.
   G. Cost target
   H. Timing to introduction
   I. Human factors engineering
   J. Safety
   K. Longevity, product life
   L. Service plan
   M. Field replacement parts
   N. Corporate standards
10. Product Release Procedure
11. Design Review Results
12. Packaging Optimization
13. Criticality Analysis
14. The Responsibilities of the Design Engineer
    A. Design verification
15. Matrix the Marketing Spec and Design Spec: Resolve Changes
16. Commitment to What Will Be Achieved

Drawings and Documentation

1. Design Documentation
   A. Design review results
2. Manufacturing Documentation
3. Repair Documentation
4. Electronic Medium Plan
5. Retrieval and Modifications
6. Portability
7. Structure and Format
8. Methods Documentation
   A. Release procedure
9. Improvement Plans
10. Deliverables

Engineering Operations

1. Product Configuration System
2. Morphing to a New Product through Engineering Change
3. Engineering Change Request / Engineering Change Notice Systems
   A. Sample engineering change procedure
   B. Product engineering change format
4. Engineering Change Request / Engineering Change Notice Database
5. Master Log and History
6. Cost/Impact Analysis, Factory Cost, and Activity-Based Costs
   A. Engineering time log
      1. Narrative Development Time Management

Significant Performance and Applications Features Document

1. Operating Envelope
2. Degree of Derating
3. Margin to Advertised Spec
4. Corporate Derating Guideline
5. Spider Web Chart of Parameters
6. Spreadsheet of Market Requirements/Design Objectives/Actual Parameters and Percentages of Each

Literature Documentation

1. Matrix of Product Information
   - A. Type A information: effecting the purchasing decision
   - B. Type B information: product usage information
   - C. Type C information: training on the product
   - D. Type D information: application of the product
   - E. Type E information: effecting the repair of the product
   - F. Type F information: prosecuting the warranty

Manufacturing

1. Manufacturing Plan
   - A. Lead time summary
   - B. Inventory analysis and allocation
2. Manufacturing Processes/Procedures
   - A. Capital procurement and setup plan
3. Layout of Line/Flow
4. Formula for Logistics
5. Manufacturing Documentation
   - A. Vendor profile
     1. Linkages
     2. Lead Time Summary
   - B. Bill of material
   - C. Part card
6. Training Plan
   - A. Employee certification
7. Failure/Fault Repair System for Production
8. Testing Plan
9. Recordkeeping
   - A. Serialization
   - B. Histogram of modules
   - C. Test data
   - D. Failure data
     1. Manufacturing Yield Data
   - E. Critical components
10. Quality Plan
    - A. Model for corrective action

Field Feedback

1. Problems
2. Complaints
3. Suggestions
4. What's Right/Needed
   A. Satisfaction assessment
5. Value Analysis
   A. Beta test feedback
   B. Beta test form
   C. Beta test summary
      1. Narrative of Program
6. Next Generation or Product Iteration

Product Liability

1. Scope Analysis
2. Scope of Product Line/Exposure
3. Financial Exposure
4. Risk Analysis
5. Recoverability Index
6. Problem Identification
7. Problem Engagement
8. Solution
9. Communication
10. Mitigation/Containment
11. Recall
    A. Product recall format
12. Definition of Completeness

Quality Report

1. Product Birth Record
2. Serial Number Database
3. Spares Repairs Tie-in
4. Field Incidents
   A. Incident information
5. Quality Database System
6. Procedures
7. Reports Data Entry
8. System Platform/Management Information System Tie-in
9. Product Life Cycle Database
   A. Quality database integration

Promotion

1. Marketing Manual
   A. Matrix of product information
   B. Quotation database
   C. Sales order entry analysis
   D. Sales representative management
   E. Sales report
   F. Deal profile with Appendices A, B, C, D
   G. Contract development quote form template
   H. Quote template
   I. Sales progress tracking
2. Promotional Literature
3. Direct Mail Pieces
4. Telemarketing
5. PowerPoint™ Presentation
6. Computer Multimedia/Video
7. Product Release
8. Promotional Plans
   A. Promotional plans and contingencies
   B. Prepare launch plans

Customer Support

1. Tools for Management
    A. Customer support database
    B. Customer service evaluation tool
    C. Format for sales compensation system
    D. Customer support inquiry management log
    E. Sales team evaluation tool

Program Management

1. Business Plan
    A. Active list of deliverables
        1. Go/No-Go Decision
2. Program Management—Narrative
3. Management Reporting
    A. Management report format
    B. Program costs summary (Includes personnel and building)
    C. Program evaluation
    D. Product development listing
4. Risk Analysis
    A. Risk assessment worksheet
    B. Impact of development cost overruns versus time
    C. Safety review
    D. Final product cost evaluation
    E. Final product forecast
5. New Product Evaluation
    A. Comparative investments (Comparative ROI worksheet)
    B. Capital requirements plan
6. Life Cycle Management
    A. Aftermarket pricing and repair
    B. Aftermarket plans
    C. Pricing policy aftermarket
    D. New product release
    E. Product obsolescence plan
    F. Value management completed
7. Intellectual / Property Protection
    A. Patent disclosure statement
    B. Patent disclosure procedure
    C. Patent review process
8. Personnel Training

PERSPECTIVES ON BUSINESS DEVELOPMENT

Like any endeavor involving people, there are four requirements for business development success:

A. Communication/accuracy. This is a basic requirement among all of the group members and the customers offering product insight and input. Accurate, timely, and unbiased communication is a fundamental requirement. Do not accept anything less.

B. Commitment. This requirement ensures some degree of momentum and sustainability in the program. A project involving people who do not have commitment will languish and fail. A program with committed people will succeed. Do not assemble a group that lacks this trait. You cannot add it in the middle of a program.

C. Agreement. It is difficult for any two people to agree on everything all of the time. It is even more difficult to get a group of people to agree with every decision in a program, especially when judgment and more than one solution are involved. The important thing to remember, however, is to allow everyone to express their opinion and make a suggestion. In the end, the manager must take all the input and make the best decision possible. It might not necessarily be the most popular decision but the manager must have enough leadership skills to make the decision and see it through. Once the decision is made, the entire product group must support it without exception or reservation. Consider the facts, consider the recommendations, and make the decision and implement it with everyone supporting it.

D. Organization. Every project has a multitude of details. The challenge is to organize them for usage and recall. Take the required time to organize the program in a format that is usable. Do not put the task off until later; you will never organize the program and, in the meantime, you will suffer from lack of organization. Lack of organization will degrade the decision-making process. (It is human nature to make an attempt to look for information, but people soon forgo the attempt when it is inconvenient and they make decisions from memory or loose facts.) A well-organized information system will aid in decision making by improving access when it is needed.

INDEX

Printed and bound by CPI Group (UK) Ltd, Croydon, CR0 4YY

08/05/2025

01864823-0005